Innovations in Landscape Research

Series Editor

Lothar Mueller, Leibniz Centre for Agricultural Landscape Research (ZALF),
Müncheberg, Germany

Aims & Scope

The Springer series "Innovations in Landscape Research" presents novel methodologies and technologies to understand, monitor and manage landscapes of the Anthropocene. The aim is to achieve landscape sustainability at high productivity. This includes halting degradation of landscapes and their compartments, developing cultural landscapes, and preserving semi-natural landscapes. Clean water and air, fertile and healthy soils for food and other ecosystem services, and a green and bio-diverse environment are attributes of landscapes for the survival and well-being of humans who inhabit them.

How do landscapes function? How do future landscapes look like? How can we sustainably develop intensively used and stressed kinds of landscapes? Scientific innovations and decision tools are key to answer and solve those challenging questions. The series will inform about advanced methods and results of disciplinary, interdisciplinary and transdisciplinary work in landscape research. It presents a broad array of methods to measure, assess, forecast, utilize and control landscapes and their compartments. These include field and laboratory measurement methods, methods of resource evaluation, functional mapping and risk assessment, and sensing methods for landscape monitoring, advanced methods for data analysis and ecosystem modeling, methods and technologies for optimizing the use of multi-functional landscapes, for the bioremediation of soil and water, and basics and procedures of landscape planning. The series provides a new view on landscapes with some focus on scientific and technological innovations, on soils and problems of optimizing agricultural landscapes under conditions of progressive urbanization. Landscape research in a globalized world of the Anthropocene is based on gathering big data and scenario modeling. International long-term experiments and agri-environmental monitoring systems will deliver data for ecosystem models and decision support systems.

Edited volumes of this series will address the following topics at high priority: Status and Trends of Landscape Research; Understanding Key Landscape Processes; Landscape Services, Functions and Biodiversity; Assessing Soil Resources and Quality; Water Resource and Quality Monitoring; Landscape Monitoring Concepts and Studies; Landscape Sensor and Monitoring Technologies; Landscape Modeling and Decision Support; Agricultural Soil and Plant Management; Basics and Tools for Landscape Planning; Tools for Water and Wetland Management; Forest Management and Agroforestry; Rehabilitation of Degraded Landscapes.

The books of this series are a source of information for researchers, teachers, students, and stakeholders interested in the topics of landscape science and related disciplines. They present status analyses, methodical chapters and case studies showing the practical relevance and feasibility of novel decision tools and technologies. Thus, the books of this series will be a particular valuable information basis for managers and decision makers at various levels, from local up to international decision bodies.

An author/editor questionnaire, instructions for authors and a book proposal form can be obtained by contacting the Publisher.

Maik Veste · Elmi Lötze · Thomas Littmann
Editors

Trees in South African Agriculture

Integration of Windbreaks in Agricultural Landscapes

Springer

Editors
Maik Veste
CEBra—Centrum für Energietechnologie
Brandenburg e.V.
Cottbus, Germany

Department of Molecular Botany
Institute of Botany
University of Hohenheim
Stuttgart, Germany

Thomas Littmann
DLC Dr. Littmann Consulting
Ennepetal, Germany

Elmi Lötze
Department of Horticulture
Stellenbosch University
Stellenbosch, South Africa

ISSN 2524-5155 ISSN 2524-5163 (electronic)
Innovations in Landscape Research
ISBN 978-3-032-00992-0 ISBN 978-3-032-00993-7 (eBook)
https://doi.org/10.1007/978-3-032-00993-7

This Springer imprint is published by the registered company Springer Nature Switzerland AG
The registered company address is: Gewerbestrasse 11, 6330 Cham, Switzerland

The book is dedicated to

Prof. Dr. Dr. h.c. Wolfgang Haber
to his 100th birthday for his outstanding scientific lifework and future-oriented research to link agriculture and landscape ecology

Preface

Over the centuries, trees have been important parts for farmers in southern Africa: they provide a variety of products such as timber, material for poles and firewood, as well as important foodstuffs such as fruit and nuts, and are also bee forage for honey production. Therefore, they fulfil a multifunctional role on the farm. In the man-made cultural landscape, they are an integral part of the landscape and increase its structural diversity. With the monotony and increasing size of fields, negative environmental impacts such as soil erosion caused by wind and water are increasing. Especially in South Africa's climatically exposed location between the two oceans, this is a challenge for land use, a factor that is becoming increasingly relevant with climate change.

Already during "Day Zero" drought in southwestern South Africa, the region experienced one of the most severe droughts in many decades, highlighting its sensitivity to climate change. This event clearly demonstrated to farmers the need to increase the resilience of their systems. Based on the increased interest of the farmers in climate-smart agricultural management, the South African German research project "FarmImpact—Development of Sustainable Water and Energy Solutions for Farms in South Africa" was initiated to learn more about the integration of trees as wind and evaporation protection and their climatic feedback. The project combines professional experience with scientific partners. The aim of the project was to increase the protection provided by windbreaks through eco-engineering measures.

The multifunctionality of the agricultural landscape is also an important element of the stabilizing diversity discussion. Structural diversity with near-natural or man-made growth forms of plant cover should be included in land use.

Based on this knowledge, the landscape ecologist Wolfgang Haber developed the concept of "differentiated soil and land use" back in the early 1970s. Wolfgang Haber, born in Datteln in 1925, studied biology, chemistry and geography at the Universities of Münster, München, Basel and Hohenheim. He received his Ph.D. from the University of Hohenheim. Furthermore, he deepen his work in botany and ecology as an assistant to Heinrich Walter at the University of Hohenheim. After working at the State Museum of Natural History in Münster, he moved to the newly founded Institute of Landscape Conservation at the Technical University of

Munich in Freising-Weihenstephan, which he later renamed the Chair of Landscape Conservation and headed until his retirement in 1993. The combination of agricultural landscape, near-natural cultivation and sustainable land use is particularly close to his heart, which he also focussed on in his decades of research work.

His scientific concept also inspired the present volume and the scientific discussion in many ways. We take this opportunity to acknowledge Wolfgang Haber's great scientific work in the field of agriculture and nature in cultural landscapes.

We do hope that this volume on windbreaks in agriculture in South Africa is useful for the readers interested in the field of ecology, agriculture, horticultural and climate change research and other related fields. We also hope to encourage farmers, policymakers, conservationist and landscapes planers to support the integration of farm trees into the landscape.

Cottbus/Stuttgart, Germany — Maik Veste
Stellenbosch, South Africa — Elmi Lötze
Ennepetal, Germany — Thomas Littmann
March 2025

Acknowledgements

The research conducted in South Africa was accompanied over many years by joint discussions with the farmers in the region. Our sincere thanks go to Johan and Gideon Roos (Paul Roos Wine, Rust en Vrede, Stellenbosch), Ian de Villiers (Babylonstoren, Paarl), Michael Back (Backsberg Wine, Paarl), Victor Sperling (Delheim Wines, Stellenbosch) and Willem van Wyk (Papkuilsfontein, Niewouldtville). Their inside views from the farming world have contributed in various ways to the scientific discussions in this book.

The perspective on water, agriculture and trade was sharpened by Vera Massie from the German Chamber of Commerce and Industry for Southern Africa in Cape Town.

Thanks for the financial support of the Federal Ministry of Education and Research (BMBF) and project management by PT-DLR for "FarmImpact—Development of Sustainable Water and Energy Solutions for Farms in South Africa" (grant no. FKZ 01LZ1711A-F) as part of "CLIENT II—International Partnerships for Sustainable Innovations". We thank Roland Keil (PT-DLR), Susanne Kiefer (BMBF) and Vera Stercken (BMBF) for their interest and endorsement of the FarmImpact project.

We would like to thank all involved research organizations, namely the National Research Foundation (NRF), the Department of Agriculture, Forestry and Fisheries (DAFF) of South Africa; German Academic Exchange Service (DAAD) and Volkswagen Foundation for their support over many years.

The editors gratefully acknowledge Doris Bleier, Saranya Kalidoss and Amudha Vijayarangan at Springer Nature for their steady interest, patience and help during the preparation of the book. Without all their efforts, we would not have been able to perform the research presented in this volume.

Contents

Editors and Contributors

About the Editors

Dr. Maik Veste is Biologist by education. He studied biology, zoology, botany and biochemistry at the University of Münster. He received his Diploma in biology from the Department of Ecology at Münster University and his Ph.D. in plant ecophysiology from the University of Hohenheim. His research fields include botany, ecophysiology, plant biogeochemistry and vegetation-climate interactions of the biogeosphere. He conducted various extensive ecological field studies in the South African agricultural landscapes, fynbos and Karoo in the last 35 years. He was Guest Scientist at various departments of Stellenbosch University for many years. He is Editor of various Springer books on agroforestry, drylands, land use and climate change.

Dr. Elmi Lötze was Senior Lecturer at the Faculty of AgriSciences, Stellenbosch University in South Africa. Her research focussed on the role of nutrition, root growth dynamics and climate on fruit yield and quality, resulting in several post graduate students and peer-reviewed papers. She also served as Chairman of the ISHS Workgroup on Mineral Nutrition of Fruit Crops. At present, she is Head of ITEST™ CARBOHYDRATES and ITEST™ LEAF at Agri Technovation and responsible for the development of carbohydrate analysis as Physiological Indicator of plant performance in various crops.

Privatdozent Dr. Thomas Littmann studied geography completed his habilitation in the field of physical geography at the Ruhr University of Bochum, and is affiliated as Privatdozent with the Martin-Luther University Halle-Wittenberg. He has led numerous research projects on climatology and environmental research during his academic career. Before founding DLC, he was Head of site assessment at a leading German company in the wind energy sector. Presently he is CEO at DLC—Dr. Littmann Consulting, an international consultancy for energy yield studies and scientific applications in the fields of wind and solar energy.

Contributors

Xiaohong Duan Helmholtz Zentrum München GmbH–German Research Center for Environmental Health, Neuherberg, Germany

Ben du Toit Department of Forest and Wood Science, Stellenbosch University, Stellenbosch, South Africa;
Verdure Consultation Services, Stellenbosch, South Africa

Jacques F. du Toit Verdure Consultation Services, Stellenbosch, South Africa

Roger Funk Leibniz Centre for Agricultural Landscape Research (ZALF E.V.), Müncheberg, Germany

Heinrich Geldenhuys Department of Horticultural Science, Stellenbosch University, Stellenbosch, South Africa;
Citrus Research International, Stellenbosch, South Africa

Tobias Hahn Orchard Research Station Müncheberg (OBVS), Müncheberg, Germany

Florian Heinlein Helmholtz Zentrum München GmbH–German Research Center for Environmental Health, Neuherberg, Germany

Florian Kestel Leibniz Centre for Agricultural Landscape Research (ZALF E.V.), Müncheberg, Germany

Anton Kunneke Department of Forestry and Wood Sciences, Stellenbosch University, Stellenbosch, South Africa

Daniel Littmann DLC–Dr. Littmann Consulting, Ennepetal, Germany

Thomas Littmann DLC–Dr. Littmann Consulting, Ennepetal, Germany

Elmi Lötze Department of Horticultural Science, Stellenbosch University, Stellenbosch, South Africa;
Agri Technovation South Africa, Wellington, South Africa

Gideon F. Malherbe Department of Forest and Wood Science, Stellenbosch University, Stellenbosch, South Africa

Lutz-Heiner Otto University of Hohenheim, Land Use Economics in the Tropics and Subtropics, Stuttgart, Germany;
Ecovest Landuse Consulting, Windhoek, Namibia

Eckart Priesack Helmholtz Zentrum München GmbH–German Research Center for Environmental Health, Neuherberg, Germany;
Institute of Soil Science, University of Hohenheim, Stuttgart, Germany

Tobias Recke DLC–Dr. Littmann Consulting, Ennepetal, Germany

Erich Seifert Scientes Mondium UG, Altomünster, Germany;
Department of Forest and Wood Science, Stellenbosch University, Matieland, South Africa

Stefan Seifert Scientes Mondium UG, Altomünster, Germany

Thomas Seifert Department of Forest and Wood Science, Stellenbosch University, Matieland, South Africa;
Chair of Forest Growth and Dendroecology, Albert-Ludwigs-University Freiburg, Freiburg, Germany

Maik Veste CEBra–Center for Energy Technology Brandenburg e.V., Cottbus, Germany;
Department Molecular Botany, Institute of Biology, University of Hohenheim, Stuttgart, Germany

Lidia Völker Leibniz Centre for Agricultural Landscape Research (ZALF E.V.), Müncheberg, Germany

Chapter 1
Agriculture and Agroforestry in South Africa: An Introduction to FarmImpact and the Book

Maik Veste, Thomas Littmann, and Elmi Lötze

Abstract The special soil and climate conditions at the south-western tip of South Africa pose major challenges for local agriculture. The region is important for the cultivation of grain, but also wine and fruit. To reduce the high local wind speeds, trees have been planted as windbreaks for many years. Especially non-native tree species such as *Populus simonii*, *Casuarina cunninghamiana*, and *Alnus cordata*, but also *Eucalyptus* species, are often planted as farm trees in the region. These trees also contribute in many ways to ecosystem services and protect agricultural crops. Based on many years of field measurements, FarmImpact develops practical recommendations for action for regional agriculture. On the one hand, this includes the design of water-efficient agricultural practices to improve the water consumption of crops through windbreak hedges. In addition, the aim is to increase soil fertility, reduce soil erosion, and provide additional ecosystem services through integrated measures in the landscape (windbreaks, ecological support zones, land management). Furthermore, the project analyses an increase in economic resilience at the farm level through wind protection measures.

Keywords Crop production · Orchards · Farms · Wind protection · Winter rain region · Cape winelands

---

M. Veste (✉)
CEBra—Center for Energy Technology Brandenburg e.V., Cottbus, Germany
e-mail: maik.veste@cebra-cottbus.de

Institute of Biology, Department Molecular Botany, University of Hohenheim, Stuttgart, Germany

T. Littmann
DLC—Dr. Littmann Consulting GmbH, Ennepetal, Germany

E. Lötze
Department of Horticulture, Stellenbosch University, Stellenbosch, South Africa

AgriTechnovation South Africa, Wellington, South Africa

M. Veste et al. (eds.), *Trees in South African Agriculture*, Innovations in Landscape Research, https://doi.org/10.1007/978-3-032-00993-7_1

1.1 Agricultural Production

As an important regional producer, South Africa makes a significant contribution to food security in the region and is also an important exporter of surplus production to Western Europe, North America, and the world market. South Africa has a market-oriented agricultural economy that is highly diversified, and includes the production of most major grains, oilseeds, deciduous and subtropical fruits, sugar, citrus, wine, and most vegetables (Vink 2004). There is also a competitive amount of livestock production. As the South African agricultural sector is not supported by state subsidies and is therefore heavily dependent on exports to the world market, there is a need to increase production and quality. Water scarcity is one of the biggest challenges of agriculture in South Africa, along with the effects of climate change, human vulnerability, and the loss of biodiversity and ecosystem functions.

Agricultural land use in the form of rangelands, fields, or orchards is an important component in the different biomes of Southern Africa. Meanwhile, South Africa's farmland makes up around 82% of the country land area. Whilst production conditions in South Africa are far from ideal, only 16.7 million hectares are considered potential arable land and 83.3% is grazing land. 1.35 million hectares of the country's potential arable land is available for irrigation and accounts for more than one-third of total output in the agriculture sector (Vink and Kirsten 2003). However, due to unfavourable climatic and soil conditions, only 12 per cent of the country is suitable for the production of rain-fed crops, with only 3 per cent considered genuinely fertile land. The majority of South Africa's land surface (69%) is suitable for grazing, and therefore, livestock farming is the largest agricultural sector in the country. The rain-fed agriculture prevails with reliable crop production typically found in the higher rainfall areas with summer rainfall (Rötter et al. 2024). Even the Western Cape with mainly winter rainfall is often referred as the 'breadbasket' of South Africa and is a centre for wheat production and therefore of high national importance for food security in the region.

According to the Western Cape Department of Agriculture (Morokong et al. 2022), approximately 2 million hectares of land were recorded as being under crop production in 2017; of this, 338.588 hectares (17%) were used for wheat and 86.670 ha (4.3%) for barley. Other important field crops are canola (90 523 ha) and lupines (72 99 ha), respectively. Most of the province's grain crops, oilseeds, and lupines are grown in the West Coast (35%) and the Overberg (30%). The West Coast also has the largest recorded area used to grow vegetables (36%), and almost the entire area used for producing tobacco, tea, and hops (99%) is mainly due to it being the major rooibos growing region (in total 58,996 ha).

Important for Western Cape agribusiness are the wine and fruit production. Wine grapes (91,221 ha) and table grapes (13,095 ha) make up the largest proportion, whilst fruit production is dominated by apples (21,512 ha), pears (10,711 ha), and oranges (7,704 ha), whilst berries production increased.

This different crop production reflects the diverse soil and microclimate conditions: high mountain regions, valleys with lush vegetation, coastal regions, semi-desert areas, and agriculture dominate large parts of the landscapes of the Western Cape. Here, the Cape Winelands District is the heart of the provincial wine industry, and more than half of the province's orchards (57%) are located here. The three distinct regions of Stellenbosch, Franschhoek, and Paarl make up most part of the Cape Winelands.

The complex interplay between the climate, soil, geology, and geomorphology makes the region highly suitable for the cultivation of different grape varieties (Bargmann 2003). The specific climate condition with the typical Mediterranean climate with summer drought and the regional wind regime is of a great importance for the wine production. It requires good management of the cultivation systems Due to the location between the ocean and the escarpment, wind protection of the crop is important to reduce soil erosion, mechanical damages, and evapotranspiration (Veste et al. 2020). On the other hand, sea breezes with the associated increase in wind speed in the afternoon and concomitant increase in relative humidity and reduction in temperature are of interest in the improvement of wine character and quality (Bonnardot et al. 2005), and therefore, it is from economic relevance for the wine industry. High radiation promotes sunburn of fruits. Nowadays, netting of orchards of citrus and berries (Fig. 1.2b) reduces the radiation effects.

However, land use changed over time due to the requirements of demand on the agricultural market and policy and regularisation (Halpern and Meadows 2013; Rötter et al. 2024). Land pressure was increased by higher utilisation due to population growth and urbanisation, whilst especially the natural vegetation of Renosterveld decreased between 1960 and 2010 (Halpern and Meadows 2013). The authors also show a shift from Renosterveld or from grain farmland to vineyards in the Swartland (Fig. 1.1b). Those land cover changes had also implication for the agricultural water use. Land cover changes in the unprotected agricultural areas in combination with droughts and strong El Niño events will also increase erosion risk in the region and a loss of fertile topsoil (Kestel et al. 2023; Baade et al. 2024).

1.2 Trees on Farmlands

Natural forests are only of minor importance in South Africa, although there are many useful tree species for timber production (van Wyk and Gericke 2018). Furthermore, wood traditionally plays an important role as firewood. To expand the wood resources, many tree species were introduced on the farms and traditionally cultivated for local wood utilisation, and non-wood forest products and ornamental purposes. Nowadays, these planted farm trees determine the visual appearance of the cultivated landscape and differ from the typical natural vegetation of the fynbos and Renosterveld. Depending on soil and water resources and their use, the trees can be arranged in different spatial settings on the farm (Fig. 1.3):

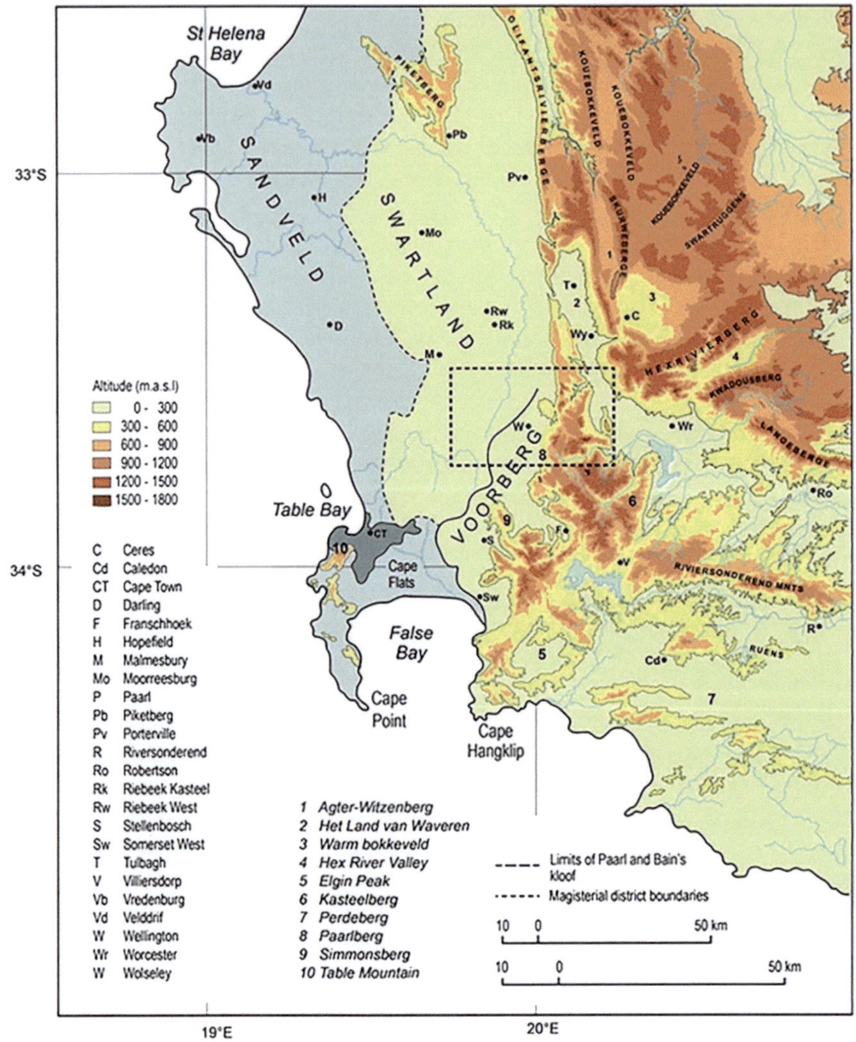

Fig. 1.1 The Cape Winelands located in the southern Western Cape Province (after Meadows 2015)

- **Single or small groups** are often use for ornamental use or for shade protection on farms.
- **Woodlots and farm forests** of small-scale production of a few hectare and arranged in mosaic pattern in appropriate landscape position.
- **Agroforestry systems** integrate trees as windbreaks with crops or pasture.

Only a few common indigenous tree species occur in the landscapes of the Western Cape and can be found on farmlands (i.e. *Olea europaea subsp. africana, Kiggelaria africana, Maytenus oleoides, Widdringtonia nodiflora, Cunonia capensis, Cassine*

Fig. 1.2 Agricultural landscapes in the Western Cape: **a** Cape Winelands: vineyards with view on Table Mountain, **b** Swartland: wheat and grape production, **c** Overberg: wheat production, **d** Cederberg: grapes and fruit trees farming along the Olifants River (Photographs: Maik Veste)

peragua, Curtisia dentata, Apodytes dimidiata, and several species of *Searsia* and *Podocarpus* (van Wyk and van Wyk 2013). However, the many trees a slow growing. Due to their higher growth rates and productivity, many tree species were introduced to Southern Africa. An overview is given by Van Wyk and van Wyk P (2013) and Glen and van Wyk (2016). The most important introduced trees on farmlands in the Western Cape are listed in Table 1.1 along with their most common uses.

Trees used for shelterbelts are often pioneer tree species (e.g. *Populus*, *Salix*), which are able to established on disturbed areas (Veste et al. 2006; Veste and Böhm 2018). They are characterised by rapid growth and high biomass production in the juvenile phase. Their ability to from new shoots from adventitious buds on the roots (root spawn) and the stem (stock spouts) also enables rapid spread at the site or re-sprouting after possible mechanical damages. As the sites of pioneer communities are often characterised by extreme and sometimes changing site conditions, these tree species must also be highly adaptable and stress-tolerant regarding local soil conditions and climatic factors and the associated water and nutrient supply (Veste et al. 2018). What they all have in common is a high light requirement.

Fast-growing trees for shelterbelts and windbreaks (Fig. 1.3) in the South African landscapes include *Casuarina cunninghamiana*, *Populus simonii*, and *Alnus cordata.* Very prominent in the different South African agricultural landscapes are the fast-growing Eucalyptus species (Fig. 1.3), which were introduced from Australia. Nearly 150 eucalypt species were introduced, mainly for forestry trials, between 1828 and 1940; nowadays, it is estimated that more than 200 species were cultivated (Hirsch

Fig. 1.3 Typical farm trees in the Western Cape: **a** eucalyptus trees, **b** wild olive (*Olea europaea* subsp. *africana*), **c** *Acacia* trees as shade cover and wind protection around the farm house on Papkuilsfontein Guest Farm, Bokkeveld Mountains, Nieuwouldtville **d** woodlot of eucalyptus in the Overberg region near Caledon, **e** woodlot of eucalyptus on Backsberg wine farm, Paarl, **f** eucalyptus and other trees on a farm in the Klein Karoo, **g** windbreak of *Populus simonii* on Babylonstoren wine farm, Paarl, **h** windbreak of *Casuarina cunninghamiana* at Rust en Vrede, Stellenbosch, **i** vineyard with shelterbelts on Delheim Wines, Stellenbosch. Photographs: Maik Veste

et al. 2019). Selected species can grow even in moderately dry areas and extent cultivation into regions where commercial forestry are limited. Thus, commercial plantation forestry is located mainly in the eastern and southern parts of South Africa with summer rainfall. On farms, they were often planted as woodlots and farm forests for timber for the pole market or sawtimber (du Toit et al. 2018). Positive features of these fast-growing trees are the provision of timber, paper, poles, firewood, shelter, ornamental value, and nectar and pollen for bees. However, their impacts on ecosystems are discussed controversially (see Chap. 14), and meanwhile, six eucalypts (i.e. *E. camaldulensis*, *E. cladocalyx, E. conferruminata, E. diversicolor, E. grandis,* and *E. tereticornis*) are listed as invasive in South Africa (Hirsch et al. 2019). Even though the trees are highly important as living windbreak hedges and as farm trees, *Alnus glutinosa*, *Casuarina cunninghamiana*, and the various oaks species are considered highly invasive and naturalised in riparian corridors and compete for water resources

Table 1.1 Overview of introduced useful trees and shrubs found on farmlands in the winter rain area of the Western Cape and Northern Cape along with their most important uses (excluding invasive thickets and commercial fruit orchards). (Armstrong and Gibbs 2001; Van Wyk, van Wyk 2013, Glen and Wyk 2016)

Species	Common name	Origan	Growth	Height	Windbreak	Timber	Firewood	Honeybee forage	Livestock fodder	Others
Introduced trees and shrubs										
Alnus cordata	Italian Alder	South Italy, Corsica			+					
Atriplex nummuralis/A. halimus	Salt bush	Australia		Shrub	(+)				(+)	
Calliandra calothyrsus	Calliandera	Central America, Mexico	Fast	Small		+	+			
Casuarina cunninghamiana	Beefwood	Australia	Fast	Medium-sized to tall			+			
Ceratonia siliqua	Locust tree	Mediterranean region		Small					+	Ornamental
Eucalyptus camaldulensis	River red gum	Australia	Fast	Tall	+		+	+		
Eucalyptus citriodora	Lemon-scented gum	Australia	Moderate			+	+	+		
Eucalyptus cladocalyx	Sugar gum	Australia	Fast			+	+	+		
Eucalyptus globulus	Blue gum	Australia	Fast	Tall		+	+	+		
Eucalyptus gomphocephala	Tuart	Australia					+	+		

(continued)

Table 1.1 (continued)

Species	Common name	Origan	Growth	Height	Windbreak	Timber	Firewood	Honeybee forage	Livestock fodder	Others
Eucalyptus melliodora	Honey-scented gum	Australia	Fast		+		+	+		
Eucalyptus polyanthemas	Red box	Australia	Fast				+	+		
Eucalyptus tereticornis	Queensland blue gum	Australia	Fast		+	+	+			Essential oils from leaves
Pinus halepensis	Aleppo pine	Eastern Mediteranean		Medium-sized to large						Shade and shelter
Pinus pinaster (=P.maritima)	Cluster pine, Maritime pine	Western Mediteranean		Medium-sized to large		+				
Pinus pinea	Stone pine	Northern Mediteranean		Medium-sized to large						Ornamental, edible nuts
Pinus radiata	Monterey pine	California		Medium-sized to large	+	+				
Populus canescens	White poplar					+	+			
Populus simonii	Chinese poplar				+					
Psidium guajava	Guava		Fast						+	Fruits
Quercus robur	European Oak	Europe		Medium-sized to large					+	Ornamental
Quercus suber	Cork oak			Medium-sized					+	Ornamental, cork

(continued)

Table 1.1 (continued)

Species	Common name	Origan	Growth	Height	Windbreak	Timber	Firewood	Honeybee forage	Livestock fodder	Others
Quercus cerris	Turkey oak	South-eastern Europe		Large					+	Ornamental
Robinia pseudoacacia	Black locust	North America	Fast	Medium-sized		+	+	+	+	

(Potgieter et al. 2014; Keet et al. 2020; Gildenhuys et al. 2024). In principle, biological invasion is an important factor for biodiversity and ecosystem functioning which needs to be under consideration for the use of trees on farmlands, natural ecosystems, and water bodies and needs special ecosystem management plans (Esler et al. 2014; van Wilgen et al. 2020). Furthermore, several species of Australian acacia were also introduced into the province, primarily for production of vegetable tannins and for stabilisation of drift sand areas, and that these species have formed invasive thickets in some regions such as the West Coast Plain, in some riverine zones and on the Agulhas Plains. The most common species in these invasive thickets are *Acacia saligna, A. cyclops, A. mearnsii,* and *A. longifolia.* They are not included in Table 1.1.

In general farm trees are important parts for the bioeconomy of the farmlands (Veste and Böhm 2018) and can be included into agricultural practice:

- Wood production for timber and poles
- Firewood and bioenergy
- Fruit and nuts production
- Essential oils from leaves
- Cork
- Honeybee forage
- Livestock fodder.

Traditionally, tree and shrub shelterbelts and windbreaks are a common measure to reduce wind speed and to protect the adjacent crops and soils. The established agroforestry systems provide various ecosystem services in the context with crop production (Sheppard et al. 2022; Veste et al. 2023):

- Modification of microclimate (windbreaks, shelterbelts, agroforest systems)
- Increase agricultural production
- Supporting pollinators and biodiversity
- Soil development
- Reduction of soil erosion
- Fertilisation
- Carbon sequestrations
- Reducing greenhouse gas emissions
- Agricultural landscape resilience.

1.3 Climatic Challenges for Agriculture

The south-western parts of South Africa receive rainfall mainly in winter or at all seasons (Fig. 1.4). Average rainfall ranges from well over 1,000 mm per annum in some mountainous areas in the south and eastern regions, to less than 150 mm over parts of the Karoo and the north-west regions (Fig. 1.4). The different variation in rainfall in combination with the different soil types contributes to the wide variety of crops grown in the province. The winter rainfall—and therefore the annual water input—is highly depending on the strength and position of the controlling pressure centres (South Atlantic Anticyclone and either dynamic mid-latitude or thermal inland Lows). In winter, the active pressure centres are either mid-latitude cyclones with frontal passage and cold air influx, strong wind, and high rainfall, or cyclones further south in the mid-latitude west wind drift with frontal systems typically not reaching the Southwestern Cape. In summer, the typical synoptic pattern represents strong inland thermal lows, a seasonally strong South Atlantic Anticyclone and following high wind speeds and temperatures in the southern Western Cape (Chap. 3).

Wind speed is also an important microclimatic driver for potential evapotranspiration (Veste et al. 2020), soil erosion, and the mechanical damages in fruit orchards. Even if agriculture is currently still taking place under more or less favourable climatic conditions, the effects of the present climate dynamic are already influencing production and yields in the region. Key challenges in the context of agriculture production in the Western Cape are water resources, climate change, coupled with and loss of biodiversity and ecosystem functions. The high interannual variations of rainfall in the last decade demonstrated already the massive impacts of climate changes. According to current scenarios, the region will be particularly affected already by the on-going climate change with high fluctuation in rainfall and an increase in extreme droughts (von Maltitz et al. 2024). One of these extreme weather events in

Fig. 1.4 Climate diagrams after Heinrich Walter showing different climate types and seasonal precipitation patterns in southern Africa (Adapted from Breckle and Rafiqpoor (2022))

the southern Western Cape was the so called 'Day Zero' drought of 2015–2018, which influenced the water supply of cities and for agriculture in the Western Cape (Theron et al. 2021). On the other hand, cut-off lows which occur typically in the transition months October and April can bring heavy rain events into the southern Western Cape with flooding affecting the agricultural production. Climate change increases the pressure to develop more resilient agricultural production systems (Talanow et al. 2021).

1.4 The Unique Flora of the Cape Floristic Region

The climatic and soil characteristics, which are also important for agriculture, are also reflected in the vegetation of the Western Cape. Typical for the Mediterranean climate zone with its winter rainfall is the Fynbos biome (Breckle and Rafiqpoor 2022; Mucina and Rutherford 2024). The Cape Floristic Region is recognised as a one world's biodiversity hotspots. In the fynbos biome, plant diversity is the highest in the world with more than 7,000 species (Esler et al. 2016). The fynbos is characterised by three different components: (i) the restioid component, belonging to the Restionaceae, and also sedges (Cyperaceae) and many grasses have the 'restoid' characteristics, (ii) the ericoid or heath component with Heaths (Ericaceae), the Daisy (Asteraceae), Blacktip (Bruniaceae), Pea (Fabaceae), Jujube (Rhamnaceae), and Thyme (Thymelaeaceae), and (iii) the proteoid component belonging to the Proteaceae. Fynbos vegetation types occur predominantly on well leached, infertile soils, where the *Protea* species are well adapted to the low nutrient soil conditions (Griebenow et al. 2022; Muller et al. 2023). In addition to Fynbos, the region features other vegetation types such as Strandveld and Renosterveld. The Strandveld Mediterranean ecosystem of the west coast of South Africa supports floristically diverse vegetation growing on mostly nutrient-poor aeolian sands and extending from the Atlantic Ocean tens of kilometres inland (Fig. 1.1). It has a diverse range of growth forms including broad-leaved and small-leaved shrubs, succulents, grasses and restios, annuals, and geophytes. The Renosterveld is characterised by the dominance of members of the Daisy Family (Asteraceae), specifically one species—Renosterbos *Elytropappus rhinocerotis*. Although Renosterbos is the characteristic dominant, many other plants are also prominent—for instance, in the Daisy Family (Asteraceae) and grasses. Furthermore, the high species richness of geophytes and orchids are also typical for the Renosterveld. Restios, ericas and proteas—typical of Fynbos—tend to be absent in Renosterveld, or are present at very low abundances, respectively. The fertile soils are largely confined to fine-grained soils, mainly clays and silts. This makes the Renosterveld interesting for agricultural land uses, and the natural vegetation is mainly transformed (Table 1.2).

Towards the north and east of the Western Cape, where the rainfall is less than 250 mm, the Renosterveld is replaced by the vegetation of Succulent Karoo. The vegetation here is dominated by dwarf, succulent shrubs, of which the Vygies (Mesembryanthemaceae) and Stonecrops (Crassulaceae) are particularly prominent.

Table 1.2 Transformation and conservation of the major fynbos vegetation (after Esler et al. 2016)

Vegetation type	Area transformed (%)	Area conserve (%)	Agricultural land-use pressure
Foothill fynbos	39	10	Vineyards, fruit orchards, pine plantations
Grassy fynbos	9	19	Pasture, pine plantations, invasive trees (*Acacia*)
Limestone fynbos	9	10	Invasive trees (*Acacia*)
Mountain fynbos (arid)	1	6	Rooibos tea
Mountain fynbos (on granite)	13	29	Rooibos and protea plantations invasive alien trees (pines, Hakea)
Renosterveld (on granite)	77	0.5	Vineyards, invasive alien plants
Renosterveld (on shale)	48	1	Cereal crops, canola, fruit orchards, vineyards invasive alien grasses
Sandveld/dune fynbos	42	2	Potatoes, rooibos tea, invasive alien plants (*Acacia*)

Mass flowering displays of annuals (mainly Daisies Asteraceae) occur in spring, often on degraded or fallow lands (Veste and Jürgens 2004). Rangelands are here the major agricultural land use in the region on farms (Fig. 1.5).

Fig. 1.5 Planted farm trees on rangelands in the Karoo, Papkuilsfontein Guest Farm, Nieuwoudtville. Photograph: Maik Veste

1.5 Objectives of the FarmImpact Project and the Book

For the integrated FarmImpact approach, South African and German research and practice partners combined technical solutions with ecological approaches to investigate the effects of windbreak hedges on the microclimate and the productivity of wine and fruits grown in the Cape Winelands (Figs. 1.1. and 1.6). Improving water use in agriculture is essential to successfully adapt to climate change. The problem will be intensified to decreased precipitation along with increased temperature as predicted consistently for the Western Cape by several climate models. The use of tree shelterbelts is a suitable eco-engineering approach to reduce water consumption and to enhance water saving in vineyards, orchards, and crop fields. One of the key objectives of creating windbreaks is to reduce the water requirements of agricultural crops. These protection strips create different field areas with different evapotranspiration and correspondingly lower irrigation requirements. In order to adapt irrigation to the reduced water requirement, an application-related basis will be created as part of FarmImpact (Fig. 1.7), which will provide agricultural enterprises with information on actual water consumption and future water requirements.

Windbreaks of various types are widely used in the fruit orchards and vineyards of the Western Cape Province. The structure of the windbreak is essential for crop protection. On various farms, the structure, growth performance, and survival of

Fig. 1.6 Experimental sites: **a** Babylonstoren, Paarl, north of the Simonsberg, **b** Farm Rust en Vrede, Stellenbosch, west of the Helderberg mountain, **c** Welgevallen experimental farm, Stellenbosch, **d** Farm Slangkop, Yzerfontein, Sandveld West Coast. Photographs: Heinrich Geldenhuys **c**, Maik Veste **a, b, d**

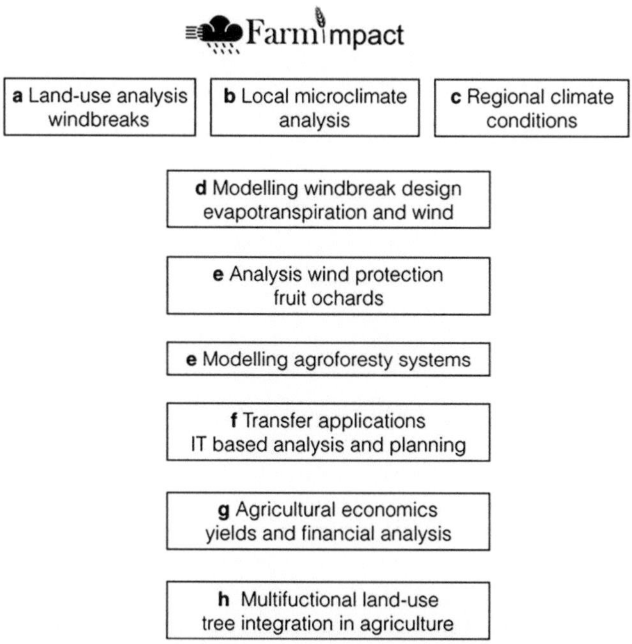

Fig. 1.7 Structure of farmimpact and the book chapters (details in the text)

planted windbreak trees were analysed (Chaps. 2 and 5). To determine the evapotranspiration of the crops, microclimate stations were installed downwind of the windbreak. These climate stations recorded wind speed, air temperature, relative humidity, global radiation, soil heat flux, and soil water content (Chap. 3). The seasonal changes of weather conditions have significant implication on the evapotranspiration (Chap. 5) in the orchards and especially the wind regime, which in turn largely controls fruit quality (Chap. 6). Interrelations of rainfall and evapotranspiration with weather types were classified for the southern Western Cape and a basis for future forecasting. The site-specific analysis forms the basis for modelling the wind protection effects of different windbreak systems (Chap. 5). In the Cape Winelands, strong winds have direct and indirect effects on the production performance of citrus trees. The extensive windbreaks in South Africa have positive effects in protecting citrus orchards from damage and thus avoiding major losses in export quality. In the first trial, the microclimate (wind speed and air temperature), production performance (yield, fruit size), and fruit quality (external and internal) were quantified at different distances from established *Casuarina* and poplar windbreaks on two mandarin orchards (Rust en Vrede and Babylonstoren, Welgevallen) (Chap. 6). Drone-based remote sensing complements the analysis of the spatial analysis of the wind protection effects on the citrus trees (Chap. 7). Sensors in the visible (RGB), multispectral, and thermal imaging, infrared ranges were combined. The RGB data provide the highest resolution, whilst the combination of different spectra in multispectral data allows information on tree vitality, and the thermal data allow estimates of water loss. The aim of the simulation of the agroforestry system (Chap. 8) was to model transpiration, the water cycle, and growth in a combined stand of hedges or tree rows and agricultural or horticultural plants using the Expert-N ecophysiological process model. Based on this, the benefits of windbreak hedges were evaluated using scenario simulations. The focus here was on the changes in water utilisation by agricultural or horticultural crops in the presence of windbreak hedges.

Transfer applications show the potential of windbreaks and agroforestry systems in other climatic conditions and are the basis for their planning and design (Chaps. 9 and 10). A tool allows a better analysis and design of landscape elements and shelterbelts (Chap. 11). A modelling system for calculating the economic profitability of investments in fast-growing windbreaks in South Africa did not exist at the time. Important for the implementation of shelterbelts and windbreaks are the financial returns. Therefore, a corresponding economic analysis was carried out using citrus as an example (Chap. 12). Finally, Chap. 13 synthesises contemporary research on the role of windbreaks in enhancing the productivity of annual and perennial crops, whilst little attention was paid to the actual calculation of the economic profitability of windbreaks. With trees, a multifunctional agriculture landscape can be developed with positive aspects for the farm development, but economy, natural resilience, and beneficial ecosystem service need to be in balance. The integrated approach for sustainable land management on farms and the integration of farm trees for climate-smart landscapes are discussed (Chap. 14).

Acknowledgements We thank our partners Paul Roos Wine, Rust en Vrede farm, Stellenbosch, Babylonstoren, Paarl, and Backsberg, Paarl for their support and permission for the field experiments. The project *Development of Sustainable Water and Energy Solutions for Farms in South Africa* was supported by the Federal Ministry of Education and Research (BMBF) and managed by PT-DLR (FKZ 01LZ1711A-F) as part of 'CLIENT II—International Partnerships for Sustainable Innovations' funding initiative.

References

Armstrong G, Gibbs L (2001) Handbook of useful trees and shrubs for rural areas of the winter rainfall region. Department of Forest Science, University of Stellenbosch, Stellenbosch

Baade J, Aucamp I, Collett A, Eckardt F, Funk R, Glotzbach C, von Holdt J, Kestel F, Knot J, Lombard A, Morgenthal T, Msipa A, Le Roux JJ et al (2024) Soil erosion research and soil conservation policy in South Africa. In: von Maltitz GP et al (ed) Sustainability of Southern African ecosystems under global change. Ecological studies 248. Springer, Cham, pp 335–368. https://doi.org/10.1007/978-3-031-10948-5_13

Bargmann CJ (2003) Geology and Wine 7. Geology and wine production in the coastal region, Western Cape Province, South Africa. Geoscience Canada 30(4). https://journals.lib.unb.ca/index.php/GC/article/view/4155

Bonnardot V, Planchon O, Cautenet S (2005) Sea breeze development under an offshore synoptic wind in the South-Western cape and implications for the Stellenbosch wine-producing area. T Heoretical Appl Climatol 81:203–218

Breckle S-W, Rafiqpoor MD (2022) Vegetation and Climate. Springer, Heidelberg. https://doi.org/10.1007/978-3-662-64036-4

du Toit B, Malherbe GF, Lambrechts H, Naidoo S, Eatwell K (2018) Market analysis to assess timber products from dryland woodlots and farm forests in South Africa. In: Revermann R, Krewenka KM, Schmiedel U, Olwoch JM, Helmschrot J, Jürgens N (eds) Climate change and adaptive land management in southern Africa—assessments, changes, challenges, and solutions. Klaus Hess Publishers, Göttingen and Windhoek, pp 336–342

Esler, K Pierce SM, de Villiers C (2016) Fynbos. Ecology and management. Briza Publisher, Pretoria

Gildenhuys CP, Potgieter LJ, Richardson DM (2024) The genus *Quercus* (Fagaceae) in South Africa: introduction history, current status, and invasion ecology. S Afr J Bot 167:150–165. https://doi.org/10.1016/j.sajb.2024.01.066

Glen H, Van Wyk B (2016) Guide to trees of southern Africa. Introduced into Southern Africa. Struik Nature, Cape Town

Griebenow S, Makunga NP, Privett S, Strauss P, Veste M, Kleinert A, Valentine A (2022) Soil pH influences the organic acid metabolism and exudation in cluster roots of Protea species from the Mediterranean-type Fynbos ecosystem, Western Cape. South Africa. Rhizosphere 21:100486. https://doi.org/10.1016/j.rhisph.2022.100486

Halpern ABW, Meadows ME (2013) Fifty years of land use change in the Swartland, Western Cape, South Africa: characteristics, causes and consequences. S Afr Geogr J = Suid-Afrikaanse Geografiese Tydskrif 95(1):38–49

Hirsch H, Allsopp MH, Canavan S, Cheek M, Geerts S, Geldenhuys CJ, Harding G, Hurley BP, Jones W, Keet J-H, Klein H, Ruwanza S, van Wilgen BW, Wingfield MJ, Richardson DM (2019) Eucalyptus camaldulensis in South Africa—past, present, future. Trans R Soc S Afr 75:1–22. https://doi.org/10.1080/0035919X.2019.1669732

Keet JH, Robertson MP, Richardson DM (2020) *Alnus glutinosa* (Betulaceae) in South Africa: invasive potential and management options. S Afr J Bot 135:280–293. https://doi.org/10.1016/j.sajb.2020.09.009

Kestel F, Wulf M, Funk R (2023) Spatiotemporal variability of the potential wind erosion risk in Southern Africa between 2005 and 2019. Land Degrad Dev 34:2945. https://doi.org/10.1002/ldr.4659

Meadows ME (2015) The Cape Winelands. In: Grab S, Knight J (eds) Landscapes and landforms of South Africa, world geomorphological landscapes. Springer, Heidelberg, pp 103–109

Morokong M, Sibulali A, Murdock J (2022) Western cape agricultural sector profile 2022 Western cape department of agriculture division for macro and resource economics, Elsenburg, pp 1–77

Mucina L, Rutherford MC (2024) Zonal biomes of Southern Africa. Springer Cham. https://doi.org/10.1007/978-3-031-63731-5

Muller K, Veste M, Privett S, Strauss P, Makunga N, Griebenow S, Valentine A (2023) Proteaceae species show different strategies for phosphorus acquisition and utilisation in P poor soils in the mediterranean-type Fynbos ecosystem. Flora 306 (2023). Article 152362. https://doi.org/10.1016/j.flora.2023.152362

Potgieter LJ, Richardson DM, Wilson JRU (2014) *Casuarina cunninghamiana* in the Western cape, South Africa: determinants of naturalisation and invasion, and options for management. S Afr J Bot 92:134–146

Rötter RP, Nkomo, M, Meyer zu Drewer J, Veste M (2024) Agricultural land-use systems and management challenges. In: von Maltitz GP, Midgley GF, Veitch J, Brümmer C, Rötter RP, Viehberg, FA, Veste M (eds) Sustainability of southern African ecosystems under global change: science for management and policy interventions, vol 248. Ecological studies. Springer Nature, Cham, pp 551–586. https://doi.org/10.1007/978-3-031-10948-5_20

Sheppard JP, Bohn Reckziegel R, Borrass L, Chirwa PW, Cuaranhua CJ, Hassler SK, Hoffmeister S, Kestel F, Maier R, Mälicke M, Morhart C, Ndlovu NP, Veste M, Funk R, Lang Seifert F, du

Toit TB, Kahle HP (2020) Agroforestry: an appropriate and sustainable response to a changing climate in Southern Africa? Sustainability 12(17):6796. https://doi.org/10.3390/su12176796

Talanow K, Topp NP, Loos J, Martín-López B (2021) Farmers' perceptions of climate change and adaptation strategies in South Africa's Western cape. J Rural Stud 81:203–219

Theron SN, Archer E, Midgley SJE, Walker S (2021) Agricultural perspectives on the 2015–2018 Western cape drought, South Africa: characteristics and spatial variability in the core wheat growing regions. Agric for Meteorol 304–305:108405. https://doi.org/10.1016/j.agrformet.2021.108405

Van Wyk B, Gericke N (2018) People's Plants: a guide to useful plants of Southern Africa. Briza Publications, Pretoria, South Africa

van Wilgen BW, Measey J, Richardson DM, Wilson JR, Zengeya TA (2020). Biological invasions in South Africa: an overview. In: Measey J (ed) Biological invasions in South Africa. Invading nature, Springer series in invasion ecology. Springer, Cham

Van Wyk B, van Wyk P (2013) Field guide to trees of Southern Africa. Struik Nature, Cape Town

Veste, M, Böhm C (2018) Agrarholz—Schnellwachsende Bäume in der Landwirtschaft: Biologie—Ökologie—Management. Springer Spektrum, Heidelberg. https://doi.org/10.1007/978-3-662-49931-3

Veste M, Gao J, Sun B, Breckle S-W (2006) The green great wall—combating desertification in China. Geogr Rundsch Int Ed 2:14–20

Veste M, Jürgens N (2004) Zonobiom III: die karoo. In Walter H, Breckle S-W, Ökologie der Erde B (eds) Spezielle Ökologie der tropischen und subtropischen Zonen, Spektrum Akademischer Verlag, pp 514–537

Veste M, Littmann T, Kunneke A, Du Toit B, Seifert T (2020) Windbreaks as part of climate-smart landscapes reduce evapotranspiration in vineyards, Western Cape Province, South Africa. Plant Soil Environ 66:119–127. https://doi.org/10.17221/616/2019-pse

Veste M, Pflugmacher C, Hartmann H, Schlepphorst R, Murach D (2018) Ökophysiologie der Agrargehölze – vom Blatt zum Bestand. In: Veste M, Böhm C (eds) Agrarholz – Schnellwachsende Bäume in der Landwirtschaft Biologie - Ökologie - Management. Springer, Berlin, Heidelberg, pp 169–227. https://doi.org/10.1007/978-3-662-49931-3_7

Veste M, Sheppard JP, Abdulai I, Ayisi KK, Borrass L, Chirwa PW, Funk R, Kapinga K, Morhart K, Mwale SE, Ndlovu NP, Nyamadzaw G, Nyoka BI, Sebola P, Seifert T, Senyolo MP, Sileshi GW, Syampungani S, Kahle H-P (2023) The need for sustainable agricultural land-use systems: benefits from integrated agroforestry systems. In: von Maltitz GP, Midgley GF, Veitch J, Brümmer C, Rötter RP, Viehberg, FA, Veste M (eds) Sustainability of Southern African ecosystems under global change: science for management and policy interventions, vol 248. Ecological studies. Springer Nature, Heidelberg, pp 587–562. https://doi.org/10.1007/978-3-031-10948-5_21

Vink N (2004) The influence of policy on the roles of agriculture in South Africa. Dev S Afr 21(1) 155–177. https://doi.org/10.1080/0376835042000181462

Vink N, Kirsten JF (2003) National report: South Africa. Paper presented at the Roles of Agriculture International Conference, Rome

Von Maltitz GP, Midgley GF, Veitch J, Brümmer C, Rötter RP, Viehberg FA, Veste M (eds) (2024) Sustainability of Southern African ecosystems under global change, Ecological studies 248, Springer, Cham. https://doi.org/10.1007/978-3-031-10948-5

Chapter 2
Windbreak Characteristics in the Cape Winelands

Ben du Toit, Gideon F. Malherbe, Anton Kunneke, Jacques F. du Toit, and Maik Veste

Abstract The chapter introduces the tree species planted on farmlands of the Cape Winelands as shelterbelts and focuses specifically on the growth rate and survival of newly established windbreaks. In existing windbreaks, *Pinus radiata* and *Casuarina cunninghamiana* showed the fastest growth rates, but the latter also had a strong competitive effect on orchard trees in their proximity. Both species produced considerable biomass (with associated carbon sequestration) but may grow too tall for a windbreak application within 30 years of age. In contrast, *Alnus cordata* displayed slower growth rates and maintained a manageable windbreak height for a longer period, with minimal competition on orchard trees in close proximity to it.

Keywords Windbreak · Shelterbelts · Farm forestry agroforestry · Italian alder · Beefwood · Chinese poplar

2.1 Introduction

Tree shelterbelts are widely used on farms of the Western Cape Province to protect fruit orchards and vineyards against the harsh wind regime in the area. They have been shown to significantly decrease wind speed and evaporation on the leeward side of prevailing winds (Veste et al. 2020; Hassler et al. 2024; Hoffmeister et al. 2024; Chap. 3). Windbreaks also afford protection against abrasive actions on fruits, thus

B. du Toit (✉) · G. F. Malherbe · A. Kunneke
Department of Forest and Wood Science, Stellenbosch University, Stellenbosch, South Africa
e-mail: ben@sun.ac.za

B. du Toit · J. F. du Toit
Verdure Consultation Services, Stellenbosch, South Africa

M. Veste (✉)
CEBra—Center for Energy Technology Brandenburg e.V., Cottbus, Germany
e-mail: veste@cebra-cottbus.de

Institute of Botany, Department of Molecular Botany, University of Hohenheim, Stuttgart, Germany

M. Veste et al. (eds.), *Trees in South African Agriculture*, Innovations in Landscape Research, https://doi.org/10.1007/978-3-032-00993-7_2

contributing to fruit quality (Geldenhuys et al 2023, Chap. 6), and are important to reduce soil erosion (Veste et al. 2024, Chap. 11). Furthermore, the integrated farm trees influence water redistribution pattern and the carbon sequestration in below-ground and aboveground biomass (Sheppard et al. 2020, 2024). The contribution of windbreaks to the productivity and resilience of the farming system may become even more important in future if current climate change predictions materialize over time (von Malitz et al. 2024). However, windbreak trees may also compete with fruit trees or vines in the rows adjacent to the windbreak (Fig. 2.1, Huth et al. 2010). This is most commonly due to competition for resources, namely nutrients, water, and solar radiation, and the utilisation of avialable resources between windbreak species and adjacent orchards/vineyards is clearly not constant among all windbreak species and tree sizes.

One of the main aims of the project is to deliver sound, scientifically based information to help South African farmers to adapt to climate change and increase their economic resilience. For management of windbreaks, the selection of the tree species is an import factor. To construct a suitable windbreak, farmers need to have information on tree species characteristics such as growth habit, growth rate, and canopy dynamics. Tree size development over time and potential longevity of the species planted are also useful data when planning the productive life of the windbreak and the spatial placement of windbreaks. One of the main elements in the calculation of

Fig. 2.1 Larger beefwood trees (*Casuarina cunninghamiana*) compete with the nearby irrigated citrus trees for water resources. Photograph: Maik Veste

the protection of any orchard behind a windbreak is the height of the windbreak as the spatial extent of crop protection afforded by the windbreak is primarily a function of tree height. Furthermore, windbreak porosity is also an important variable. Nowadays, terrestrial LiDAR allow better structural analysis of windbreak hedgerows and porosity in poplar and blackwood windbreaks on experimental farms in the Western Cape is presented in Chap. 5. Finally, it is also useful to have information on the spacing and the stem volumes that are grown in windbreaks over time, as it can be used to estimate the carbon sequestration and/or the volume of wood that will become available when old windbreak trees are cut to be replaced (Amichev et al. 2016). Basically, the evergreen and dense beefwood has the lower permeability than the two deciduous species poplar and Italian alder.

There is currently no information on the growth rate and the development of stem, height, and crown dimensions for most commonly used windbreak species in the Western Cape Province, South Africa. For a *Populus simonii* windbreak, the biomass and carbon storage were investigated at the farm Babylonstoren (Sheppard et al. 2024); furthermore, species abundance and the density of windbreaks per unit area of land is currently undocumented. This chapter reports on a survey that was conducted to determine the status quo of windbreaks commonly used in the Cape Winelands. It provides details on (a) the stem and height development of commonly used species in windbreaks, (b) windbreak species abundance and windbreak densities commonly implemented in the region, and (c) the survival of newly established windbreaks.

2.2 The Study Area in the Cape Winelands

A total of ten farms in the Cape Winelands district (Chap. 1) were selected to study the abundance, growth rate, and development of windbreak species. The climate of the study area is classified as class Csb by the Köppen system (Mediterranean climate, warm, with dry summers (see Chap. 4)). The long-term mean annual precipitation and temperature as well as the distance of the weather station from the farms is shown in Table 2.1, and the study site locations are shown on a map of the area in Fig. 2.2.

2.3 Windbreaks in the Cape Winelands

2.3.1 Tree Species

All indigenous tree species in the region are slow growing, and this characteristic, combined by the small area occupied by indigenous trees in the province, has led to the introduction of exotic fast-growing species for use on farmlands. To be used in windbreaks, however, it requires rapid growth and high biomass production in the juvenile phase to fill the available free space (Veste et al. 2018). By far, the most

Table 2.1 Farms used in the windbreak study with their climatic conditions. MAP: mean annual precipitation, MAT: mean annual temperature. See Fig. 2.1 for farm locations

Study sites/farm name	MAP (mm)	MAT (°C)	Distance from weather station (km)
Rust en Vrede, Roos Winery	615	17.4	0.2
Babylonstoren	800	17.5	2.1
Slaley	522	17.4	3.3
Delheim	522	17.4	0.6
Boschendal	944	18.1	0.5
Nooitgedacht	647	17.9	4.5
Hillcrest	808	17.1	2.5
Mariendahl	647	17.9	1.9
Welgevallen	735	17.3	1.5
Backsberg	800	17.5	0.9

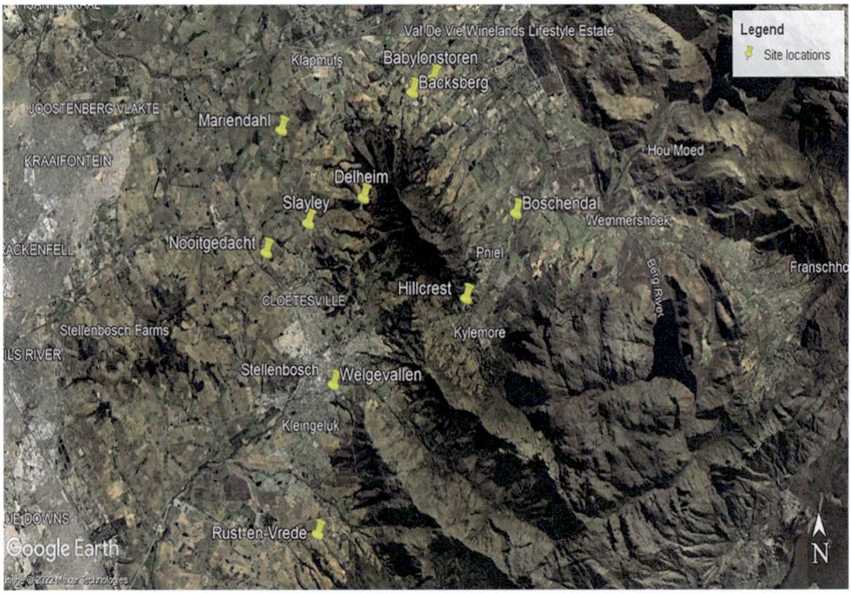

Fig. 2.2 The location map of farms around Stellenbosch where windbreak characteristics were measured

common windbreak species in the Cape Winelands are the following tree species (Van Wyk and van Wyk 2013).

- ***Casuarina cunninghamiana*, Beefwood**

Beefwood (Fig. 2.3a) is a fast-growing tree in the family Casuarinaceae and typically grows to a height of 15–35 m with DBH 0.5–1.5 m. Originally, the tree is native to Australia and New Guinea and introduced to Southern Africa between 1902 and 1907 mostly from Australia and predominantly cultivated for shade, shelter, and ornament (Potgieter et al. 2014). As an evergreen species with a dense grown, the tree provides a good shelter during the entire year. Casuarinas form symbiotic N-fixing associations with soil actinomycetes from the genus *Frankia*. These symbiotic associations allow casuarinas to fix nitrogen at rates similar to nodulated legumes, while mycorrhizal fungi promote phosphate uptake in P-poor soils. Additionally, the formation of cluster roots may be advantageous for nutrient poor by providing a reliable mechanism to enhance nutrient uptake. Casuarina may pose an invasive threat in some climate zones/provinces of South Africa, and a good management of the trees is recommended. Furthermore, high competition with orchard trees is another factor influencing the establishment of beefwood windbreaks.

- ***Populus simonii, Simon's poplar or Chinese poplar***

Various poplar species from the Salicaceae family have been introduced to South Africa as fast-growing tree species. Primarily, *Populus simonii* (Fig. 2.3b) is used as a windbreak, as it is not considered to be highly competitive. The species is native to northeastern China and Mongolia. Furthermore, it is commonly utilized for low-quality timber or as energy wood and for phytoremediation purposes. It reaches a height of 12–15 m.

Fig. 2.3 Common tree species for windbreaks in the Cape Winelands: **a** *Casuarina cunninghamiana,* beefwood; **b** *Populus simonii,* Simon's poplar or Chinese poplar; **c** *Alnus cordata*, Italian alder; **d** *Pinus radiata,* Monterey pine or Radiata pine. Photographs: Maik Veste (**a–c**) Elmi Lötze (**d**)

- ***Alnus cordata*, Italian alder**

The Italian alder (Fig. 2.3c) belongs to the family Betulaceae and is originally distributed in Italy in the southern Apennines Mountains but has also been cultivated in other countries in Europe, Chile, and New Zealand. As a deciduous tree, it can grow up to 25 m with a stem up to 70–100 cm. Typical for the genus Alnus, it is able to fix nitrogen by the symbiotic Actinomyces bacteria (*Frankia alni*) (Veste et al. 2018) and grows rapidly even under nutrient-poor soils, which are widespread in the Western Cape where it is commonly used in windbreaks.

- ***Pinus radiata, Monterey pine or Radiata pine***

Monterey pine (Fig. 2.3d) is native to the Central Coast of California and Mexico as well on Guadalupe Island and Cedros Island. As an evergreen conifer, it belongs to family of Pinaceae. The species is a versatile, fast-growing, medium-density softwood, suitable for a wide range of uses and valued for rapid growth. It is a medium-size to large tree and growing to 15–30 m. In the past, it was used in windbreaks in the Western Cape.

Furthermore, the presence of eucalypts on Cape Winelands farms and in woodlots plays an important role in sustaining honeybee populations during the dry season (mid to late summer, when orchards are not flowering) (De Lange et al. 2013). It follows that the availability of healthy populations of pollinators is essential. Often the eucalyptus is also used as shelter trees on different farms. Furthermore, fast-growing tress with strong apical dominance are commonly used on farms for pole production and as windbreaks, while trees that produce edible pods or nuts and acorns for animal fodder are in demand by livestock farmers. Lastly, general utility trees for shelter, beautification, recreational use, and the production of firewood are also commonly planted on farms in the region (Chap. 1).

2.3.2 *Design of Windbreaks*

In the Western Cape, windbreaks normally consist of a single tree row (Fig. 2.4). However, they are sometimes planted as parallel lines. They may link to many lateral branches of other windbreaks, thus mostly forming roughly square or rectangular boundaries around orchards (Fig. 2.5). In the case of square blocks or cells bordered by windbreaks, the windbreak length per unit area of land varies as shown in Table 2.2. If we presume that (in the case of rectangular orchards) the shortest distance between orchard boundaries is usually orientated to be parallel to the prevailing wind, we deduce from Table 2.2. that the average distance between windbreaks at a windbreak density of 100 m ha^{-1} is probably in the order of roughly 100 to 200 m (depending on rectangle configurations). Seeing that the protection afforded by most windbreaks is still strongly significant at roughly 5 times the windbreak tree height (Veste et al. 2020), these calculations suggest that tree height of roughly 20 m tall will afford good protection for an orchard that has windbreaks spaced 100 m apart. Windbreaks

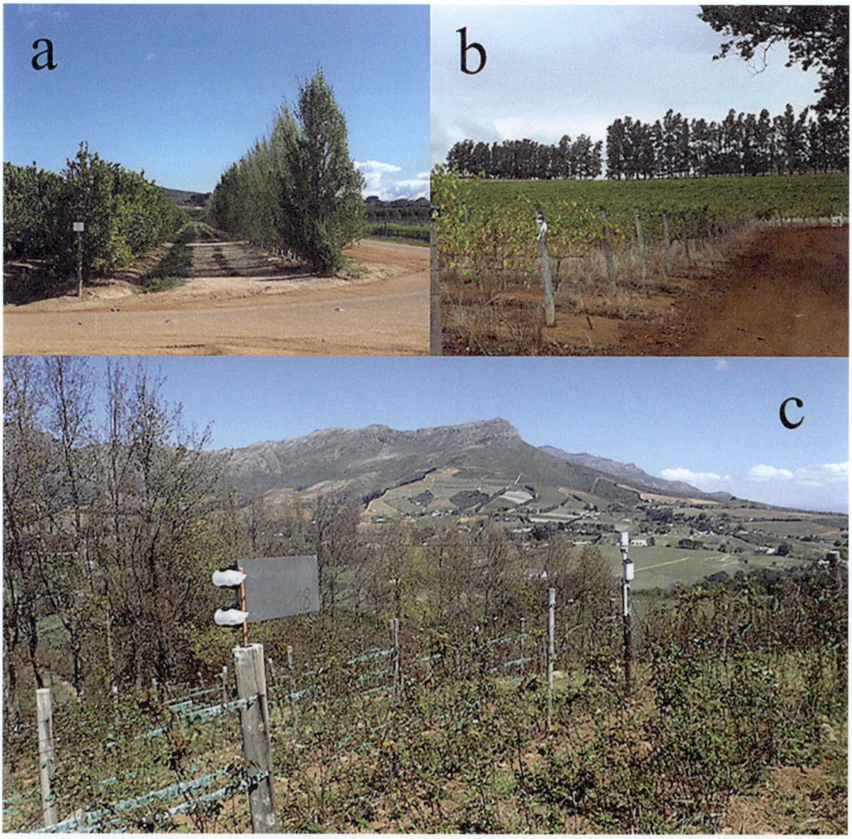

Fig. 2.4 Single row shelterbelts of **a** *Populus simonii* at Babylonstoren, Stellenbosch, **b** *Casuarina cunninghamiana* at Delheim, Stellenbosch, **c** *Alnus cordata* at Hillcrest Berry Orchards, Pniel, Stellenbosch. Photographs: Maik Veste (**a, b**), Roger Funk (**c**)

will have to be planted in denser configuration than is currently the case on several farms (i.e., approximately 50 m apart) if they are to give significant protection by the time trees are 10 m high.

2.4 Species Abundance and Ages

The total length of windbreak per farm, the species distribution, and the density of windbreak coverage across several farms in the Cape Winelands area are summarized in Table 2.3. The windbreak density ranges from approximately 48 to 143 m length per hectare, with an average of 100.3 m ha^{-1}.

Fig. 2.5 Windbreak design forming roughly square or rectangular boundaries in the Western Cape. Photograph: Maik Veste

Table 2.2 Relationship between the length of windbreak per unit area of land and the distance between windbreaks (orchard cell size dimensions) for windbreaks arranged as either square (top half of table) or rectangular cells (bottom half of table). The shaded areas correspond to windbreak densities of 100 m ha^{-1}

Square cells size (m)	Windbreak length (m)	Orchard area (ha)	Running m ha^{-1}
50 × 50	100	0.25	400
100 × 100	200	1.00	200
200 × 200	400	4.00	100
400 × 400	800	16.00	50
Rectangular cells size (m)	Windbreak length (m)	Orchard area (ha)	Running m ha^{-1}
25 × 100	125	0.25	500
50 × 200	250	1.00	250
100 × 400	500	4.00	125
125 × 500	625	6.25	100
200 × 800	1000	16.00	62.5

On most farms in the district with young or immature windbreaks, *Populus simonii* is increasingly preferred over *Casuarina cunninghamiana*, since it is clearly less competitive with adjacent orchards and vines (personal observations on Rust-en-Vrede, Babylonstoren, Delheim, and Hillcrest farms). Our age class distribution data support this observation. The fact that *P. simonii* is deciduous also makes it less competitive during the winter and early spring seasons. Furthermore, we also observed that *Alnus cordata* is less competitive than *Casuarina*, but this species is less well known to most farmers and nursery managers that grow windbreak species (telephonic interviews).

Table 2.3 Windbreak characteristics at the different locations

Location	Farm	Orchard Area (ha)	Running Meters of windbreak (m)			Total length of windbreaks (m)	Windbreaks length per unit area of land (m/ha)
			Casuarina	Populus	Alnus		
Wolseley	Verrekyker 1/338	72	6487			6487	90.1
Tulbagh	portion 5/62	37	3512			3512	94.9
Stellenbosch	Rust-en-Vrede	36	3799			3799	105.5
Stellenbosch	Nooitgedacht	86	2104	4922		7026	81.7
Stellenbosch	Slaley	33	2920		1783	4703	142.5
Wolseley	Romansrivier	76	7117			7117	93.6
Wolseley	La Plaisant	208	9901			9901	47.6
Franschhoek	Portion 10/1655	15.4	1629			1629	105.8
Franschhoek	Boschendal portion 1/1647	30	3721			3721	124.0
Franschhoek	Boschendal portion 1510	92	6994			6994	76.0
Pniel	Hillcrest portion RE/400/124	7			989	989	141.3

2.5 Windbreak Growth Rate

Species typically utilized in windbreaks, such as *Casuarina cunninghamiana*, *Populus simonii,* and *Alnus cordata* tree breaks were selected and measured. Furthermore, *Pinus radiata* was extensively used for windbreaks in the past, but it was gradually replaced first with *C. cunninghamiana* and later with *A. cordata* and *P. simonii.* For this reason, *P. radiata* data on tree height growth were sourced from the commercial forest plantation industry to serve as comparison to the other species in this study. For the three key species, windbreaks of varying ages were selected using a chronosequence approach across all ten study farms. An attempt was made to measure the widest possible age range. The ages were sourced from the landowners, or the best available historical Google Earth images where landowners' records were incomplete.

The ages of windbreaks included in our study in the Cape Winelands are indicated in Table 2.4. Ages for *P. simonii* are mostly younger than ten years, with only one site at age twenty. This is an indication that *P. simonii* was introduced as a species suitable for windbreaks in the region only during the last twenty years. On the four available sites for *A. cordata,* the ages were all twenty years and older. No windbreaks of *A. cordata* that are younger than twenty years could be identified in the study area. A search for the species in tree nurseries in the Western Cape confirmed that the species are not actively grown by nurseries anymore and were not sold from nurseries in recent years.

The tree height (Ht, in meters) and diameter at breast height (DBH, in cm) as well as tree spacing within the windbreak row were measured. The latter was estimated by measuring the distance between 11 originally planted tree positions in a windbreak row (i.e., a distance that contains 11 trees and 10 spaces among tree positions) and dividing the distance by 10. The total number of windbreaks measured were thirteen sites of *C. cunninghamiana,* six sites of *P. simonii,* and four sites of *A. cordata.* On all the sites, diameter at breast height (DBH) and total tree height (HT) were measured. The average HT per windbreak was calculated. The age/HT pairs per species were used to calculate an age/HT regression function per species. The rainfall and temperature conditions across the farms where measurements were taken varied in a fairly narrow band (Table 2.4). Seeing that there were only small variations in climate across sites, it was deemed acceptable to bulk all the height and DBH measurements together in one dataset, using the chronosequence approach.

The height growth of trees in in a windbreak is seen as a key element in the calculation of the effect and protection of the crop on the downwind side of the windbreak. The height development of a specific tree species will be the basis of models to predict at what age, and distance from the windbreak, the reduction in wind speed will peak. The quantification of indirect effects of the windbreak, i.e., the reduction in evapotranspiration, increase in temperature, conservation of soil moisture, utilization of biomass, will all rely on the height growth model per species.

This study provides the regression functions that can be used to predict height growth of *C. cunninghamiana, P. simonii* and *A. cordata* in the Western Cape. For this

Table 2.4 Windbreak age class distribution and location of sites included in the height growth and development study. See Fig. 2.2 for site locations

Age years	Alnus cordata	Casuarina cunninghamiana	Populus simonii
2		Nooitgedacht	Nooitgedacht
3			Nooitgedacht
4			Nooitgedacht
6		Rust-en-Vrede	Nooitgedacht
7		Nooitgedacht, Rust-en-Vrede	Nooitgedacht
8		Nooitgedacht	
20	Slaley		
22			Babylonstoren
26		Nooitgedacht	
29		Slaley	
30	Slaley	Boschendal, Rust-en-Vrede	
31	Slaley		
33		Slaley	

reason, the average height is presented and not the dominant height. It can be clearly seen that *C. cunninghamiana* had the fastest height growth, averaging 24 m at age 30. *P. simonii* reached 15 m at age 22 years on one site. *Alnus cordata* showed the slowest height growth rate over time and reached only 10 m at a mature age of 30 years. The average height development of *P. radiata* calculated with standard growth models for the species in plantation conditions and for an average site index applicable for the Stellenbosch area is included in the graph in Fig. 2.6 as a benchmark of height growth for a well-known commercial tree species. It is important to note that the *P. radiata* growth model was developed in plantation conditions and the effect of competition will not be the same as in a single- or double-row windbreak. Furthermore, the *P. radiata* trees were grown from seedlings approximately 0.25 m tall when planted, whereas most windbreak trees are mostly planted as 1 to 1.5 m tall trees. *Pinus radiata* showed the expected logarithmic function for height development over time, while the best fit regression lines for the three windbreak species were power functions. The height over age regression functions for *A. cordata, C. cunninghamiana,* and *P. simonii* are indicated in Table 2.5.

The time taken to reach heights of 10 m varies among species and can be estimated from Fig. 2.7 as approximately 8, 13, and 28 years for *C. cunninghamiana, P simonii,*

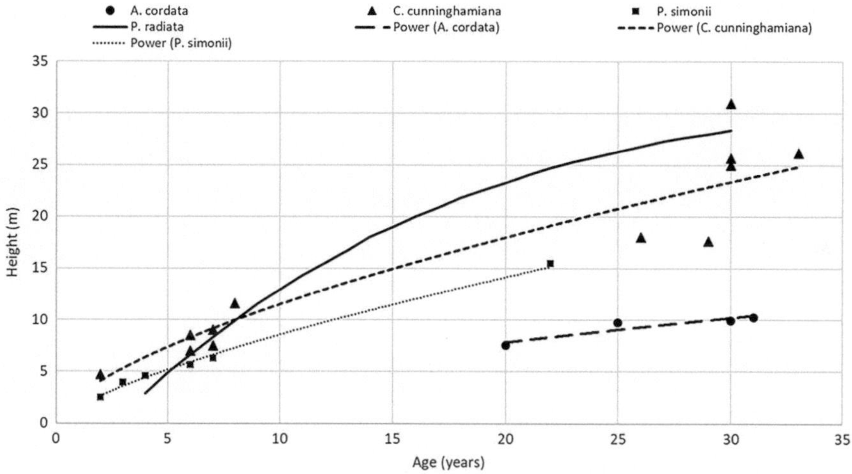

Fig. 2.6 The height growth over age for *P. radiata* as a benchmark species (solid line) and three windbreak species with measured heights and regression lines (dotted lines)

Table 2.5 The height over age regression function for the three windbreak species

Species	Alnus cordata	Casuarina cunninghamiana	Populus simonii
Regression function	HGT = $1.0845(\text{age})^{0.6594}$	HGT = $2.6086(\text{age})^{0.6445}$	HGT = $1.6116(\text{age})^{0.7251}$
R^2	0.8681	0.9321	0.9893

and *A. cordata,* respectively. It follows that *A. cordata* windbreaks will take much longer to become mature and afford good protection as far as 50 m downwind from the break, but once mature, they will endure for longer before they become too tall and large where trimming and/or replacement is required. The slow growth rate for *A. cordata* may be one of the reasons why it is less desirable for new windbreaks in fruit orchards and vineyard. However, it brings up the question whether multi-species rows of trees in windbreaks could not form a more effective windbreak barrier over time that has more flexibility to be managed as the trees age. This is a question that needs further investigation.

The development of tree dbh and tree volume over time is shown in Fig. 2.7. At 30 years of age, the casuarina trees are substantially larger than the alders and the plantation-grown pines. Note that as competition sets in for the plantation-grown pines, the rate of DBH development over age slows down. At 30 years of age, *C. cunninghamiana* trees attain a large size (dbh exceeding 40 cm from Fig. 2.7). Trees of this size (Fig. 2.8) have a very high percentage recovery when used for solid wood products, and it is also comparatively cheap per unit volume to harvest these sizes with the appropriate (heavy) equipment. This means that there is substantial carbon sequestration potential and income that can be generated for the farmer from harvested wood products when the windbreak reaches the end of its life and needs to be replaced.

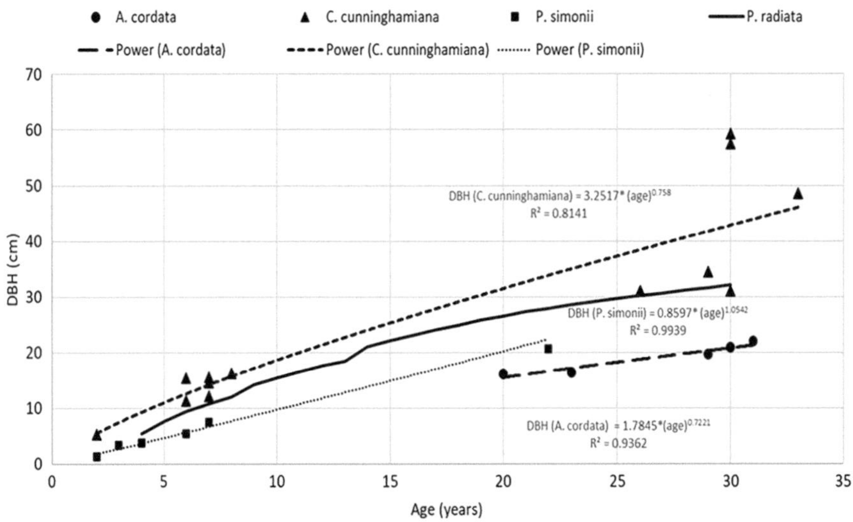

Fig. 2.7 Development of diameter at breast height (DBH) in three windbreak species contrasted to plantation-grown *Pinus radiata* on sites of comparable quality in the Western Cape. Note that the kinks in the *P. radiata* line at ages 8 and 13 years represent thinnings from below, resulting in a sudden increase in average tree diameter for the trees remaining on-site in the subsequent measurement year

Fig. 2.8 Larger *Casuarina cunninghamiana* trees can be used for solid wood products. Photograph: Maik Veste

2.6 Survival of Planted Trees

2.6.1 Experimental Approach

The survival of established windbreaks is another important factor for tree selection. To test the survival of the trees, at three sites around Stellenbosch, test trials were established:

(a) **Welgevallen Experimental Farm at Stellenbosch.** The windbreak is positioned next to fruit orchards on Welgevallen, and the objective is to protect the orchards from southeasterly winds that is expected to cause wind damage to the fruit (Chap. 7).
(b) **Backsberg Farm**. The trial is positioned next to a road and orientated to measure the protection against the northwesterly winds. The Backsberg trial was planted in three rows, with the objective to develop a management strategy, which will include the felling and replacement of one of the three lines at specific intervals, to prevent the windbreak to grow to very big tree sizes and the subsequent competition to the adjoining crop.
(c) **Mariendahl Experimental Farm**. The trial site was not suited to the typical windbreak design with two or three lines next to an orchard, but the site was used to plant the same species, which were used on Welgevallen and Backsberg, in square plots and replicated to compare the growth rate between species.

The location of the three sites can be seen in Fig. 2.2. Due to a limited number of plants per species and available area on all three trial sites, all species could not be replicated on all sites. Table 2.6 indicates how the species were distributed between the three trial sites. The Welgevallen site is the only site where indigenous species *Searsia lancea*, *Kiggelaria africana,* and *Catha edulis* were planted to see how these indigenous species will compare to Eucalypt species and traditional windbreak species like *P. simonii*. Depending on their growth rate and growth habit, it is also possible that the indigenous species can be planted in complementary fashion among traditional windbreak species. Additional exotics were also included for testing. *Eucalyptus cloeziana* is known for excellent form and the utilization as high-value poles in the Limpopo province, and it was included at Backsberg and Mariendahl to test the adaptability of the species in the Western Cape. *Eucalyptus cladocalyx* is the only eucalypt species that was traditionally planted on farms in the Western Cape and the only species that was replicated on all three sites.

The plants were sourced from different nurseries, and they were all at different sizes at planting, i.e., average height, six months after planting range from 43 cm (*E. cloeziana*) to 2 m (*C. cunninghamiana*). The height growth in this initial establishment report can therefore not be used to compare species. The survival and height *increment* will therefore be discussed under the results. The Welgevallen trial received irrigation on the same schedule as the adjoining orchard and the Backsberg trial was manually watered during the first dry season after planting. The Mariendahl trial did

Table 2.6 Detail of establishment of three windbreak species trials

Trial/location	Welgevallen	Backsberg	Mariendahl
Planting date	June 2019	June 2019	July 2019
Trial design	Two rows	Three rows	Square plots
Species	*Eucalyptus cladocalyx*	*E. cladocalyx*	*E. cladocalyx*
	E. grandis x resinifera	*E. grandis x urophylla*	*E. grandis x urophylla*
	E. grandis x tereticornis	*E. cloeziana*	*E. cloeziana*
	Populus simonii	*Corymbia henryi*	*Corymbia henryi*
	Sercia lancea	*Casuarina cunninghamiana*	*Casuarina cunninghamiana*
	Catha edulis	*Populus simonii*	
	Kiggelaria africana		
Espacement	2 m between × 1 m within rows	2 m between × 1 m within rows	3 × 2 m
Fertilizer	Superphosphate 100 g/tree	Superphosphate 100 g/tree	Superphosphate 100 g/tree
Survival and height measurement	January 2020 June 2020 March 2021	January 2020 June 2021 October 2021	January 2020 June 2021

not receive any additional watering after planting. The vegetation on all three sites was grass before planting. Site preparation before planting includes ploughing and/or strip spray at Welgevallen, strip spray and/or hoe at Mariendahl, and ploughing at Backsberg; follow-up weeding with manual hoeing was done when necessary. Blanking was done on all three sites during the next winter in 2020 with big plants in one-litre bags, grown in the DFWS nursery, for the specific reason to do a late blanking (where necessary) during the rainy season. In some cases, where the plants for blanking were not enough, blanking was done with filler plants of other species or hybrids, so as to maintain a level of competition among trees. The survival percentage for two years after planting includes the blanked plants where late blanking was done with the same species or hybrid, and excludes blanked plants when blanking was done with a different species or hybrid. Survival counts and height measurements were done on the ages as indicated in Table 2.6. The Mariendahl site received only one height measurement six months after planting. As discussed before, early height measurements are strongly influenced by tree size at planting because the different species that could be obtained for planting were of varying sizes. This is the reason why the Mariendahl site (Fig. 2.11) does not include height increment data. The height increment measurements included in Figs. 2.6 and 2.7 were calculated from two measurements, over the last 13 months (Welgevallen) or 22 months (Backsberg).

2.6.2 Results of the Trials

Survival at Welgevallen and Backsberg was at acceptable levels above 80% for all species except for E. *cloeziana* and *P. simonii.* The poor survival of P. simonii was unexpected, compared to the general survival of the species observed in commercial windbreaks in the same area. The results confirm an assumption that *P. simonii* is dependent on regular irrigation as is the case with commercial windbreaks. The poor survival of *E. cloeziana* was not unexpected. The species is a tropical species that grew well in the Limpopo province and was never before planted and tested in the Western Cape. Reasons for the poor survival of all species on the Mariendahl site (all species below 80%) might include temporarily flooding after the planting due to an impenetrable clay layer in the subsoil, a very dry first summer season with no irrigation and too much suppression from competing grass vegetation.

For the investigated tree species, we can summarize:

(a) ***Casuarina cunninghamiana*** Survival range from 69% at Mariendahl to 99% at Backsberg. Reasons for the low survival at Mariendahl were no watering after planting as well as heavy browsing of the tops by cattle that accidentally got access to the trial site. The plants were the tallest at planting (2 m), but the initial growth was substantially lower than the eucalypt species (Fig. 2.9).

(b) ***Populus simonii***: Survival was poor at Backsberg (21%) and Welgevallen (35%) despite initial watering. The poor survival is much worse than the actual survival found on farms in the same area. The only possible reason for the poor survival

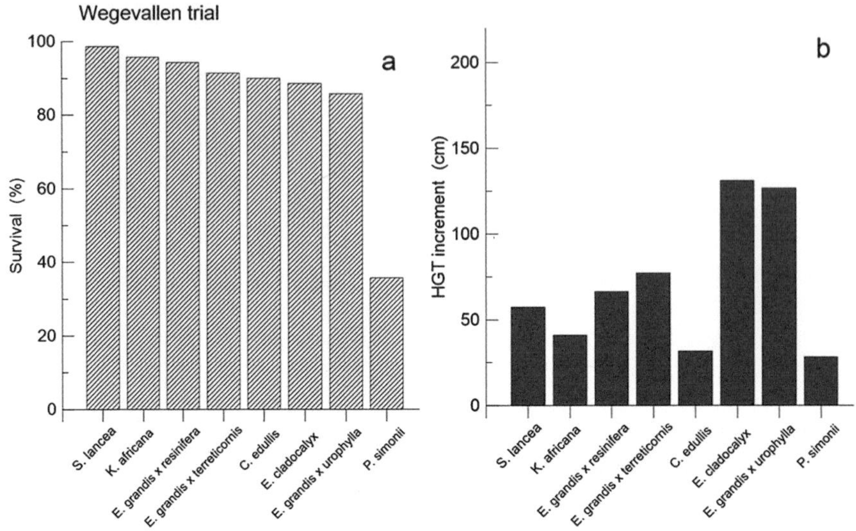

Fig. 2.9 Welgevallen trial: **a** survival % 21 months after planting and **b** increment in height growth during the last 13 months

was that watering was stopped on both trial sites after the first summer, generally compared to the continuation of watering on the same schedule as the adjoining orchard on a commercial farm. The increment in height growth of the small percentage of trees that did survive was also the lowest of all the species.

(c) ***Eucalyptus cladocalyx***: *Survival range from 50% at Mariendahl to 87% at* Welgevallen and 88% at Backsberg. The increment in height growth after planting was the highest of all the species with 204 cm after 22 months at Backsberg and 131 cm after 13 months at Welgevallen (Figs. 2.8 and 2.9).

(d) ***Eucalyptus grandis x urophylla***: Survival range from 77% at Mariendahl to 85% at Welgevallen and 99% at Backsberg. The increment in height growth after planting was the second best after *E. cladocalyx,* at Welgevallen and third best after C. henryi at Backsberg (Figs. 2.8 and 2.9).

(e) ***Eucalyptus cloeziana***: Survival range from 40% at Mariendahl to 77% at Backsberg. The height growth was the poorest of the four eucalypt species.

(f) **Corymbia** ***henryi***: Survival range from 68% at Mariendahl to 85% at Backsberg. Increment in height growth was the second best after *E. cladocalyx*, at Backsberg (Figs. 2.8 and 2.9).

(g) **Indigenous species**: The survival of the three indigenous species was good as expected from 89 to 97%, but the height growth was substantially lower than the eucalypt species at Welgevallen (Figs. 2.8, 2.10).

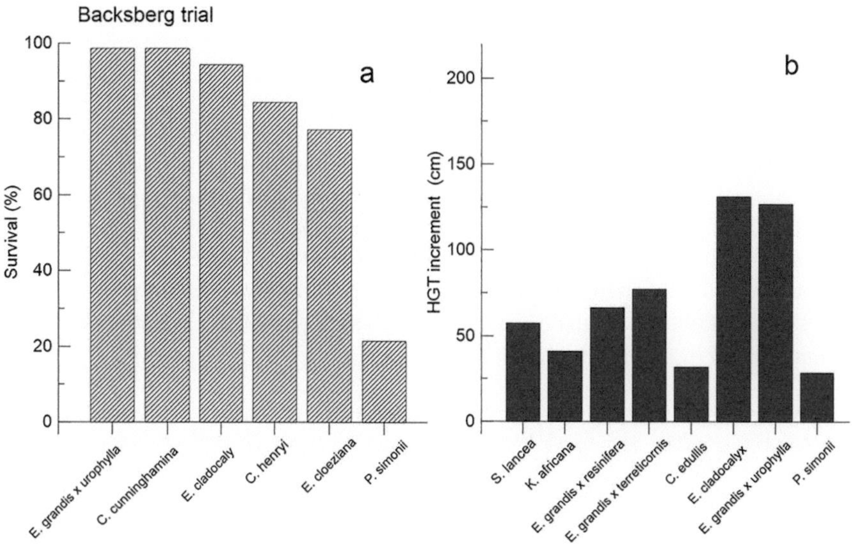

Fig. 2.10 Backsberg trial: **a** survival %, 27 months after planting and **b** increment in height growth during the last 22 months

Fig. 2.11 Mariendahl trial: survival %, 24 months after planting

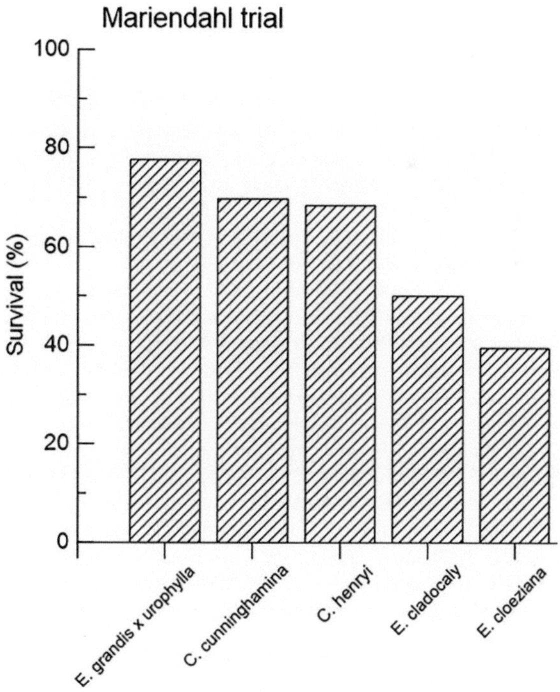

2.7 Conclusions

The height growth of trees in a windbreak is seen as a key element in the calculation of the effect and protection of the crop on the downwind side of the windbreak. The height development of a specific tree species will be the basis of models to predict at what age, and distance from the windbreak, the reduction in wind speed will peak. The quantification of indirect effects of the windbreak, i.e., the reduction in evapotranspiration, increase in temperature, protection of soil moisture, and utilization of biomass, will all rely on the height growth model per species. This study provides the regression functions that can be used to predict height growth of *C. cunninghamiana*, *P. simonii,* and *A. cordata* in the Western Cape.

Acknowledgements We thank the farmers of the different farms for their permission for the survey and, especially, Michael Back (Backsberg, Paarl), Ian de Villiers (Babylonstoren, Paarl), and Johan and Gideon Roos (Paul Roos Wine, Rust en Vrede, Stellenbosch) and the experimental farms Welgevallen and Mariendahl, Stellenbosch University for their support for the field experiments. The project was supported by the Federal Ministry of Education and Research (BMBF) and managed by PT-DLR (grant no. FKZ 01LZ1711A-F) as part of "CLIENT II—International Partnerships for Sustainable Innovations" funding initiative.

References

Amichev BY, Bentham M J, Kurz WA, Laroque C P, Kulshreshtha S, Piwowar J M, Van Rees, KC J (2016) Carbon sequestration by white spruce shelterbelts in Saskatchewan, Canada: 3PG and CBM-CFS3 model simulations. Ecol Model 325:35–46. https://doi.org/10.1016/j.ecolmodel.2016.01.003

De Lange WJ, Veldtman R, Allsop MH (2013) Valuation of pollinator forage services provided by Eucalyptus cladocalyx. J Environ Manag 125:12–18

Geldenhuys H, Lötze E, Veste M (2023) Fruit quality and yield of mandarin (*Citrus reticulata*) in orchards with different windbreaks in the Western Cape, South Africa. Erwerbs-Obstbau. https://doi.org/10.1007/s10341-022-00725-3

Hassler SK, Bohn Reckziegel R, du Toit B, Hoffmeister S, Kestel F, Kunneke A, Maier R, Sheppard JP (2024) Multivariate characterisation of a blackberry–alder agroforestry system in South Africa: hydrological, pedological, dendrological and meteorological measurements. Earth Syst Sci Data 16:3935–3948. https://doi.org/10.5194/essd-16-3935-2024

Hoffmeister, S., Bohn Reckziegel, R., du Toit, B., Hassler, S. K., Kestel, F., Maier, R., Sheppard, J. P., and Zehe, E. (2024) Hydrological and pedological effects of combining Italian alder and blackberries in an agroforestry windbreak system in South Africa, Hydrol. Earth Syst. Sci., 28, 3963–3982, https://doi.org/10.5194/hess-28-3963-2024

Huth NI, Robertson MJ, Poulton PL (2010) Regional differences in tree-crop competition due to soil, climate and management. Crop Pasture Sci 61(9):763–770. https://doi.org/10.1071/CP09254

Potgieter LJ, Richardson DM, Wilson JRU (2014) *Casuarina cunninghamiana* in the Western Cape, South Africa: determinants of naturalisation and invasion, and options for management. S Afr J Bot 92:134–146

Sheppard JP, Bohn Reckziegel R, Borrass L, Chirwa PW, Cuaranhua CJ, Hassler SK, Hoffmeister S, Kestel F, Maier R, Mälicke M, Morhart C, Ndlovu NP, Veste M, Funk R, Lang Seifert F, du

Toit TB, Kahle HP (2020) Agroforestry: an appropriate and sustainable response to a changing climate in Southern Africa? Sustainability 12(17):6796. https://doi.org/10.3390/su12176796

Sheppard JP, Larysch E, Cuaranhua CJ, Schindler Z, du Toit B, Malherbe GF, Kunneke A, Morhart C, Bohn Reckziegel R, Seifert T (2024) Kahle H-P (2024): Assessment of biomass and carbon storage of a Populus simonii windbreak located in the Western Cape Province, South Africa. Agroforest Syst 98:697–714. https://doi.org/10.1007/s10457-023-00940-1

Veste M, Schillem S, Böhm C (2018) Baumarten für die Agrarholzproduktion. In: Veste M, Böhm C (eds) Agrarholz—Schnellwachsende Bäume in der Landwirtschaft. Springer Spektrum, Heidelberg. 80–118. https://doi.org/10.1007/978-3-662-49931-3_4

Veste M, Littmann T, Kunneke A, Du Toit B, Seifert T (2020) Windbreaks as part of climate-smart landscapes reduce evapotranspiration in vineyards, Western Cape Province, South Africa. Plant Soil Environ 66:119–127. https://doi.org/10.17221/616/2019-pse

Veste M, Sheppard JP, Abdulai I, Ayisi KK, Borrass L, Chirwa PW, Funk R, Kapinga K, Morhart K, Mwale SE, Ndlovu NP, Nyamadzaw G, Nyoka BI, Sebola P, Seifert T, Senyolo MP, Sileshi GW, Syampungani S, Kahle H-P (2024) The need for sustainable agricultural land-use systems: benefits from integrated agroforestry systems. In: von Maltitz GP, Midgley GF, Veitch J, Brümmer C, Rötter RP, Viehberg, FA, Veste M (eds) Sustainability of southern African ecosystems under global change: science for management and policy interventions. Ecological studies 248, Springer Nature, Heidelberg, pp 587–623. https://doi.org/10.1007/978-3-031-10948-5_2

Von Maltitz GP, Midgley GF, Veitch J, Brümmer C, Rötter RP, Viehberg FA, Veste M (eds) (2024) Sustainability of Southern African ecosystems under global change, ecological studies 248, Springer, Cham. https://doi.org/10.1007/978-3-031-10948-5

Van Wyk B, van Wyk P (2013) Field guide to trees of southern Africa. Struik Nature, Cape Town

Chapter 3
Local Wind Field on Farms in the Cape Winelands

Thomas Littmann, Tobias Recke, and Maik Veste

Abstract The establishment of hedges and tree strips are important measures to minimize wind erosion and protect agricultural crops. In the Western Cape region, strong winds influence evapotranspiration during the growing season of different crops. The effects of windbreaks on the local wind field were investigated on two farms in the Cape Winelands: on Babylonstoren wine farm, Chinese poplars (*Populus simonii*) are planted as windbreaks for citrus plantations and vineyards, while evergreen beefwood (*Casuarina cunninghamiana*) hedges are used on Rust en Vrede to protect citrus plantations from wind damage. The analysis showed clear seasonal differences in wind speed and wind direction. Strong winds are most common in the Western Cape during the summer months, especially from November to February. The main wind direction in the summer months is primarily east-southeast to south-southeast, while the winds in winter come from north-northwest to west-southwest. In the vineyards, the wind reduction is almost 22% and particularly in the summer months with the strong winds from the south 24% to 29%. Within citrus orchards with beefwood windbreaks, local differences in the average wind speed between the measuring stations of 18% were observed during the summer months. The seasonal changes of the microclimate within the citrus plantations were attributed to the different weather conditions.

Keywords Windbreaks · Microclimate · Citrus orchard · Vineyard · *Casuarina* · *Populus*

T. Littmann (✉) · T. Recke
DLC—Dr. Littmann Consulting GmbH, Ennepetal, Germany
e-mail: littmann@littmann-consulting.de

M. Veste
CEBra—Center for Energy Technology Brandenburg e.V., Cottbus, Germany

Department of Molecular Botany, University of Hohenheim, Institute of Biology, Hohenheim, Germany

M. Veste et al. (eds.), *Trees in South African Agriculture*, Innovations in Landscape Research, https://doi.org/10.1007/978-3-032-00993-7_3

3.1 Introduction

The reduced wind speed and evaporation and transpiration from soils and plants are the main ecological benefits of tree shelterbelts and agroforestry systems (Kanzler et al. 2019; Veste et al. 2024). The reduction of wind speed on the downwind side of a shelterbelt is a function of distance, aerodynamic porosity and tree height (Littmann et al. 2004). Furthermore, the topography of the terrain and the surrounding landscape structures play an important role in shaping the microclimate. The local wind field is controlled by the regional wind conditions and the seasonally changing large-scale weather conditions (Chap. 4). To assess the effects of the different windbreaks (Chap. 5), spatially detailed measurements in the different stands are necessary. For the analysis of the effects of windbreaks on the local wind fields, experimental sites on two farms in the Cape Winelands (Meadows 2015). In both regions, tree belts are traditionally used to protect crops (Chap. 2).

3.2 Research Sites in the Cape Winelands

(a) Babylonstoren in the Paarl Wine Region

The commercial farm Babylonstoren (33°49′27.69”S, 18°55′19.38”E, altitude 194 m a.s.l.) is located in the Paarl wine district of the Cape Winelands Western Cape, South Africa. The vineyards are found on the lower slopes of Paarl Rock, on the northern side of Simonsberg (peak 1,399 m) (Chap. 1), which is part of the Cape Fold Belt and in the valley of the Berg River. The two mountains provide well-drained granite and shale soils for the vines, while the soil on the valley floor is more sandstone-based. Paarl's climate is relatively hot compared to the more coastal areas of the Western Cape. Long, hot summers and cold, rainy winters are typical of the area with an average annual rainfall of 770 mm. In a first approach, microclimatic measurements were installed within a 4.2 ha trickle irrigated vineyard (Fig. 3.1) with an NW–SE row orientation and a spacing of 2 m. The masts were located (i) 18 m north (hedge station) of the existing Chinese poplar windbreak (*Populus simonii*) parallel to the vineyard rows and (ii) as a reference in approximately 100 m distance in the open field (Fig. 3.2). The height of the tree shelterbeld was 6 m (Veste et al. 2020). Horizontal wind speed was recorded at 2 m and 6 m and wind direction at 6 m (RM Young Wind Sentry, USA). All microclimatic data were sampled at 1 Hz and were recorded as 5 min averages by a data-logger (Campbell Scientific CR1000, Logan, USA). The data were transferred from the data-logger to a server at Stellenbosch University by GSM-data-modem. Data availability over the measurement period was 100% in all cases. The first measurement period was April 2015 to March 2019 and the second from 2020 to 2021 for comparison with the nearby citrus orchard.

For second experiment, automatic microclimate stations were installed on 6 March 2020 and collecting data until January 2021 in the citrus orchard located (Fig. 3.2). The citrus plantation borders directly to the south of the poplar windbreak with a

Fig. 3.1 Location map of the wind measurements on the farms Babylonstoren (**a**) and Rust en Vrede (**b**)

height of 6 m and north of a 11–13 m height poplar tree strip (Fig. 3.1a). In the citrus orchard, wind speed was measured at 3 m height (Farm 1) and in the center of the orchard at 2 m height (together with other micoclimate data for the calculation of evapotranspiration, Chap. 5, Geldenhuys et al. 2023) and recorded as 10-min averages by a data-logger (ML- × 17, YDOC, The Netherlands) and transferred onto an online cloud server daily.

(b) Rust En Vrede Farm, Paul Roos Wine in the Stellenbosch Wine Region

In the Stellenbosch region, vineyards cover the gently rolling hills of Stellenbosch, from Helderberg in the south to the lower slopes of Simonsberg Mountain in the north. The region's climate is relatively hot and dry, although a maritime influence comes from False Bay in the south. Cooling southeasterly breezes wash through the vineyards in the afternoons, Mean annual in Stellenbosch 742 mm, respectively (Chap. 4).

Fig. 3.2 Set-up of microclimate measurements and wind speed and direction at the Farm Babylonstoren. **a, b** measurements near the hedge, **c** in the center of the vineyard, **d** in the citrus orchard

Wind measurements at the farm Rust en Vrede, Paul Ross Wine, Somerset West, were installed within a citrus orchard (34°00′42″S, 18°48′35″E, Fig. 3.1) surrounded in all directions by 8–10 m high beefwood (*Casuarina cunninghamiana*) hedgerows (higher hedges on the western and eastern sides; lower ones north and south). Wind speed was measured with an anemometer (Thies compact, Göttingen, Germany) at 3 m above canopy (Farm 3 North, Farm 4 South), and at 2 m within canopy (Farm 5) (Fig. 3.1, Fig. 3.3). Together with other microclimate parameters (Chap. 5), wind speed was recorded by a data-logger (ML- × 17, YDOC, The Netherlands) and transferred onto an online cloud server daily. The measurement period was March 2020 and January 2022. Additionally, a reference station (Fig. 3.3d) was installed north of the orchard in an open field (Fig. 3.1b). Wind direction and wind speed were measured with a all-in-one weather station (Atmos 41, Meter, Pullmann, USA, München, Germany).

Fig. 3.3 Set-up of microclimate measurements and wind speed and direction in the citrus orchard at the Farm Rust en Vrede: **a, b** Farm 3 and Farm 4 measurements near the windbreak, **c** Farm 5 in the center of the orchard, **d** reference station outside the orchard. For location see Fig. 3.1

3.3 Wind Patterns at the Farm Babylonstoren

3.3.1 Vineyard

Figure 3.4 shows the daily mean wind speeds at the measuring points Field (US 1; Reference measurement in the center of the field) and Hedge (18 m north of the poplar row; US 2), which were determined from 10 min raw data.

The time series show a relatively homogeneous seasonal structure over the entire period and do not show a significant trend. To check the validity of the measurement data, correlation calculations were carried out with the nearest node for re-analysis data (MERRA 2-daily means for 10 m height). Figure 3.5 shows the result for the monthly averages of the Field US 1 wind measurement and the long-term monthly averages of the MERRA 2 data; extrapolated for the period 1994–2019 (26 years). As expected, the correlation shows systematic deviations, but is highly significant at 89.3%. The representativeness of the in situ wind measurements can therefore be assumed.

In the case of daily values, the correlations are expected to be smaller due to the higher dispersions due to the location (Table 3.1). The very significant connection between the wind speeds of the two measuring points at a height of 6 m and the

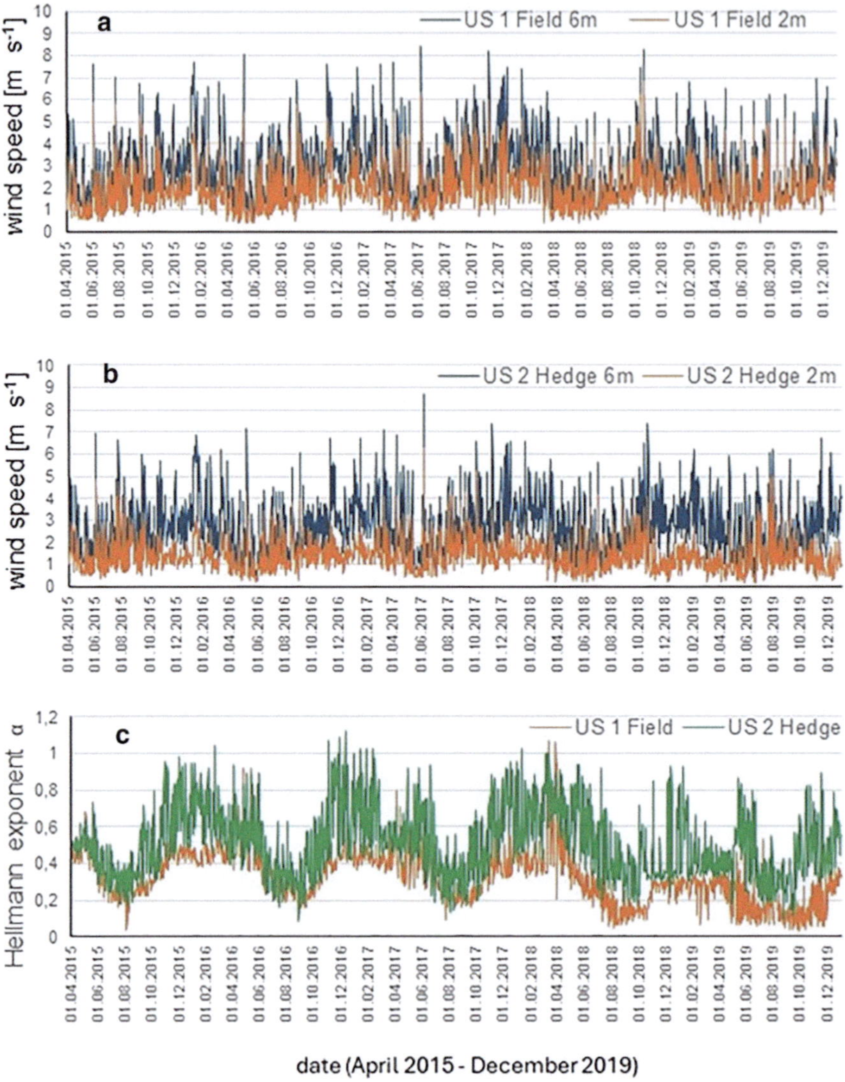

Fig. 3.4 Average daily wind speed at the measurement point at 2 m and 6 m height in the vineyard **a** in field center US 1, **b** near the hedge US 2 and, **c** the calculated daily Hellmann exponent in the vineyard at Babylonstoren April 2015–December 2019

2 m reference measurement (Field US 1) with the 6 m measurement near the hedge is striking. This speaks for a relatively undisturbed structure of the local wind field at a height of 6 m and for no influence on the 2 m measurement of the reference point by the hedge. However, the correlations with that of the 2 m measurement near the hedge are much smaller, even with the 6 m measurement at the same measuring

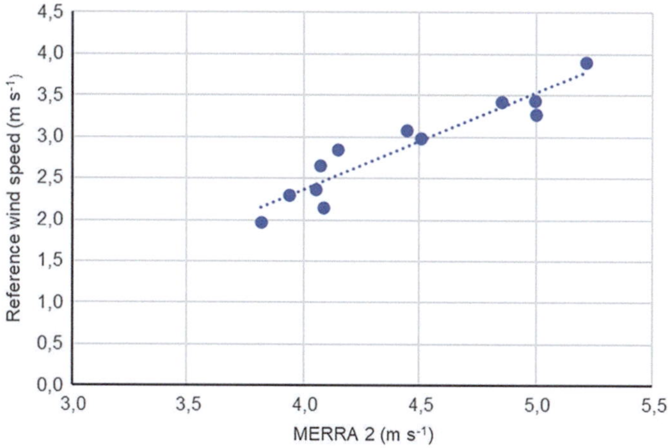

Fig. 3.5 Correlation of the mean reference wind speed at 6 m (Field US 1) with mean MERRA 2 at 10 m-re-analysis data (1994–2019) at the vineyard in Babylonstoren ($R^2 = 0.8934$)

point. This cannot be explained by the direct reduction of the wind speed by the canopy but must be due to the decelerating effect of the windbreak hedge.

This finding was confirmed by calculations of the Hellmann exponent from the power law of the vertical increase in wind speed above ground (Fig. 3.4c). The Hellmann exponent (wind gradient) is used to determine the increase in wind with height and is dependent on the roughness of the ground.

The exponents are generally very high immediately above the canopy of the vines and even reach values > 1.0 on days with very low wind speeds (< 1 m s^{-1}), which otherwise apply to surface roughness of an inner-city development. However, the following figure shows the very clear differences between the two measuring points. Due to the strong reduction in wind speed at a height of 2 m in the hedge measurement, there is a significantly higher wind gradient up to the measurement height of 6 m compared to the reference measurement (Table 3.2).

The results also reflect the seasonal wind regime of the Western Cape region (Chap. 4). The maximum wind speeds are in the southern summer (October to March; especially from November to February) and the minimum in the southern winter (May to August). This seasonal structure is controlled by the seasonal change of the pressure centers over the Kalahari in the north (heat low in summer, anticyclones

Table 3.1 Correlation coefficients (r^2) of wind measurement series at the vineyard at Babylonstoren

r^2	Field 2 m	Hedge 6 m	Hedge 2 m	MERRA 2
Field 6 m	0.93	0.98	0.73	0.47
Field 2 m		0.94	0.82	0.43
Hedge 6 m			0.79	0.50
Hedge 2 m				0.38

Table 3.2 Mean monthly wind speeds differences between 6 and 2 m height and Hellmann values at the reference in the field center (US 1) and near the hedge (US 2) of the long-term measurements (2015–2019) in the vineyard at Babylonstoren

	US 1 Field		US 2 Hedge		
Month	Wind Gradient 6 m–2 m [m s^{-1}]	Hellmann exponent	Wind Gradient 6 m–2 m [m s^{-1}]	Hellmann exponent	MERRA 2 1994–2019
Jan	1.37	0.40	1.86	0.65	5.22
Feb	1.19	0.40	1.59	0.61	5.00
Mar	1.10	0.45	1.32	0.58	4.51
Apr	0.96	0.44	1.11	0.55	4.07
May	0.65	0.40	0.85	0.59	3.82
Jun	0.55	0.29	0.74	0.45	4.09
Jul	0.49	0.23	0.65	0.35	3.94
Aug	0.47	0.20	0.66	0.32	4.06
Sep	0.55	0.20	0.92	0.39	4.15
Oct	0.76	0.26	1.21	0.50	4.45
Nov	1.09	0.35	1.52	0.58	4.86
Dec	1.13	0.39	1.54	0.63	5.00
2015–2019	0.86	0.33	1.17	0.52	4.43

in winter) and the winter cyclogenesis in the South Atlantic. Depending on the characteristics of these controlling pressure structures, this leads to a bidirectional wind regime with southerly wind directions in summer and northerly wind directions in winter (Chap. 4) (Figs. 3.6 and 3.7).

However, the summer half of the year is also the growing season with foliage of the vines from October to April (irrigation is already stopped for the harvest in March). The increased aerodynamic roughness during the foliage phase has no direct influence on the reduction of the wind speed directly above the canopy (2 m measurements), but on the gradient of the wind profile: the Hellmann exponents in the summer half of the year are considerably higher than those of the winter half of the year (Table 3.2).

In any case, the influence of the wind speed directly above the stand height by the windbreak hedge, which is independent of the stand, has been proven by the measurements taken so far. Even at a height of 6 m, slight reductions in wind speed at the hedge site can still be observed. However, the reduction of the wind speed at a height of 2 m is quite considerable at almost 22%, whereby especially in the summer half year with the predominance of the strongly drying southerly winds, there are reductions of 24% to 29%. To the same extent, the increase in wind speed is reduced with height.

The annual wind direction distributions (Fig. 3.8) did not show any significant differences over the measurement period. At the reference measurement in the center

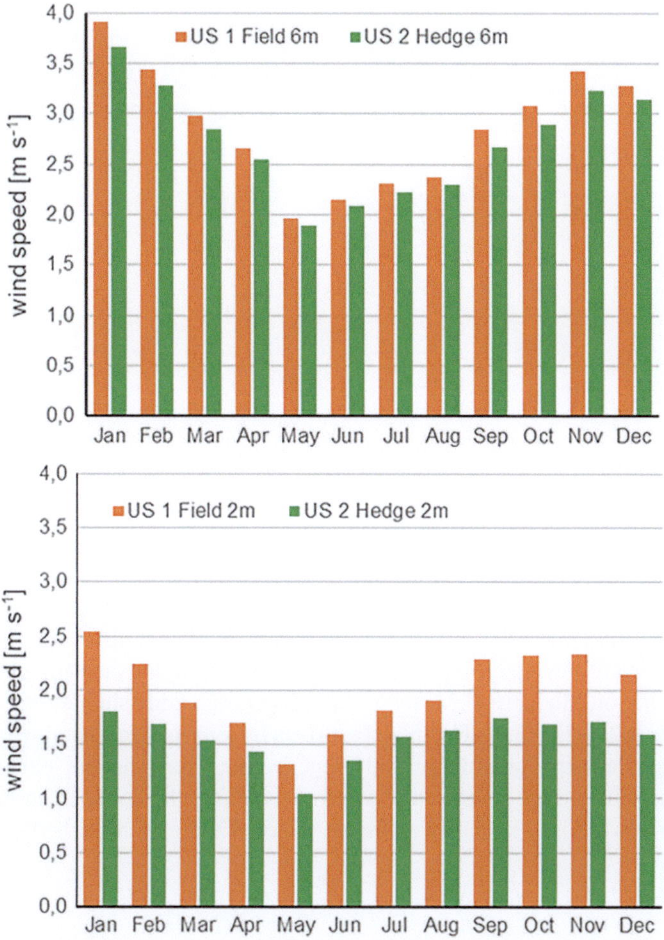

Fig. 3.6 Annual course of mean wind speed at 2 m in the vineyard (US1 center, US2 near the hedge) 2015–2019

of the field, the main wind direction in the summer half year is east-southeast to south-southeast, secondarily in the winter half year in the sectors from north-northwest to west-southwest. However, this does not quite correspond to the long-term distributions of the regional weather stations (e.g., Paarl with the main wind directions south and southwest, north and northeast). Local terrain effects or the influence of the existing building on the measurement at a height of 6 m can play a role here. A clear indication of the turbulent influence of the wind direction measurement at the hedge station is given by the deviation of the frequency distributions with a summer maximum at east-northeast and north-northeast, followed by the secondary maximum from west-southwest to south-southeast. This is a twisting of the wind

Fig. 3.7 Relative difference of wind speed at the reference measurement (field) and at the hedge measurement. Monthly means 2015–2019

when flowing over or toward the windbreak hedge (typical right-hand rotation when flowing over an obstacle).

The measured effects of windbreak hedges were also confirmed by numerical modeling. The following example shows a model run applying the Parallelized Large Eddy Simulation Model (PALM) that is based on Large Eddy simulations (Maronga et al. 2015, 2020) (Fig. 3.9). The model is driven by long-term climatological parameters and a parametrization of crop and hedge features (stand height, leaf area index, root depth). The leeside reduction of wind speed along the hedge in case of south-easterly winds is visualized as well as the effect of the gap in the hedge along the road.

3.3.2 Citrus Orchard

The second wind measurements implemented in the citrus stand at Babylonstoren (Fig. 3.2d) make it possible to compare the wind conditions directly above the canopy of the citrus orchard and the vineyard. The measurements US 1 and US 2 are above the stand height of the wine (2 m); the Farm 1 and Farm 6 measurements above the stand height of the Citrus (3 m). Farm 1 is located about 20 m north of the hedge on the south side of the citrus field; Farm 6 is in the middle of the stand (Fig. 3.1a). For this evaluation, the period April 2020 to March 2021 was considered to analyze two complete meteorological half-years (winter: April to September 2020; Summer: October 2021 to March 2021) (Fig. 3.10).

The mean wind speeds (01.04.2020–31.03.2021) show significant differences (Table 3.3): US 2 is about 30% below the comparative speed of the reference measurement US 1; the wind speeds in the citrus stand are considerably lower than in the

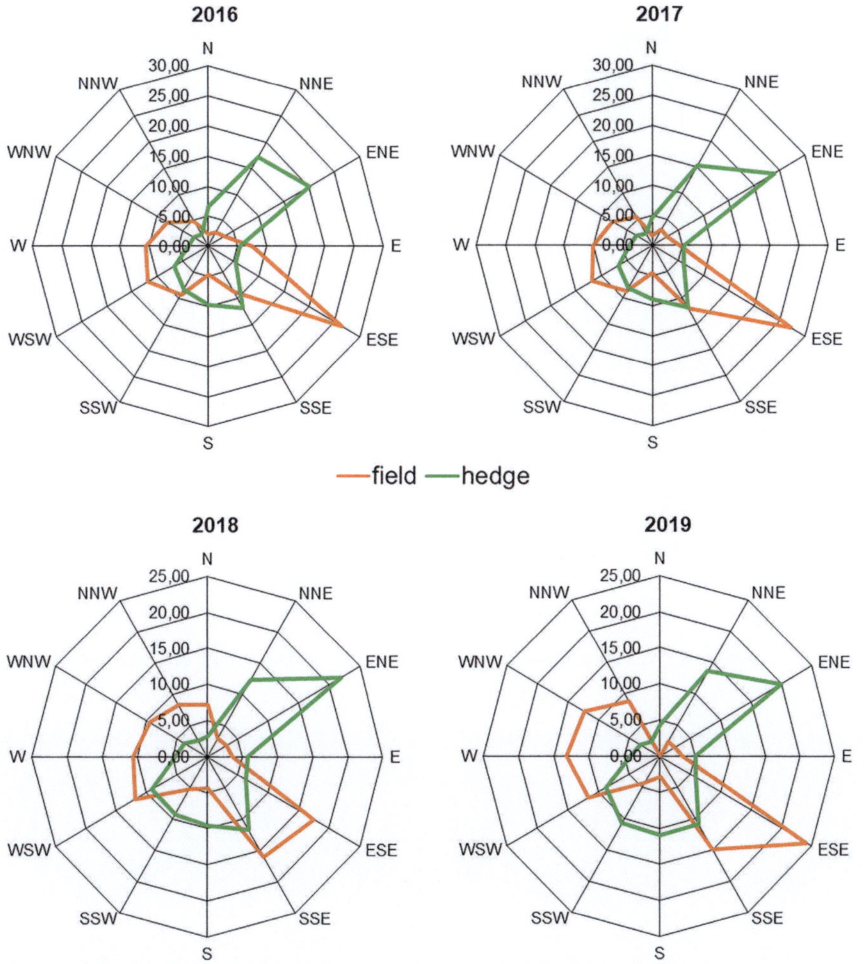

Fig. 3.8 Annual wind direction frequencies at the measurement stations in the center of the field site and near the hedge in the vineyard at Babylonstoren between 2016 until 2018

relatively open vineyard and obviously depend more on the position within the stand than on the distance to the nearest hedge (farm 6 in the middle of the stand shows the lowest wind speed). Seasonally, there are significantly higher wind speeds in summer compared to winter; only in the center of the citrus orchard are there no differences in the generally very low wind speeds. Mean maximum wind speeds (10 min average) show no clear seasonal differentiation; they are in the vineyard and in the citrus orchard in an order of magnitude in which fruit damage may be expected.

Despite the clear differences in mean wind speed, it is striking that the correlative interrelation of the measuring points is significant with the only exception of station US 2 (Table 3.3). From this, it can be deduced that the flow through and over the

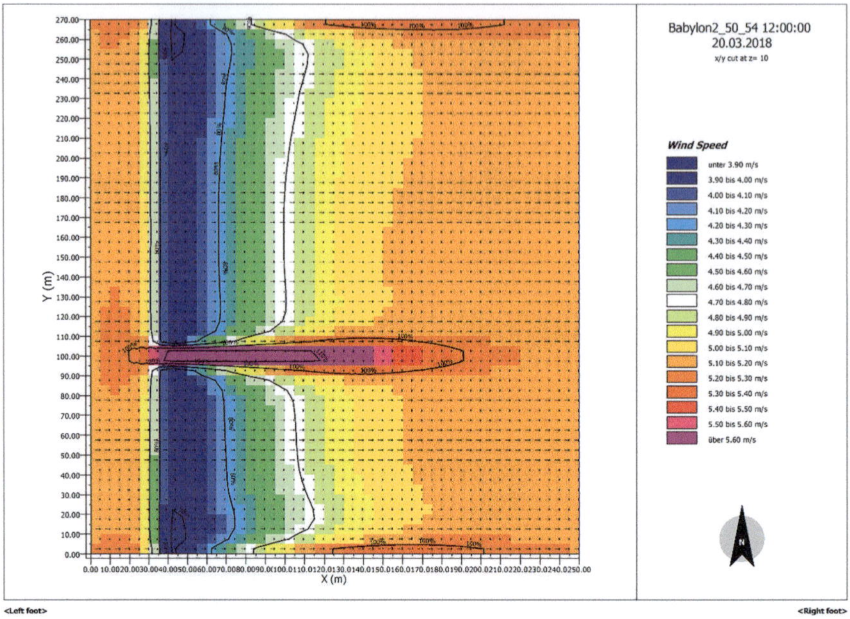

Fig. 3.9 Example of PALM model run of leeside wind speed at 10 m in case of southeasterly wind at Babylonstoren

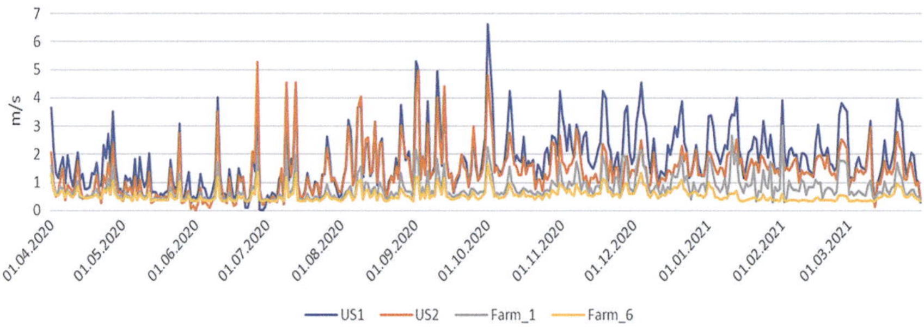

Fig. 3.10 Time series of wind speed in the vineyard (US 1 and US 2) and in the citrus orchard (Farm 1 and Farm 6) at Babylonstoren

Table 3.3 Mean wind speeds in the vineyard and the citrus orchard at Babylonstoren 2020–2021

Measurement	Crop	Crop height [m]	Measurement height [m]	Mean [m s^{-1}]	Summer [m s^{-1}]	Winter [m s^{-1}]	Max winter [m s^{-1}*10 min]	Max summer [m s^{-1}*10 min]
US 1	Vine	2	2	1.82	2.20	1.44	9.54	9.34
US 2	Vine	2	2	1.40	1.57	1.23	9.04	7.3
Farm 1	Citrus	3	3	0.81	0.96	0.65	4.07	4.76
Farm 6	Citrus	3	2	0.55	0.57	0.52	2.88	2.88

Table 3.4 Correlation of wind measurements at Babylonstoren (see Table 3.2)

r^2	US 1	US 2	Farm 1	Farm 6
US 1		0.7828	0.8334	0.9049
US 2			0.6315	0.7170
Farm 1				0.8824

stands is relatively homogeneous and only the higher poplar hedge at the southern edge of the vineyard causes a significant disturbance of the wind field at the canopy level and a corresponding reduction in wind speed. In the citrus stand, the effect of the southern hedge is only indirectly visible through a comparison with US 1; the actual reduction of the inflow from the main wind direction southeast cannot be quantified because of the additional effect of the citrus stand (cf. Farm 6) (Table 3.4).

3.4 Citrus Orchard at Rust En Vrede Farm

Over the summer months with prevailing winds from Southeast in the Cape Flats area, average wind speeds were about 18% lower at Farm 4 on the leeside of the southern hedgerow as compared to Farm 3 (Table 3.5). In the winter half year with prevailing winds from northerly sectors the effect is reversed but significantly weaker (Farm 3 shows 4% lower mean wind speeds than Farm 4). 10-min mean maximum wind speeds are also highest during summer and are 3% lower at Farm 4. A clear reduction of mean wind speed was measured within the citrus canopy at Farm 5.

Correlation coefficients of the measurements show differing interrelation (Table 3.6). While Farm 4 and Farm 5 interrelate significantly and Farm 5 and Farm 3 show a moderate interrelation, the interrelation of Farm 4 and Farm is not significant. This points to clearly different effects of the bordering hedges and the orchard itself on the canopy wind field. Because of wind direction measurements missing over the 2020 period, further analysis had to rely on local weather forecast wind directions data base provided weather forecast (YR- Norwegian Weather Service, www.yr.no.

The results (Fig. 3.11) show Farm 3 wind speeds to be double as high as at Farm 4 in case of southerly winds while it was only half as high in case of northerly directions.

Table 3.5 Mean wind speeds in the citrus orchard (citrus tree height 3 m) at Rust en Vrede during the measuring period 2020/2021

Measurement	Measuring height [m]	Annual mean [m s^{-1}]	Summer [m s^{-1}]	Winter [m s^{-1}]	Max winter [m s^{-1}*10 min]	Max summer [m s^{-1}* 10 min]
Farm 3	3	1.18	1.48	0.99	7.04	7.27
Farm 4	3	1.14	1.25	1.03	6.88	7.07
Farm 5	2	0.55	–	–	3.61	

Table 3.6 Correlation coefficients of wind measurements in the citrus orchard at the Farm Rust en Vrede (see Table 3.5

r^2	Farm 3	Farm 4	Farm 5
Farm 3		0.4311	0.7599
Farm 4			0.8507

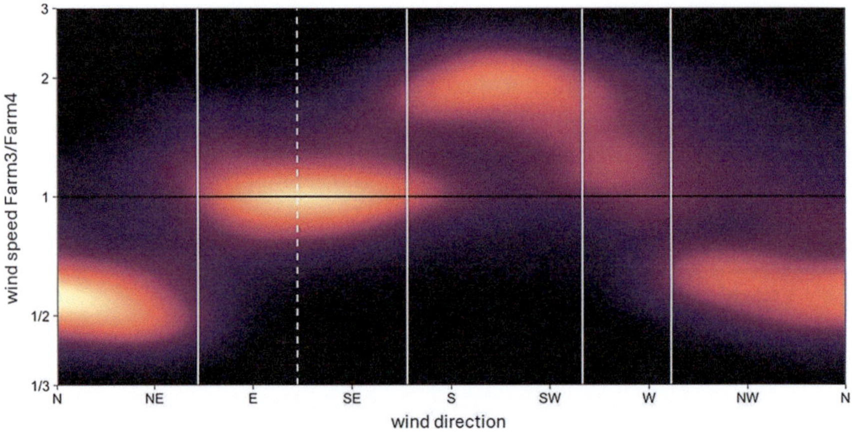

Fig. 3.11 Wind speed ratios Farm 3/Farm 4 as a function of wind direction weather forecast wind direction data base provided by YR- Norwegian Weather Service www.yr.no

Although these results obtained over 12 months of measurements indicate a general pattern, they are still too rough to explain the obvious statistical independency of Farm 3 wind speeds near the northern bordering hedgerow.

Data from the 2 m reference wind measurement located outside the orchard north of the citrus were available as from 22 April 2021 (analyzed until 1 June 2021). The measurement includes 10-min values of mean wind speed, gust (maximum wind speed value), and wind direction (Fig. 3.12).

Wind speeds > 2 m s^{-1} by 30° directions showed highest frequencies from N and NNE and secondarily from S and SSW at the reference measurement. This differed except of N and S in terms of a clockwise 60° offset to the weather forecast distribution for the same period, pointing to normal local effects of terrain and surface roughness on site wind direction distributions. However, Farm 3 and 4 wind speed distributions were fairly like the weather forecast pattern. For the northerly sectors, Farm 4 shows wind speeds up to 4 m s^{-1} while reference station wind speeds were up to 7.4 m s^{-1}. Farm 3 showed very low wind speeds from northerly sectors (an effect of the windbreak) but peaks in the southwesterly and southeasterly sectors. This pattern is not found at Farm 4.

These patterns seem to be triggered by gusts (maximum 10-min value) up to 15 m s^{-1} sweeping across the orchard area (Fig. 3.13). Farm 3 is fairly shielded in case of gusts from the northern sectors and whereas Farm 4 is not hit by higher

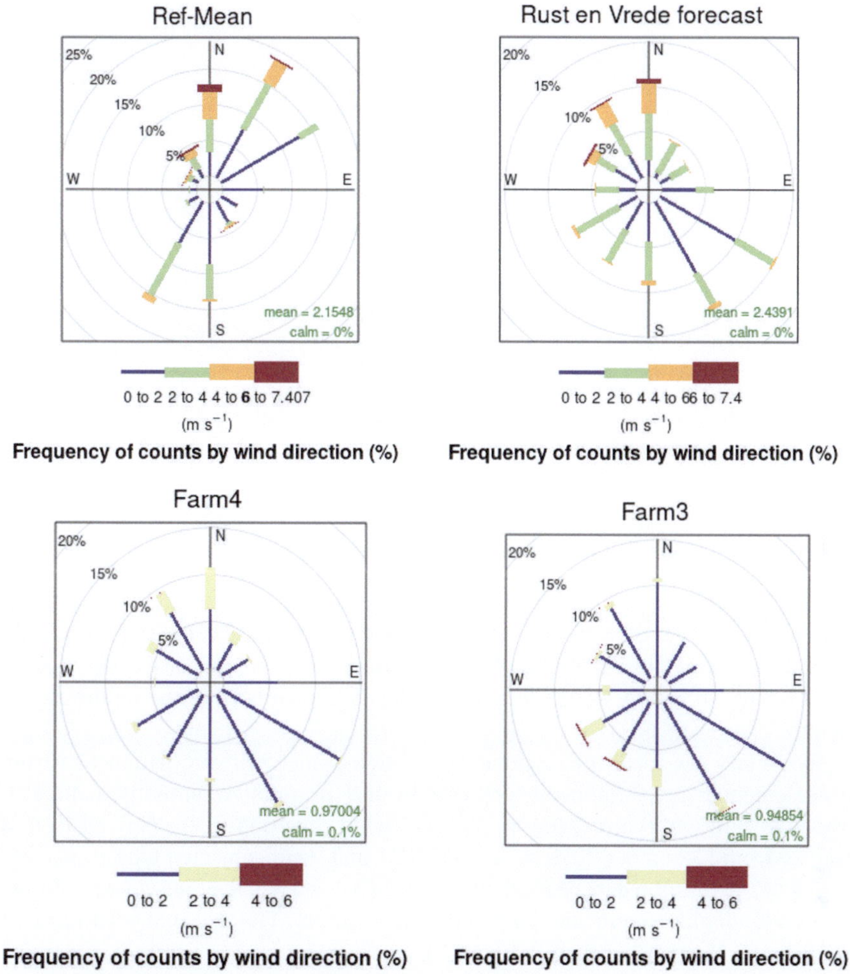

Fig. 3.12 Frequency of wind speed intervals by wind direction measured at the reference station, Farm 3 and Farm 4 and weather forecast for the period 2020/21. Weather forecast data provided by YR- Norwegian Weather Service, www.yr.no

wind speeds. Vice versa, gusts from southerly sectors show no effect at Farm 4 but Farm 3 is affected by considerably higher speeds in this situation. This may be an explanation why Farm 3 and Farm 4 show a weak correlation in the wind speed time series.

Mean maximum wind speeds from both northerly and southerly sectors were highest mainly in the afternoon and evening hours; a finding that should be further observed in separate evapotranspiration computations.

Table 3.7 gives sector wise mean and mean maximum wind speeds. The patterns indicated above can be seen in the distributions of mean wind speeds. The higher

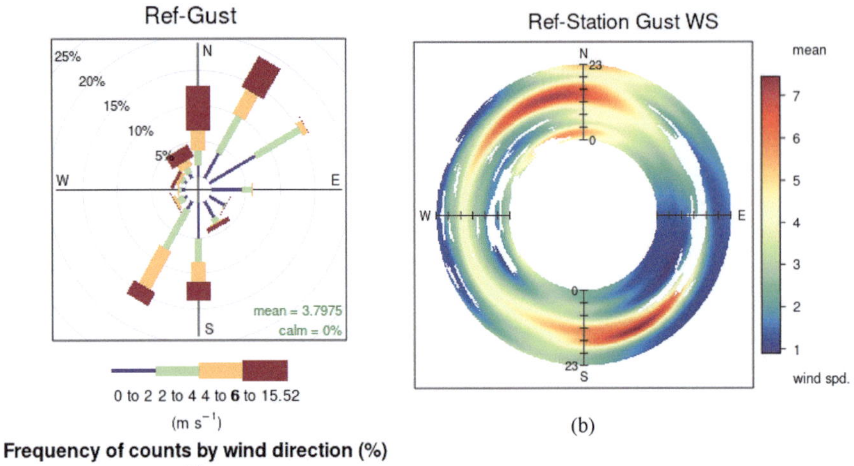

(a)

Fig. 3.13 **a** Frequency of wind gust intervals by wind direction and **b** diurnal course (0..23 h) of guest wind by wind direction at the reference station outside the citrus orchard

wind speeds in the sectors N, NNE, WNW and NNW at the reference station are only found in the Farm 4 distributions while the southeasterly maxima also occur at Farm 4 and Farm 3. It is only Farm 3 that shows the additional peak in the southwesterly sectors.

To classify the local microclimatic conditions, microclimatic parameters from measurement Farm 4 (ambient temperature at 2 m, relative humidity, computed evapotranspiration, wind speed at 3 m and wind direction) were derived from a data set from November 2020 to August 2021 and were subjected cluster analysis (Fig. 3.14). With silhouette coefficients between 0.4 and 0.5, the four-cluster solution shows not a highly significant but practicable structure. The seasonal occurrence of the different clusters is shown in Fig. 3.15.

- **Cluster 1** represents warmer days with lower relative humidity, low wind speed, and high evapotranspiration rates mainly in the transitional season from March to May, as well as some cooler days in summer. Evapotranspiration rates seem to be driven mainly by a higher saturation deficit.
- **Cluster 2** shows cool and moist days with high wind speed and high evapotranspiration rates occurring as short periods at all seasons.
- **Cluster 3** is the typical winter case with low temperatures, high relative humidity, very low wind speeds and low evapotranspiration rates.
- **Cluster 4** represents the typical summer case with high temperatures, dry air, average wind speed and highest evapotranspiration rates.

Table 3.7 Sector wise mean and maximum wind speeds [m s^{-1}] at the reference microclimate station and Farm 3 and Farm and the wind data from the weather forecast provided by YR- Norwegian Weather Service, www.yr.no

Sector	345–15	15–45	45–75	75–105	105–135	135–165	165–195	195–225	225–255	255–285	285–315	315–345	Mean
mean_Ref_WS_avg	3.37	2.63	1.54	1.09	0.76	1.37	1.75	2.26	1.51	1.25	2.56	3.09	1.93
mean_Farm3_WS	0.77	0.6	0.53	0.42	0.49	1.1	1.54	1.73	0.87	0.74	0.7	0.8	0.86
mean_Farm4_WS	1.71	1.15	0.47	0.43	0.41	0.77	0.82	0.96	0.84	0.81	1.27	1.61	0.94
mean_RV_Forecast_WS	3.33	2.5	1.78	1.83	1.95	2.25	2.43	2.48	2.08	2.24	2.7	3.26	2.40
max_Ref_WS_avg	7.41	6.69	4.02	3.41	5.81	6.94	4.86	5.41	3.41	3.76	6.2	7.3	5.44
max_Farm3_WS	2.12	1.76	1.45	0.87	4.42	5.59	4.73	4.81	2.08	1.84	1.77	2.1	2.80
max_Farm4_WS	3.6	3.35	1.47	1.09	3.96	4.59	2.59	2.74	2.27	2.11	3.74	4.04	2.96
max_RV_Forecast_WS	7.4	7.2	6.97	6.85	6.62	6.5	6.35	6.3	6.4	6.27	5.93	6.4	6.60
max_Ref_WS_gust	12.62	12.16	6.72	6.24	15.52	13.7	10.49	11.33	7.16	8.52	12.46	12.07	10.75

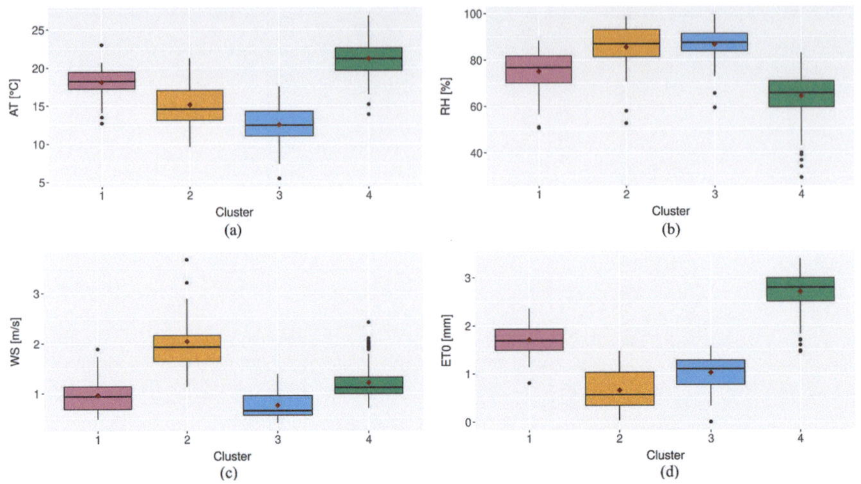

Fig. 3.14 Boxplot of microclimate clusters Rust en Vrede (2020/21). Microclimate data ambient temperature at 2 m (**a**), relative humidity (**b**), wind speed at 3 m (**c**) and computed evapotranspiration (**d**) from the Farm 4 station

3.5 Conclusions

Knowledge of local wind field patterns and associated microclimatic parameters is essential for the consideration of windbreaks to shelter crops from wind damage and high evapotranspiration especially over the summer months with dry and heavy southeasterly winds in the Western Cape. The investigations on two farmlands with different crops and different arrangements of windbreaks showed comparable seasonal wind conditions but different microclimatic response depending on the arrangements of windbreaks. While a single line windbreak at Babylonstoren Farm reduces wind speeds in summer very effectively over a part of the vineyard, the four side windbreak arrangement at Rust en Vrede Farm reduces wind speeds at all seasons but also leads to increased turbulence within the citrus canopy layer. Cluster analysis of microclimatic parameters at Rust en Vrede Farm showed typical seasonal patterns closely linked to weather types (Chap. 4). Wind speed as a main driver of increased water loss by evapotranspiration is typical for the summer case but may also occur in a different microclimatic pattern in short intervals all over the year. In the transitional season from March to May, higher evapotranspiration loss is mainly controlled by the saturation deficit, not by wind speeds.

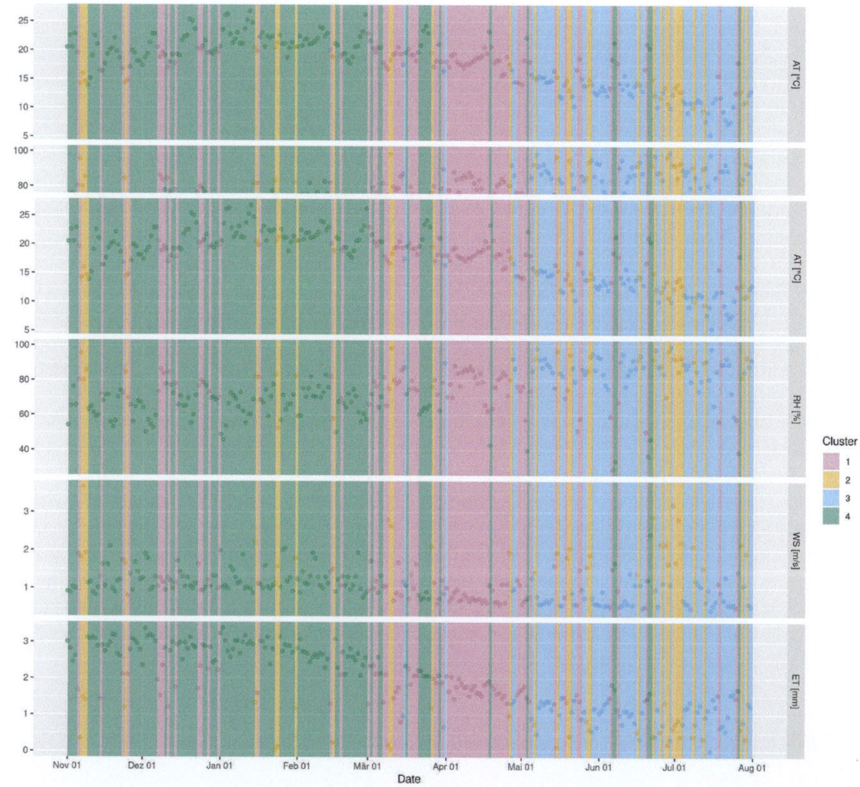

Fig. 3.15 Seasonal occurrence of microclimate clusters (for explanation of the cluster 1–4 see text)

Acknowledgements We thank our partners Paul Roos Wine, Rust en Vrede farm, Somerset West, Babylonstoren, Simondium, and for their support and permission for the field experiments. The wind measurements in the vineyard at Babylonstoren are supported by the project AgroSolarSystems, funded by the Department of Agriculture, Forestry and Fisheries, South Africa. FarmImpact project was supported by the Federal Ministry of Education and Research (BMBF) and managed by PT-DLR (FKZ 01LZ1711A-F) as part of "CLIENT II—International Partnerships for Sustainable Innovations" funding initiative.

References

Geldenhuys H, Lötze E, Veste M (2023) Fruit quality and yield of mandarin (*Citrus reticulata*) in orchards with different windbreaks in the Western Cape, South Africa. Erwerbs-Obstbau 65:959–369. https://doi.org/10.1007/s10341-022-00725-3

Kanzler M, Böhm C, Mirck J, Schmitt D, Veste M (2019) Microclimate effects on evaporation and winter wheat (Triticum aestivum L.) yield within a temperate agroforestry system. Agrofor Syst 93(5):1821–1841. https://doi.org/10.1007/s10457-018-0289-4

Littmann T, Steinrücke J, Bürger M (2004) Elemente des Klimas. Grundlagen und Anwendungen. Gotha: Klett-Perthes, 208 S

Maronga B, Banzhaf S, Burmeister C, Esch T, Forkel R, Fröhlich D, Fuka V, Gehrke KF, Geletič J, Giersch S, Gronemeier T, Groß G, Heldens W, Hellsten A, Hoffmann F, Inagaki A, Kadasch E, Kanani-Sühring F, Ketelsen K, Khan BA, Knigge C, Knoop H, Krč P, Kurppa M, Maamari H, Matzarakis A, Mauder M, Pallasch M, Pavlik D, Pfafferott J, Resler J, Rissmann S, Russo E, Salim M, Schrempf M, Schwenkel J, Seckmeyer G, Schubert S, Sühring M, von Tils R, Vollmer L, Ward S, Witha B, Wurps H, Zeidler J, Raasch S (2020) Overview of the PALM model system 6.0. Geosci Model Dev 13:1335–1372

Maronga B, Gryschka M, Heinze R, Hoffmann F, Kanani-Sühring F, Keck M, Ketelsen K, Letzel MO, Sühring M, Raasch S (2015) The parallelized large-eddy simulation model (PALM) version 4.0 for atmospheric and oceanic flows: model formulation, recent developments, and future perspectives. Geosci Model Dev 8(8):2515–2551

Meadows ME (2015) The cape winelands. In: Grab S, Knight J (eds) Landscapes and landforms of South Africa. World geomorphological landscapes. Springer, Heidelberg, 103–109

Veste M, Littmann T, Kunneke A, Du Toit B, Seifert T (2020) Windbreaks as part of climate-smart landscapes reduce evapotranspiration in vineyards, Western Cape Province, South Africa. Plant Soil Environ 66:119–127. https://doi.org/10.17221/616/2019-pse

Veste M, Sheppard JP, Abdulai I, Ayisi KK, Borrass L, Chirwa PW, Funk R, Kapinga K, Morhart K, Mwale SE, Ndlovu NP, Nyamadzaw G, Nyoka BI, Sebola P, Seifert T, Senyolo MP, Sileshi GW, Syampungani S, Kahle H-P (2024) The need for sustainable agricultural land-use systems: benefits from integrated agroforestry systems. In: von Maltitz GP, Midgley GF, Veitch J, Brümmer C, Rötter RP, Viehberg, FA, Veste M (eds) Sustainability of southern African ecosystems under global change: Science for management and policy interventions, ecological studies 248, Springer Nature, Heidelberg, 587–62. https://doi.org/10.1007/978-3-031-10948-5_21

Chapter 4
An Empirical Classification of Weather Types and Interrelations with Rainfall and Evapotranspiration in the Southern Western Cape, South Africa

Thomas Littmann, Tobias Recke, and Maik Veste

Abstract Daily data from various sources (aerological soundings for three geopotential heights, reanalysis, weather station, FarmImpact measurements) were analyzed for the 2010–2022 period by k-means cluster analysis. The basic pattern of the classification of 5 weather types derived from cluster analysis is the seasonal change of near-ground circulation depending on the strength and position of the controlling pressure centers (South Atlantic Anticyclone and either dynamic mid-latitude or thermal inland lows). In winter, the active pressure centers are either mid-latitude cyclones with frontal passage and cold air influx, strong wind and high rainfall (Type 2 with overall minor frequencies) or cyclones further south in the mid-latitude west wind drift with frontal systems typically not reaching the Southwestern Cape (Type 4 as the more frequent type). In summer, Type 3 represents the typical synoptic pattern of strong inland thermal Lows, a seasonally strong South Atlantic Anticyclone and following high wind speeds and temperatures in the southern Western Cape. Microclimatic processes like rainfall and evapotranspiration and the resulting water balance on farmlands of the southern Western Cape are largely driven by the mesoscale atmospheric dynamics as could be shown for the Rust en Vrede Farm site. Irrigation water consumption or mechanical fruit damage caused by high wind speeds is defined by frequencies of dry and windy weather types in the growing season (Types 3 and 1) as well as of rain-bearing types (Types 2, 4 and 5) over the winter half year, determining soil, groundwater and reservoir water storage.

Keywords Weather types · Circulation patterns · Cluster analysis · Drought · Water consumption · Fruit damage

T. Littmann (✉) · T. Recke
DLC—Dr. Littmann Consulting, Ennepetal, Germany
e-mail: littmann@littmann-consulting.de

M. Veste
CEBra—Center for Energy Technology Brandenburg e.V., Cottbus, Germany

Department of Molecular Botany, University of Hohenheim, Institute of Biology, Hohenheim, Germany

M. Veste et al. (eds.), *Trees in South African Agriculture*, Innovations in Landscape Research, https://doi.org/10.1007/978-3-032-00993-7_4

4.1 Introduction

4.1.1 *Agricultural Production and Weather in the South Western Cape*

The Western Cape is one of the important agricultural centers of South Africa and traditionally the winter wheat production area. In the recent decades fruit and wine production increased, which highly dependent on sufficient water supply and therefore from the precipitation in the wintertime. The characteristics of the different weather types over the year have different effects on agricultural production and differ in terms of the effects on the different agricultural crops (Chaps. 5, 6). The last multi-year droughts of 2015–2018 have led to drastic crop losses in the Western Cape region, likely to be exacerbated under future climate change (Theron et al. 2021). Beside the seasonal rainfall pattern and temperature, the wind direction and wind speed in one of the drivers affecting the agricultural production. Especially wind speed influencing the evapotranspiration and agricultural water use (Veste et al. 2020, Chap. 5) and can lead to fruit damage and economical losses (Geldenhuys et al. 2023, Chaps. 7, 12).

Despite many local factors determining the climatic water balance at the measurement sites, rainfall and evapotranspiration are predominantly controlled by larger-scale weather types. For simplification, we may refer in our analysis to winter (April to September) and summer (October to March) half years while an empirical definition of seasons in South Africa based on temperature data (Van der Walt and Fitchett 2020) confined winter to June to September and summer to December to March.

Although prior research on major synoptic types led to a deeper understanding of the interrelation of circulation patterns, synoptic types, teleconnections, rainfall and wind in South Africa, empirical work explicitly covering the southern Western Cape region is still missing. Within the specific framework of the FarmImpact project, we will focus on the area of the southern Western Cape between Cape Town, Somerset West, Stellenbosch up to Paarl with its vineyards and fruit orchards (Fig. 4.1).

4.1.2 *Climate Background of the Western Cape*

South African climate is characterized by seasonal variations of the intertropical convergence Zone (ITCZ), the subtropical branch of the Hadley circulation with its high-pressure cells, and the mid-latitude westerlies south of the continent. Furthermore, the Agulhas Current and the Benguela upwelling systems are important drivers for oceanic and terrestrial climate in the region (Biastoch et al. 2024; Brandt et al. 2024). While the descending branch of the Hadley cell is mainly visible as the quasi-permanent South Atlantic Anticyclone west of the Western Cape the South Indian Ocean Anticyclone on the eastern side of the continent is highly volatile and is often replaced by low pressure systems originating in the mid-latitude westerly

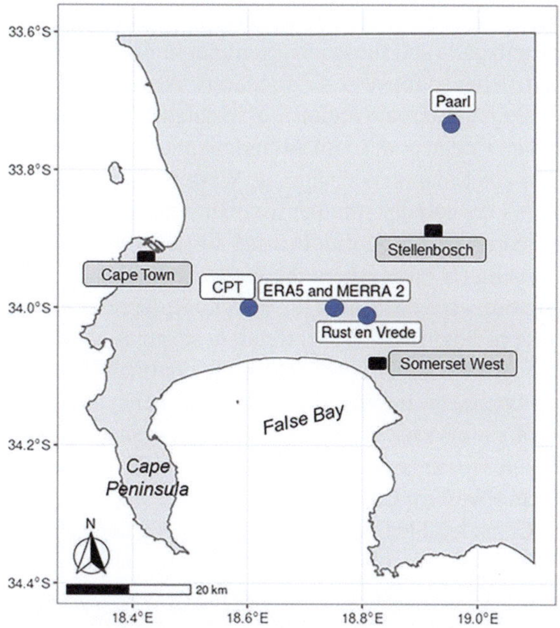

Fig. 4.1 Location map of data sets (blue dots) obtained in the southern Western Cape, South Africa. CPT: Cape Town international airport, explanation see Table 4.1

jet. As the stable Atlantic Anticyclone shifts southward in summer (around 31° S), it may lead to strong southeasterly winds over the southern Western Cape, being reinforced further north as offshore trade winds leading to cold water upwelling of the Benguela current (Weischet and Endlicher 2000). Further inland the shallow subtropical Kalahari Low may well extend into coastal areas. In winter, the South Atlantic Anticyclone is located further north (around 25–28° S), giving space to the strengthened mid-latitude westerly jet leading to various forms of cyclogenesis depending on the wave and circulation type (Weischet and Endlicher 2000). This basic pattern is confined to the western parts of the Cape region; the eastern parts of South Africa show a different seasonal climate primarily controlled by subtropical low pressure and rainfall in summer and the absence of mid-latitude frontal systems in winter (Weischet and Endlicher 2000).

4.1.3 Classification of Weather Types

Earlier definitions of weather types in South Africa (Longley 1976; Hurry and van Heerden 1982; Taljaard 1995) were mainly derived from synoptic map analysis and identified 6 main types that are based on the seasonal pressure systems. This qualitative approach was formally applied by Littmann (2000) for the definition of weather types in the Mediterranean Basin. Two main types are associated with the extension ("ridging") of the South Atlantic Anticyclone over most parts of South

Africa in summer, either leading to moisture flux from the Indian Ocean into the eastern parts and subsequent summer rainfall or with cool moist air flow in the eastern parts from southwesterly directions. Also in summer, a second rain-bearing pattern for the summer rain region is associated with either easterly waves between the South Indian Ocean and South Atlantic Anticyclones or a stronger thermal low extending from the Kalahari deep into the Western Cape region. In both cases the Western Cape shows hot and dry summer weather with southeasterly flow. The main winter pattern is characterized by rain-bearing westerly depressions originating along troughs of the westerly jet reaching the Western Cape as the South Atlantic Anticyclone is in a far northward position. Those cyclonic depressions do not normally reach the eastern parts of South Africa as a frequent strong trough in the westerly belt leads to a very strong low further southeast of the continent and easterly frontal systems reaching the continent are rare occasions. Further types such as cutoff lows originating in weak flow waves of the westerlies and tropical cyclones (Taljaard 1995) are relevant features for the eastern parts and the southern Cape but do not show relevant influence for the Western Cape area.

Considerable empirical work has been dedicated to interrelations of regional and hemispheric circulation pattern and rainfall in South Africa. Mahlobo et al. (2019) showed dry winters to be linked to above-normal southward extensions of the ITCZ and Hadley circulation leading to an easterly position of a strong South Atlantic Anticyclone blocking effectively the migration of frontal systems toward the Western Cape while dry summers occur when the South Indian Anticyclone is absent and water vapor flux from the Indian Ocean is decreased. Mason and Jury (1997) also derived such contrary patterns of dry and wet years for the summer and winter rainfall areas in South Africa. They showed seasonal circulation anomalies linked to larger-scale anomalies such as El Niño events when sea surface temperatures (SST) are lower in the southern Indian Ocean, thus decreasing moisture influx and summer rainfall in the eastern parts of South Africa, while higher SST off the Western Cape enhance momentum flux and cyclogenesis in the westerlies and winter rainfall in the Western Cape. At the same time, the South Atlantic Anticyclone is much weaker and in a westerly or northerly position giving space for the penetration of strong mid-latitude cyclones and high to extreme rainfall, e.g., in 2001 (Reason 2002). In this case moisture influx originates in the western equatorial Atlantic, the South Atlantic jet is extremely strong with high cyclonic vorticity and low-level convergence near the Western Cape, both effects being triggered by high SST anomalies and instability of the lower atmosphere (Reason et al. 2003). However, storm events do not imply spatially uniform rainfall patterns in the Western Cape as storm cell characteristics and orography may vary over short distances (Hughes and Wright 1988). Engelbrecht and Landman (2016) defined 35 different nodes as 850 hPa-level synoptic types for the southern Cape coast from Ward's minimum variance cluster analysis of daily values 1979–2011. Summer rainfall anomalies in this region were also linked to southwesterly ridging of the South Atlantic Anticyclone and moisture influx from the Indian Ocean while above-normal frequencies of frontal systems in June occurred during El Niño years, below normal frequencies during La Niña years. Above-normal rainfall along the Southern Cape may also occur when cutoff lows of a weak mid-latitude jet are frequent.

Strong winds in the Western Cape are always related to either a southerly position of the Atlantic Anticyclone with strong divergence and eastward ridging resulting in strong southeasterly winds or to frontal passages in winter (Kruger et al. 2010). In Cape Town, most annual maximum wind gusts 1993–2008 occurred during frontal passage events in winter while gusts in the eastern parts of South Africa are mainly linked to summerly thunderstorms. In this way, summerly "ridging" of high pressure is only a secondary process for the generation of severe gusts in the Western Cape (Kruger et al. 2013). In their analysis of damaging strong winds in the Western Cape, Goliger and Retief (2002) also showed mid-latitude cyclones in winter to be the predominant feature, followed by cutoff lows in the transitional seasons and shallow southeasterly winds in summer.

4.2 Methodology for Classification of Weather Types

For the classification and description of major weather conditions of the South African Western Cape region, five different types of data sources were used.

Historical, vertical profile data of the atmosphere (*sounding data*) of the SYNOP station Cape Town Airport were obtained (REF http://weather.uwyo.edu/upperair/sounding.html) for the period 01.01.2010–31.12.2021 (measurements at 00.00 h UTC). Parameters used for further analysis are air temperature (AT [°C]), relative air humidity RH [%], wind speed (WS [m s^{-1}]), and wind direction (WD [°]) for geopotential heights 500 hPa, 850 hPa, and 1000 hPa. Furthermore, the elevation of these pressure levels was considered. For the observation period, 3475 data sets were available, which corresponds to a data availability of about 79%.

Two *reanalysis data*sets of the grid point S 34.00°/E 18.75°, which is representative of the study region, were used for the same period (open source: EMD WindPro). From the hourly datasets MERRA2 (height 50 m a.g.l.; parameters air temperature AT [°C], wind direction WD [°], wind speed WS [m s^{-1}]) and ERA5 (height 100 m a.g.l., parameters WD [%], WS [m s^{-1}]) values between 00:00 and 01:00 UTC were extracted for comparison with the Cape Town International Airport soundings data. Reanalysis data have an availability of 100%.

*Weather station data*sets in daily resolution for the SYNOP stations Cape Town International Airport (ID: 68,816 WS) and Paarl (ID: 68,713 WS) were obtained from https://www.ogimet.com/ for the period indicated above (Cape Town International Airport) and for 01.01.2014 to 31.12.2021 (Paarl). Parameters used for analysis are AT [°C], relative humidity RH [%], precipitation P [mm], WD [%], and WS [m/s].

For the visualization of results and comparison to regional synoptics, *synoptic weather maps* of the South African Weather Service were used for selected days (https://www.weathersa.co.za/home/historicalsynoptic).

Finally, two *micrometeorological data*sets of FarmImpact project site measurements in a citrus orchard of Rust en Vrede farm (Chap. 3) were included in the evaluation. Data from the measuring stations F3 and F4 were available for the period 01.04.2020 to 31.12.2021, with F3 showing a data failure from January to March 2021. Parameters used are air temperature (AT [°C]), relative air humidity (RH [%]), wind speed (WS [m s^{-1}]), short wave incident irradiation [W m^{-2}], short wave reflected irradiation [W m^{-2}] and soil heat flux [W m^{-2}]. Total evapotranspiration was computed following the FAO modified Penman-Monteith approach (Allen et al.1998).

In terms of an objective statistical approach to identify weather types with clear individual characteristics and showing significant differences primarily, the Cape Town International Airport soundings data were subjected to cluster analysis, especially the k-means algorithm. Cluster analysis is a process of unsupervised learning to classify objects, where objects with similar properties are mapped to the same cluster and those with dissimilar properties to other clusters. The k-means clustering algorithm is considered a widely used tool in the scientific and industrial field. It is used in meteorology, e.g., for clustering similar weather data (Dubey and Choubey 2017). The R and RStudio software packages were used for data analysis (Czernecki et al. 2020; RStudio Team 2020; Wickham et al. 2019; Wickham 2006). The k-means algorithm as well as the variable reduction by means of the UMAP method are not scale-independent. An equilibrium of variable values with different units of measurement is carried out by a standardization, the so-called ranging, whereby the minimum of a variable (y min) is subtracted from the variable value (y i) and then divided by the interval (y max–y min) (Legendre and Legendre 2012).

Variable reduction was conducted using the UMAP method. The Uniform Manifold Approximation and Projection (UMAP) is a dimension reduction method that can be used for visualization like the t-distributed stochastic neighborhood embedding (t-SNE) for visualization, but also for general nonlinear dimension reduction. The theoretical foundations for UMAP are largely based on the theory of diversity and topological data analysis. The UMAP algorithm offers several advantages over the t-SNE, such as higher speed and better preservation of the global structure of the data (McInnes et al. 2018). The UMAP algorithm was used to reduce the dimensions of the standardized sounding data.

K-means—Cluster Analysis: the k-means algorithm divides a set of M of points in an N-dimensional space into K clusters, so that the distances within a cluster is minimized (Hartigan and Wong 1979). The number of k centroids is predetermined–then k clusters are formed by assigning each point to the next centroid. Iteratively, the centroid of each cluster is recalculated until the centroid does not change. The points in the clusters must have a high intra-class similarity and a low inter-class similarity (Dubey and Choubey 2017). In this study, the Hartigan and Wong (1979) algorithm was used for cluster analysis of the sounding data, which uses the squared Euclidean distance as a measure of dissimilarity.

Quality of cluster solutions: A suitable measure for the selection of the best parameter k of a k-means cluster analysis is the silhouette coefficient (Everitt et al. 2011). This is calculated from the silhouettes of the individual objects, which describe the distance to the two nearest clusters. The silhouette coefficient can take a maximum

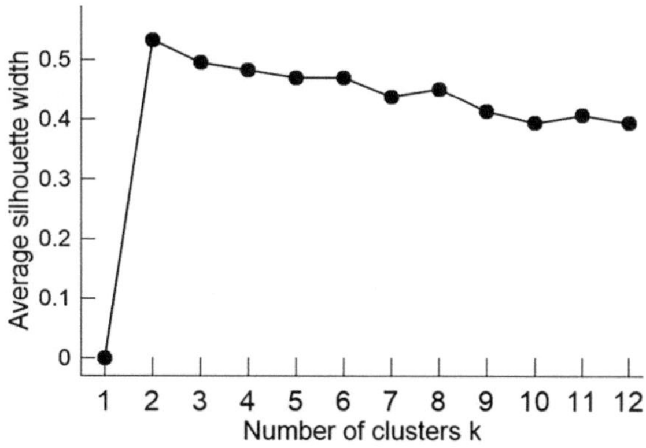

Fig. 4.2 Silhouette coefficients of k-means cluster analyses with k 2 to 1

value of 1, whereby the structure within the clusters increases as the silhouette coefficient increases. From a value of less than 0.25, no structure is assumed. The optimal number of clusters was derived from the silhouette coefficient values. The silhouette coefficients (Fig. 4.2) for $k = (4, 5)$ are 0.47 each. Iteratively, 5 clusters were found to be the best-fit solution. The very low number of negative silhouettes in the clusters were excluded from the analysis. Figure 4.3 summarizes the methodical approach of weather type definition.

Following cluster analysis of the Cape Town International Airport soundings data as the most detailed dataset, both reanalysis and weather station data were subsequently assigned to the clustered time series of the soundings data for further description of characteristic properties of the clusters. The focus was on the comparability of soundings, reanalysis and weather data by means of characteristic value calculations for the meteorological parameters AT, RH, WS, WD, and P within the clusters. On the other hand, the characteristic values also serve to investigate the seasonal and interannual variability during the study period. This analysis was mainly based on weather station data.

4.3 Classification of Weather Types

4.3.1 Results of Cluster Analysis

The following Tables and Figures show the results of cluster analysis of the soundings, reanalysis, and weather station data. Clusters 1 to 5 denominate the empirical weather types and show significant differences of parameters (Kruskal-Wallis probabilities < 0.0001, maximum group-wise Kruskal–Wallis probability < 0.05).

Results of sounding data clustering show significant differences of geopotential height, especially on the 500 hPa and 580 hPa levels (ridging or trough structures), warm and dry (types 3 to 5) or cold and wet conditions (types 1 and 2) in the lower

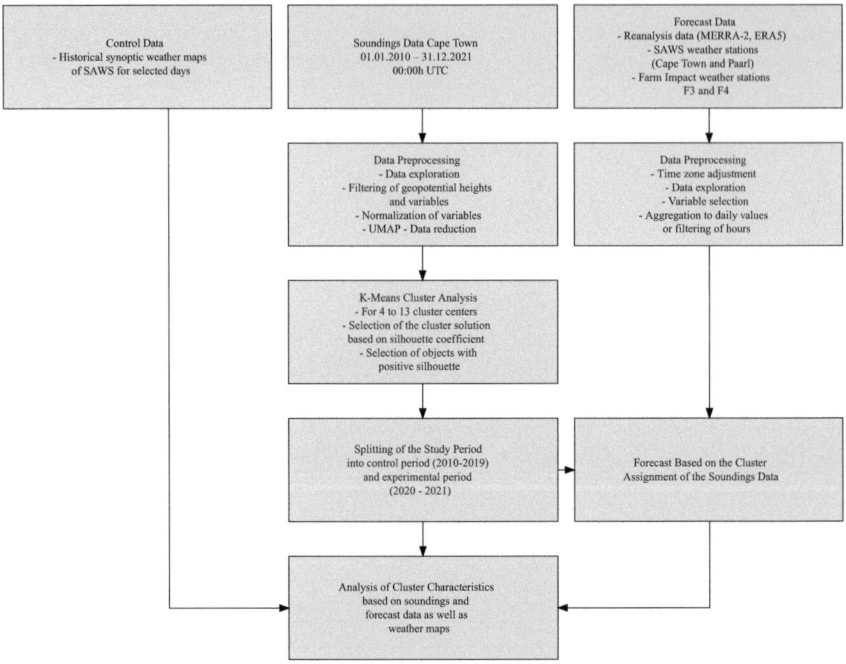

Fig. 4.3 Summary of the methodology for classification and work flow of weather types

troposphere, and higher (types 1 to 3) or lower wind speed (types 4 and 5) at all three geopotential heights (Table 4.2 and Fig. 4.4).

Reanalysis data in hourly resolution also show similar differences of wind speed as found in the sounding data. The higher time resolution also allows for a better definition of mean wind directions of the clusters. Types 1, 3, and 4 are southeasterly flow patterns while types 2 and 5 show northwesterly flow (Table 4.3 and Figs. 4.5 and 4.6).

The basic characteristics of clusters indicated by sounding and reanalysis data were also found in Cape Town Airport and Paarl weather station data. At both locations, type 2 is the coldest weather type with highest precipitation, while type 3 is the warmest, windiest, and driest type of all (Table 4.4 and 4.5 and Figs. 4.7 and 4.8).

Finally, the total and seasonal frequency of clusters shows clusters 3 and 4 as the most frequent types but it is only clusters 3 (summer) and 2 (winter) that show a clear seasonal pattern except cluster 5, which is a rare case.

4.3.2 Synoptics and Description of Weather Types

Weather Type 1

Table 4.1 Overview of used datasets from the southern Western Cape, South Africa, location map see Fig. 4.1

Data type	Dataset	Location	Coordinates (geogr. WGS 84)	Elevation a.s.l ´[m]	Time period
Rawinsonde data		CPT—Cape town international airport	S -33.9648 E18.6017	46	2010-01-01–2022-12-31
Reanalysis data	MERRA 2 ERA5	Grid point western cape	S -34.00 E 18.75 S -34.00 E 18.75	50 100	2010-01-01–2022-12-31
Meteorological stations	SYNOP stations	CPT—Cape town international airport Paarl	S -33.9648 E 18.6017 S -33.7217 E 18.9719	46 109	2010-01-01–2022-12-31 2014-01-01–2022-12-31
	FarmImpact	Rust en Vrede F3 Rust en Vrede F4	S -34.0112 E 18.8098 S -34.0120 E 18.8103	126	2020-04-12–2021-12-31
Weather maps	South African weather service	South Africa	–	–	Selected days

Typical synoptic weather charts for days classified as Type 1 (Fig. 4.9) show the South Atlantic Anticyclone in a southern position but ridging over the Western Cape which results in a moderate southeasterly flow with light rainfall in the Western Cape and further east. The main controlling pressure system is in all cases the strong mid-latitude low with or without cold front systems; the South Atlantic Anticyclone is in a far southerly position.

This type shows slightly below average geopotential height of the 500 hPa and 850 hPa levels but an above-average height of the 1000 hPa level, indicating a shallow mid-tropospheric anticyclonic system with near-ground low pressure. Wind directions are typically from the southeast with average or slightly above-average wind speeds. Temperature and humidity data imply a relatively cool and moist air mass with moderate rainfall. With a frequency of 14% type 1 is a secondary weather type with no clear seasonal characteristics but the highest frequency in the summer months October to March, while the winter months May to July have the lowest frequency.

Weather Type 2

Typical synoptic weather charts for days classified as Type 2 (Fig. 4.10) show in all cases the presence of cold fronts passing the Western Cape and adjacent areas. Therefore, the anticyclonic characteristics implied by the pressure level soundings are a secondary feature of this type as the southern frontal systems dominate, bringing

Table 4.2 Cluster characteristics from Cape Town international airport soundings 2010–2019. GH: geopotential height [m], AT: ambient temperature [°C], WD: wind direction [°], WS: wind speed [m/s], RH: relative humidity [%]

	Sum	Mean	Mean	Mean	Mean	Mean	Mean	Mean	Mean	Mean	Mean	Mean	Mean
Type/ cluster	N	GH_500	AT_500	WD_500	WS_500	GH_850	AT_850	WD_850	WS_850	GH_1000	AT_1000	WD_1000	WS_1000
1	379	5.715	−14.7	242	19.7	1.515	6.4	175	7.1	161	14.6	171	7.2
2	527	5.658	−16.2	270	28.6	1.489	4.7	277	11.5	142	13.5	314	6.8
3	804	5.804	−11.1	247	12.2	1.524	15.0	115	5.1	141	17.2	166	8.2
4	808	5.781	−12.0	277	17.6	1.516	13.9	301	6.3	136	16.1	145	4.9
5	224	5.758	−11.9	288	23.4	1.498	13.3	307	11.7	126	15.0	330	5.8
Total/ average	2742	5.753	−12.9		18.9	1.511	11.4		7.5	141	15.6		6.6

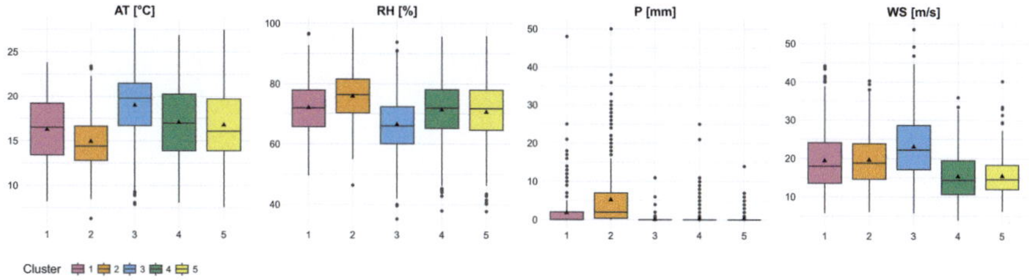

Fig. 4.4 Boxplots of the cluster solution of the sounding data of Cape Town international airport in the period 2010–2019

Table 4.3 Reanalysis data characteristics of clusters

Reanalysis data		MERRA 2			ERA5	
Type/cluster	N	Mean WS [m s^{-1}]	Mean WD [°]	Mean AT [°C]	Mean WS [m s^{-1}]	Mean WD [°]
1	379	5.6	166	12.5	5,5	167
2	527	6.7	286	11.8	6,3	299
3	804	7.1	129	14.2	7,1	130
4	808	4.8	151	13.2	4,5	153
5	224	5.4	292	12.5	5,2	299
Total/average	2742	5.9	205	12.8	5,8	188

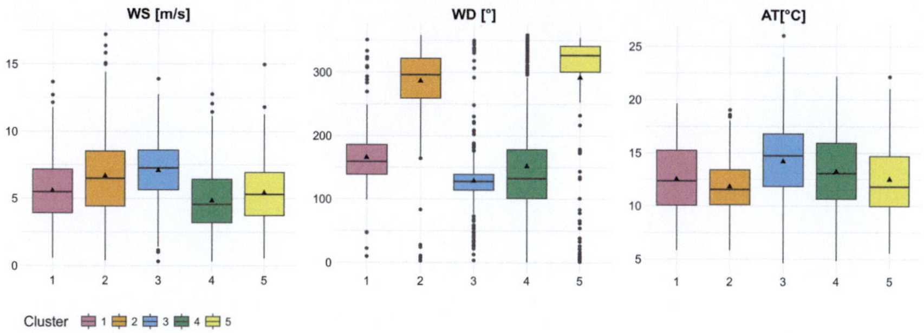

Fig. 4.5 Boxplots of MERRA 2 data (height 50 m) based on the soundings cluster solution in the period 2010–2019

cold air and precipitation (rain and snow) to the Western Cape. With a frequency of 19.4% type 2 is another secondary weather type mainly occurring in the winter half year with highest frequency May to September.

Weather Type 3

The Type 3 (Fig. 4.11) shows clearly above-average geopotential heights of all three pressure levels, indicating a high-reaching low pressure system with predominantly southeasterly wind directions, highest wind speeds of all clusters as well as

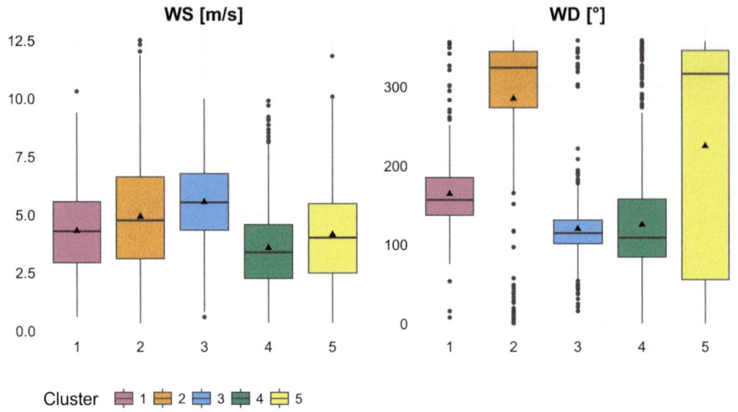

Fig. 4.6 Boxplots of ERA5 data (height 100 m) based on the soundings cluster solution in the period 2010–2019

Table 4.4 Weather station characteristics of clusters at Cape Town international airport (CPT) and Paarl. N: cases; mean AT: ambient temperature, RH: relative humidity P: precipitation WS: wind speed

Type/cluster	N	AT [°C]	RH [%]	P [mm]	P sum per day [mm]	WS [m/s]
Cape town international airport daily 2010–2019						
1	379	16.3	71.9	1.8	4.0	5.5
2	527	15.0	75.8	5.1	6.2	5.5
3	804	18.9	66.7	0.1	1.5	6.2
4	807	17.0	71.3	0.3	3.1	4.3
5	224	16.8	70.5	0.3	2.2	4.4
Average	2741	16.8	71.2	1.5	3.4	5.2
Paarl daily 2014–2019						
1	213	17.5	64.2	1.3	3.2	2.0
2	275	15.2	74.6	7.3	9.8	1.8
3	433	21.5	52.3	0.2	2.2	2.2
4	436	18.9	61.6	0.6	3.4	1.5
5	127	17.5	65.0	0.4	3.0	1.2
Average	1484	18.1	635	2.0	4.3	1.7

high temperatures. This type is the typical summer case with low-level convective troughs dominating the Western and Eastern Cape regions, sometimes connected to the shallow thermal Kalahari Low further North. The South Atlantic Anticyclone is reinforced as the descending branch of the Hadley Cell but remains in western or southern positions while convective and monsoonal summer rainfall may occur in the eastern regions; in rare cases reaching the Western Cape as in March 2022. Type

Table 4.5 Seasonal frequency of clusters. Summer: October to March; Winter: April to September

Cluster	Summer [%]	Winter [%]	Total [%]
1	7.8	6.2	14.0
2	5.4	14.0	19.4
3	21.2	7.9	29.1
4	13.4	15.8	29.2
5	2.3	6.0	8.3
Total	50.0	50.0	100

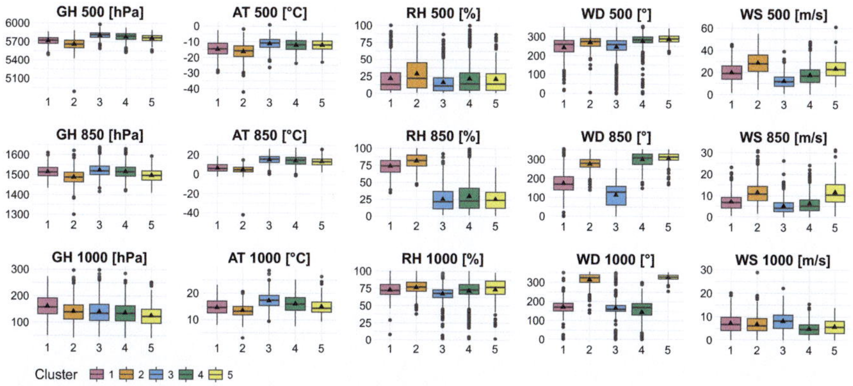

Fig. 4.7 Boxplots of the weather station data of Cape Town based on the soundings cluster solution in the period 2010–2019

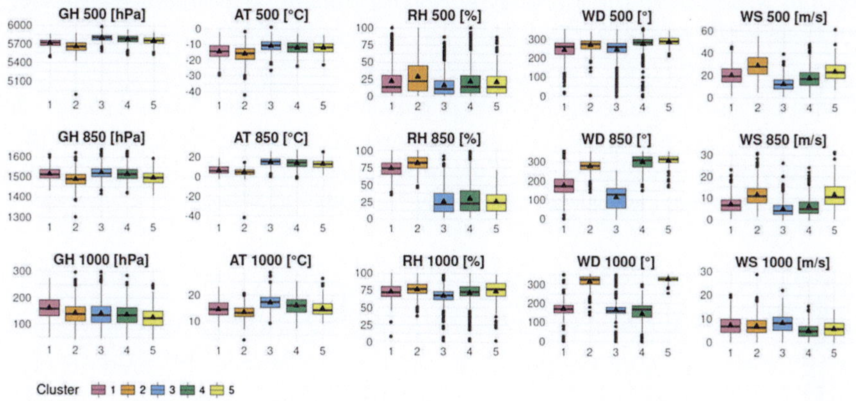

Fig. 4.8 Boxplots of the weather station data of Paarl based on the soundings cluster solution in the period 2014–2019

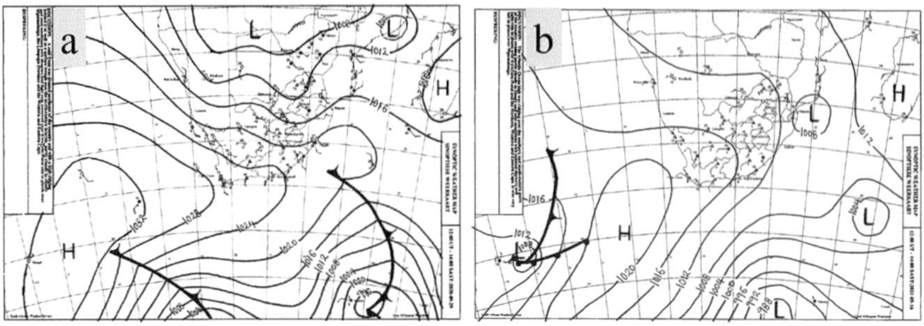

Fig. 4.9 **a** Synoptic weather map showing type 1 (example 2020-09-29). Frontal passage led to extreme rainfall at Cape Town International Airport weather station. **b** Synoptic weather map showing type 1 (example 2021-05-16). *Source* South African Weather Service

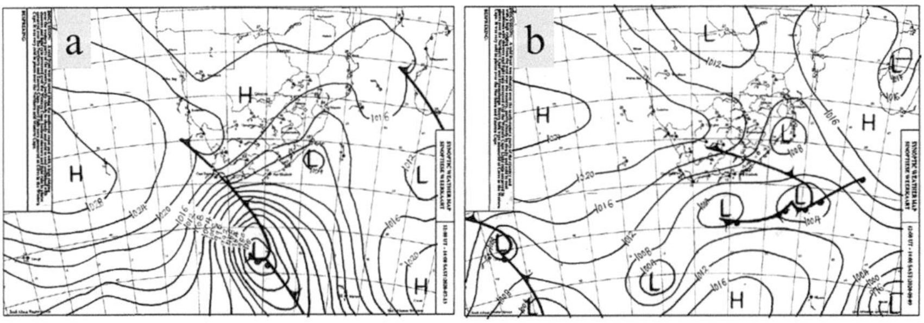

Fig. 4.10 Synoptic weather map showing type 2 **a** example 2020-07-13 and **b** example 2020-08-09. *Source* South African Weather Service

3 does not show a clear ridging of high pressure between the South Atlantic and South Indian Ocean Anticyclones; the latter may even be completely absent. With a frequency of 29,1% type 3 is one of the two main weather types and the typical hot and windy summer case (highest frequency November to March).

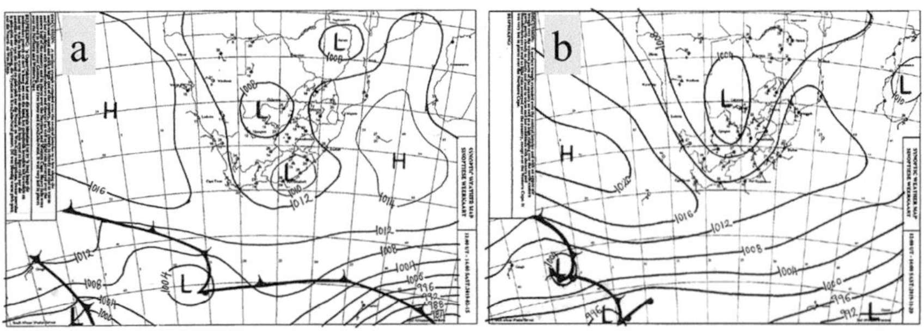

Fig. 4.11 Synoptic weather map showing type 3 **a** example 2019-02-15 and **b** example 2019-11-20. *Source* South African Weather Service

Weather Type 4

Average mid-tropospheric geopotential heights and a slightly lower level at 1000 hPa, basically indicating shallow anticyclonic flow are characteristics of the Type 4 (Fig. 4.12). Thus, predominant wind directions are southeast with light wind and a cool and moist air mass. Rainfall averages are comparably low in this situation. The typical synoptic type shows the South Atlantic Anticyclone in a southern position and cyclonic frontal systems dominating in the Eastern Cape. Type 4 is with a frequency of 29.2% the second main type but shows no clear seasonal pattern. Highest monthly frequencies are, however, in the winter half year (April to September). Different to Type 2 in winter, the southern position of the South Atlantic Anticyclone with predominant ridging toward the South Indian Ocean Anticyclone prevents a direct passage of larger frontal system with high rainfall over the Southwestern Cape. In summer, Type 4 represents a weaker flow as compared to Type 3 as the South Atlantic Anticyclone and the inland thermal low are not as strong as in case of Type 3.

Weather Type 5

The Type 5 (Fig. 4.13) shows average mid-tropospheric geopotential heights and a clearly lower level at 1000 hPa, indicating shallow anticyclonic flow. However, all data sources show predominantly northwesterly wind directions. The typical synoptics show low pressure over the Western Cape, in summer as a larger thermal Low and, predominantly, in winter a strong mid-latitude Low south of the continent. In all cases the South Atlantic Anticyclone is far west or southwest. This Type 5 is less frequent than Type 2, bringing less rainfall and cold air to the region as the mid-latitude frontal systems are too far south or frontal systems reach the continent further east. Wind speeds are generally below average. It occurs mainly in winter with highest monthly frequencies from May to August.

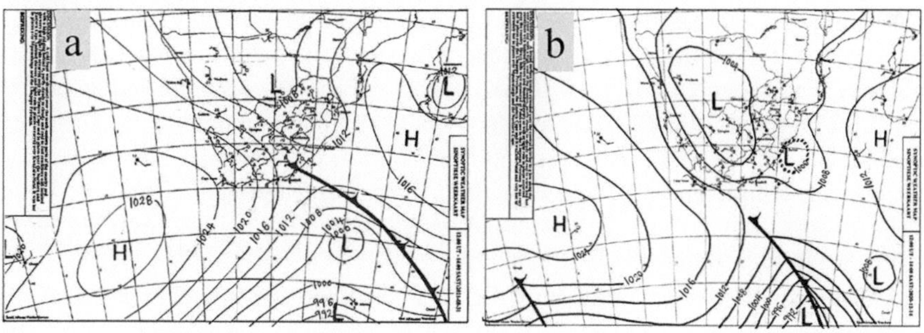

Fig. 4.12 Synoptic weather map showing type 4 **a** example 2021-08-31 and **b** example 2020-12-14. *Source* South African Weather Service

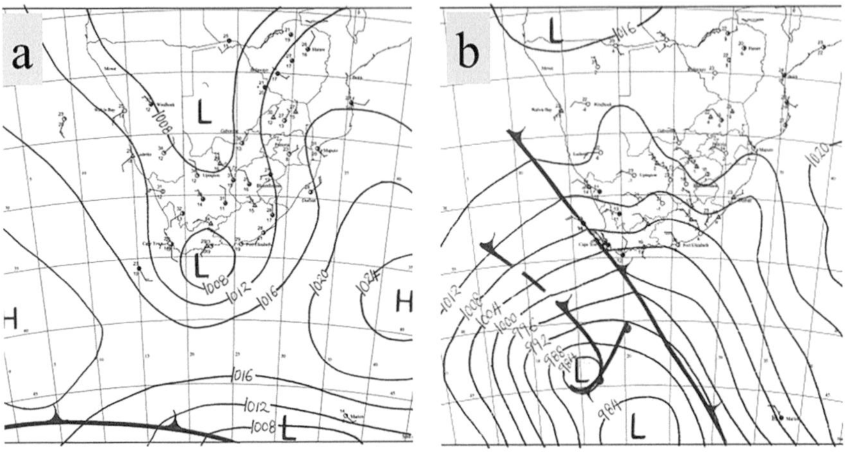

Fig. 4.13 Synoptic weather map showing type 5. **a** Example 2018-02-14 and **b** 2018-06-14. *Source* South African Weather Service

4.3.3 Weather Types During the Research Period 2020–2021

To assess the representativity of the measurement period in 2020–2021 at Rust en Vrede Farm, the same cluster analysis was applied to the 2020–2021 dataset. The following tables show the results (Tables 4.6, 4.7, 4.8 and 4.9).

The comparison of the measurement period to the 10-year definition period of weather types shows the following seasonal differences.

Main Type 3 showed a 2.9% decrease in summer while main Type 4 showed a 2.3% increase in the winter half year. Subtype 1 increased by 2.5% mainly in summer while subtype 2 decreased by 2.2% in winter. Subtype 5 showed only a slight change in frequency. The control period was generally cooler in winter (mean temperatures of Types 2 and 4 decreased by 0.4 °C and 0.6 °C at Cape Town station and by 0,8 °C and 1.2 °C at Paarl). Rainfall (and even snowfall in higher altitudes) showed a strong increase in winter (Types 2 and 5 at Cape Town, only Type 2 at Paarl), mainly because of some single high precipitation events (e.g., Type 5 with high rainfall at Cape Town on 29.09.2020; not at Paarl station). Wind speeds were slightly lower over all Types at both weather stations.

An analysis of Rust en Vrede farm daily values of the measurement period April 2020 to December 2021 (579 days) showed the frequency of weather types to be quite like the weather stations over the entire 2020 to 2021 period. Because of 3 summer months missing in the measurement series (January to March 2020), Type 3 had 3,6% less frequency compared to the entire period and winter Types 2 and 4 showed a slightly higher frequency.

Temperatures within the orchard were about 2 K lower than at Cape Town International Airport (CPT) weather station, while relative humidity was 5% higher. Wind speeds in the Cape Flats area are not expected to be lower than at CPT station but

Table 4.6 Cluster characteristics from Cape Town international airport soundings 2020–2021

	Sum	Mean	Mean	Mean	Mean	Mean	Mean	Mean	Mean	Mean	Mean	Mean	Mean
Type	N	GH_500	AT_500	WD_500	WS_500	GH_850	AT_850	WD_850	WS_850	GH_1000	AT_1000	WD_1000	WS_1000
1	107	5.727	−14.4	243	19.5	1.522	6.7	186	7.4	166	15.1	175	7.3
2	112	5.669	−16.8	267	27.3	1.504	4,7	276	11.1	156	13.5	316	6.4
3	175	5.817	−11.0	249	12.5	1.536	15,1	113	5.3	153	17.1	168	7.9
4	211	5.786	−12.4	277	18.6	1.529	13,2	302	7.0	152	16.0	139	4.8
5	56	5.782	−11.7	284	22.0	1.521	13,2	306	11.5	146	15.8	328	5.4
Average	661	5.756	−13.2		20.0	1.522	10,6		8.4	155	15.5		6.3

Table 4.7 Reanalysis data characteristics of clusters 2020–2021

Reanalysis data		Merra2			ERA5	
Type	N	Mean WS	Mean WD	Mean AT	Mean WS	Mean WD
1	107	5.5	166	12.5	5.5	167
2	112	6.6	284	11.5	6.2	293
3	175	6.8	129	13.9	6.8	137
4	211	4.9	164	12.6	4.5	167
5	56	5.1	285	12.3	4.9	290
Average	661	5.8		12.6	5.6	

Table 4.8 Weather station characteristics of clusters at Cape Town International Airport (CPT) and Paarl 2020–2021. Mean values of AT: ambient temperature, WS: wind speed, P: precipitation [mm]

Type/cluster	N	AT [°C]	RH [%]	P [mm]	P sum per day [mm]	WS [m $^{-1}$]
Cape town international airport daily 2020–2021						
1	107	16.6	71.1	2.1	4.7	5.7
2	112	14.7	78.0	8.4	10.3	5.3
3	175	18.5	68.5	0.4	1.7	6.1
4	211	16.5	73.1	0.6	4.1	4.2
5	56	16.6	72.3	2.4	10.2	4.0
Average	661	16.6	72.6	2.8	6.2	5.1
Paarl daily 2020–2021						
1	107	17.4	64.6	1.8	5.0	2.1
2	112	14.4	76.9	9.3	11.7	1.7
3	175	20.9	55.0	0.2	1.4	2.2
4	211	17.7	64.3	1.0	5.5	1.4
5	56	17.2	68.0	0.5	1.6	1.1
Average	661	17.5	65.8	2.6	5.0	1.7

Table 4.9 Seasonal frequency of clusters for 2020–21. Summer: October to March; Winter: April to September

Type	Summer [%]	Winter [%]	Total [%]
1	10.4	6.1	16.5
2	5.8	11.4	17.2
3	191	7.1	262
4	11.6	19.9	31.5
5	2.9	5.7	8.6
Total	50	50	100.0

Table 4.10 Characteristics of weather types measured at Rust en Vrede Farm (average of measurements F3 and F4). Mean of AT: ambient temperature [°C], WS: wind speed [m s^{-1}] P: precipitation [mm] and ET: Evapotranspiration within the citrus orchard [mm]

Cluster type/ type	N	N %	AT [°C]	RH [%]	P [mm]	P sum per day [mm]	WS [m s-1]	ET [mm]
1	97	16.8	15.26	75.43	1.0	3.5	1.54	2.0
2	109	18.8	13.22	86.77	6.9	8.9	1.07	1.2
3	131	22.6	17.95	69.56	0.3	2.8	1.05	2.4
4	189	32.6	15.32	77.90	1.5	5.4	0.98	1.7
5	53	9.2	15.46	80.02	1.8	5.8	0.94	1.3
Total	579	100	15.44	77.94	2.3	5.3	1.1	1.7

Table 4.11 Overall totals of precipitation (P) and evapotranspiration (ET) over the measurement period 2020–2021

Type	P [mm]	ET [mm]	Difference [mm]
1	97.6	196.9	−99.3
2	750.0	126.1	623.9
3	35.6	315.8	−280.2
4	278.6	322.3	−43.7
5	93.2	71.2	22.0
Total	1255.0	1032.3	222.7

are greatly reduced at the farm because of the presence of windbreak hedges and additionally within the citrus canopy. This is especially visible in the case of Type 3 where southeasterly strong winds in summer (6.1 m/s at Cape Town station) are effectively reduced by dense *Casuarina* hedges orientated southwest-northeast, i.e., perpendicular to the main wind direction of Type 3 (Table 4.10).

Rainfall and evapotranspiration computed following the FAO modified approach (Allen et al.1998) showed clear differences of the weather types. The dry summer Types 3 and 1 were characterized by large water deficits followed by Type 4 with its low rainfall tendency in winter. The main phases of positive water balance were days with Type 2 cyclonic rainfall in winter. Type 5 represents the cyclonic winter situation with lower rainfall. The overall climatic water balance within the measurement period was positive; mainly an effect of a lower frequency of Type 3 events and higher rainfall at Type 2 events as compared to the 10-year period 2010–2019 (Table 4.11).

4.4 Variability of Weather Types

Time series analysis of Cape Town International Airport weather station for the entire period 2010 to 2021 was conducted on daily values of the occurrence of weather types and rainfall. Data availability was poor for rainfall data (availability between 36 and

78%) for the different weather types and there was a complete data gap for November 2013 to March 2014. Therefore, we will consider relative deviations from the data set average for rainfall only (Tables 4.12 and 4.13).

The deviation of frequency of the weather types relative to the time series mean in relation to annual rainfall anomalies shows three main patterns:

1. **Drought years** with rainfall anomalies of −13% to −67% (2010–2011, 2015–2017). The typical pattern is a significant increase in frequency of Type 3 in summer combined with a strong decrease of Types 2 and 5 in winter. In those years, the frequency of rain-bearing coastal cyclones is significantly lower, while the inland summer thermal depressions dominate for longer periods. The main reason for low winter rainfall is a strong South Atlantic Anticyclone while high sea surface temperatures along the South African coast and in the Indian Ocean prevent the development of winter cyclones. 2015 and 2016 was a strong El Niño event (October to December 2015) followed by La Niña later in 2016 while 2010/2011 was a typical La Niña year. While El Niño and La Niña show opposite pressure patterns over the Pacific Ocean, their effect in both cases is an increase in sea surface temperatures around the southern African continent in winter (Haeseler and Ziese 2016), effectively preventing regional cyclogenesis and forcing the mid-latitude lows further south.
2. **Wet years** with positive rainfall anomalies of + 26% to + 33% (2012–2013). This pattern is characterized by a strong increase of Type 2 in winter while Type 3 frequencies are rather low in summer. Additionally, Types 5 or 1 are more frequent, bringing additional winter rainfall to the region. When Types 1 and 5 do not show a simultaneous increase with Type 2, rainfall anomalies remain moderately negative (2018–2019). This implies that regional winter cyclogenesis

Table 4.12 Relative frequency of weather types (T1…T5) for the years 2010 to 2021 in % at Cape Town international airport (CPT station)

Year	T1	T2	T3	T4	T5
2010	10.6	17.1	35.5	31.0	5.7
2011	13.9	19.7	30.0	29.4	7.0
2012	11.0	21.8	29.1	28.8	9.3
2013	15.0	22.5	23.6	30.7	8.2
2014	13.1	20.9	30.1	27.7	8.3
2015	16.7	12.5	31.0	30.3	9.4
2016	12.7	17.5	31.3	31.0	7.5
2017	15.4	15.4	31.3	28.4	9.6
2018	14.4	20.9	26.4	30.3	7.9
2019	13.9	22.0	26.1	29.8	8.1
2020	14.5	18.1	25.9	33.4	8.1
2021	17.7	15.9	26.9	30.9	8.6
Mean	14.1	18.7	28.9	30.1	8.2

Table 4.13 Annual differences to the time series average in %. Cape Town International Airport station. Dark gray: significantly negative, light gray: significantly positive differences

Year	T1	T2	T3	T4	T5	Rainfall
2010	−3.5	−1.5	6.6	0,9	−2.4	−12.9
2011	−0.1	1.0	1.1	−0,7	−1.2	−23.9
2012	−3.1	3.1	0.2	−1,3	1.2	26.1
2013	0.9	3.8	−5.3	0,6	0.1	33.0
2014	−1.0	2.2	1.2	−2,5	0.1	−3.6
2015	2.6	−6.1	2.1	0,2	1.3	−34.1
2016	−1.4	−1.2	2.4	0,8	−0.6	−26.4
2017	1.3	−3.3	2.3	−1,8	1.5	−67.4
2018	0.4	2.2	−2.6	0,2	−0.2	−16.8
2019	−0.2	3.3	−2.8	−0,3	0.0	−13.4
2020	0.4	−0.6	−3.0	3,3	0.0	107.1
2021	3.7	−2.8	−2.0	0,7	0.4	32.2

(Type 2) alone does not imply clearly above-normal rainfall years; passages of larger-scale mid-latitude low pressure systems are additionally required.

3. **Single event years** occurred when both Type 2 and Type 3 were below average, but rainfall anomalies were exceptionally high (2020–2021). In 2020, Cape Town International Airport station recorded three heavy winter half year rainfall events on 12.04.2020 (230 mm; Type 2), 14.07.2020 (78 mm; Type 2) and on 29.09.2020 (99 mm; Type 1). Type 5 synoptics led to untypical summer rainfall events on 07.11.2020 (10 mm) and on 11.12.2021 (99 mm). Regarding rainfall anomalies, 2020 and 2021 were untypical years with several single events exceeding the overall seasonal synoptic patterns: Type 3 and Type 2 frequencies were lower than average while Type 4 (2020) and Type 1 (2021) showed the highest relative increase over the time series. Both years were affected by another La Niña event; in this case leading to some exceptional events triggered by sea surface warming, such as the extension of eastern summer rainfall deep into the Western Cape area. Further inland, the single events may not show the same rainfall anomalies as at Cape Town International Airport station. Paarl station showed much lower to no rainfall for the three winter rainfall events and especially no rainfall for the two summer rainfall events. Contrary to the larger-scale synoptic patterns of drought

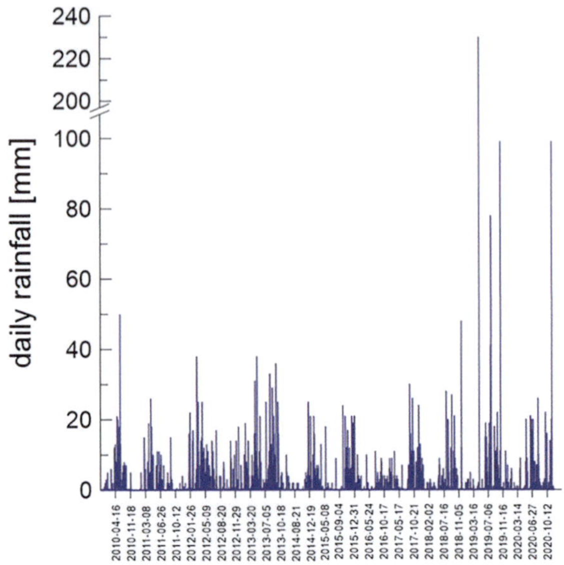

Fig. 4.14 Daily rainfall 2010–2021 recorded at weather station Cape Town international airport

and wet years, this implies more local characteristics of rainfall anomalies during single event years (Fig. 4.14).

4.4.1 Implications for Agriculture and Horticulture

Microclimatic processes like rainfall and evapotranspiration and the resulting water balance on farmlands of the southern Western Cape are largely driven by the mesoscale atmospheric dynamics as could be shown for the Rust en Vrede Farm site south of Stellenbosch. Apart from the entire set of local factors effecting crop evapotranspiration, irrigation water consumption or mechanical fruit damage caused by high wind speeds the overlying pattern is defined by frequencies of dry and windy weather types in the growing season (Types 3 and 1) as well as of rain-bearing types (Types 2, 4, and 5) over the winter half year, determining soil, groundwater, and reservoir water storage.

4.5 Climate Change, Extreme Events, and the "Day Zero" Cape Town Drought

Under the influence of climate change in Southern Africa (von Maltitz et al. 2024a), current climate models predict increased drought, while the western part of the continent being particularly affected by droughts. Associated with climate change is a significant increase in average temperatures and, thus, in evapotranspiration.

According to these model approaches (Engelbrecht et al. 2024), precipitation in the region will decrease as the mid-latitude cyclones move further south. Thus, a significant increase in extreme weather events under climate are expected in the coming decades with its immense impact on the ecosystems, agriculture, and water reservoirs as well on the urban water supply in the larger Cape Town area (von Maltitz et al. 2024a). One of these extreme weather events in the southern Western Cape was the so-called Zero-day drought of 2015–2018. If we look at this drought over a longer period than the analysis period, we can see that it was the driest since 1933 and a drought return period of once in 84 years (Wolski 2018). The period 2015–2017 were characterized by an expansion of quasi-stationary anticyclones in the South Atlantic and South Indian Ocean shifted poleward (Sousa et al. 2018; Burls et al. 2019), which corresponds with the predicted circulation changes and reduction of rainfall in the region (Engelbrecht et al. 2024). In so far, our finding that severe droughts are specifically linked to a strong increase of Type 3 in summer and a simultaneous decrease of Type 2 in winter is an empirical validation of predictive regional circulation modeling. Although a significant increase in droughts and thus a reduction in winter rainfall is predicted, above-average (winter) precipitation and the importance of off-season rain must also be considered for the long-term prediction of agricultural water supply and regional hydrological processes. De Kock et al. (2021) emphasized their importance for drought recovery periods and the rainwater supply for dams in the region of southern Western Cape. Cutoff lows which occur typically in the transition months October and April also can bring heavy rain events during short bursts of intensive rainfall in the summer months in the southern Western Cape. A study Abba and Abiodun (2020) showed that the Western Cape experiences 10 cutoff lows per year and they contribute about 11% of the annual precipitation over the province. In the recent years, extreme rainfalls occur in the Western Cape Region and affecting Cape Winelands (Fig. 4.15).

In July 2024, the highest monthly rainfall amounts exceeded the historical records at Cape Town (South African Weather station Oranjezicht) with 317.6 mm. This extreme events exceeded the historical average (128.0 mm) and is also the highest

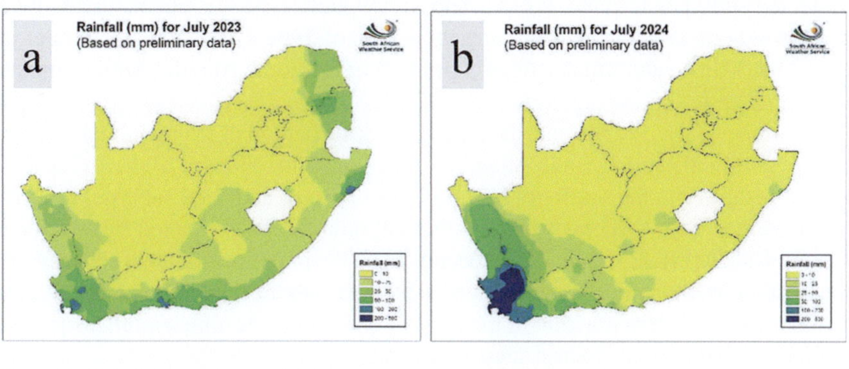

Fig. 4.15 Rainfall map for July 2023 (**a**) and July 2024 (**b**) (South African Weather Service 2024)

monthly rainfall recorded for July since comprehensive records began in 1960 (SAWS 2024). In contrast, the previous year saw only 81.8 mm of rain in July, highlighting the major difference in rainfall between these two years. In the same period the Cape Winelands stations, including, e.g., Paarl (455.6 mm), Worcester (280.6 mm), Jonkershoek (418 mm), and Cape Town International Airport (273.4 mm) have experienced substantial rainfall this winter season, with long-term monthly records being broken in these areas (South African Weather Service 2024). Furthermore, extreme strong winds and minor disruptive hail were reported on July 7, 2024, in Stellenbosch's Helshoogte Pass.

The current observations shows clearly that the interactions between ocean, atmosphere, and land are more complex and influenced at different scales by different drivers (von Maltitz et al. 2024b). Global circulation patterns have impact on local drivers of climate such as the formation of convectional storms and the strength and prevalence of offshore winds affecting the location and intensity of upwelling of the Benguela current. Climate change impacts on ocean currents could therefore have profound and poorly understood impacts on local weather and climate and finally on agriculture and land-use.

4.6 Conclusions

The basic pattern of the weather type classification derived from cluster analysis is the seasonal change of near-ground circulation depending on the strength and position of the controlling pressure centers (South Atlantic Anticyclone and either dynamic mid-latitude or thermal inland Lows). In winter, the South Atlantic Anticyclone is generally weaker and in a far western or northwestern position. The active pressure centers are either mid-latitude cyclones with frontal passage and cold air influx, strong wind, and high rainfall (Type 2 with overall minor frequencies) or cyclones further south in the mid-latitude west wind drift with frontal systems typically not reaching the Southwestern Cape (Type 4 as the more frequent type). Types 1 and 5 are mixed subtypes of main Types 2 and 4 in winter: Type 5 is characterized by a strong southern dynamic Low not present in case of Type 4 and Type 1 is similar to Type 2 but with an abnormal southerly position of the South Atlantic Anticyclone in winter.

In summer, Type 3 represents the typical synoptic pattern of strong inland thermal Lows, a seasonally strong South Atlantic Anticyclone and following high wind speeds and temperatures in the southern Western Cape. Types 1 and 4 are subtypes of Type 3 in this case: Type 1 occurs when the thermal inland Low is situated further north over the Kalahari Region and Type 4 occurs when the pressure gradients between the South Atlantic Anticyclone and the thermal inland Lows are much weaker because of an abnormally western position of the anticyclone. Thus, the high overall frequency of Type 4 is based on its hybrid seasonal characteristics.

In so far, our empirical classification of 5 different weather types characteristic for the southern Western Cape fits well into the main synoptic types defined in earlier works (e.g., Taljaard 1995) but provides a higher and specifically regional resolution based on different datasets although the investigation period of 12 years is rather limited.

When assigning a specific weather type to the daily time series 2010–2021 as derived from the k-means cluster solution the resulting time series shows drought and wet years characterized by abnormal frequencies of either the typical summer Type 3 or of all cyclonic Types (1, 2, 5), respectively. As the recent research has shown, both anomalies clearly identified in our analysis as well as the occurrence of single strong rainfall events are linked to anomalies of sea surface temperatures of the southeastern Atlantic and, thus, an effect of global anomalies of the tropical Walker circulation. Also, on a longer timescale, drought periods occurred with a clear 35 year signal (1920s–1930s and 1960s to 1970s) driven by a strengthening of the Hadley Cell and associated poleward displacement of the mid-latitude frontal systems. However, the exceptional drought in the Western Cape 2015–2017 did not correlate with long-term trends during the analyzed time period. Hemispheric drivers as revealed by a 30–35 year period indicate larger-scale shifts of circulation patterns and energy transfer from the tropical to the mid-latitudes also in the northern hemisphere (Littmann 1991).

This overall pattern does not imply uniform climatic characteristics at the southern Western Cape coastline and further inland, irrespective of orographic control. A comparison of the averages at CPT and Paarl stations showed that all weather types are warmer and less moist and windy at Paarl but have higher rainfall per event (especially Type 2) except of Type 1. This implies a higher frequency of cyclonic passage and frontal systems further inland in winter and a stronger influence of the thermal lows in summer. The southern Western Cape differs also in terms of long-tern rainfall trends from the northern and southeastern Cape regions (Wolski et al 2021).

Acknowledgements We thank Johan and Gideon Roos for their permission for the operation of the weather stations at Paul Roos Wine, Rust en Vrede Farm, Stellenbosch. FarmImpact project was supported by the Federal Ministry of Education and Research (BMBF) and managed by PT-DLR (grant no. FKZ 01LZ1711A-F) as part of "CLIENT II—International Partnerships for Sustainable Innovations" funding initiative.

References

Abba OS, Abiodun B (2020) Characteristics of cut-off lows during the 2015–2017 drought in the Western Cape, South Africa. Atmos Res 235:104772. https://doi.org/10.1016/j.atmosres.2019.104772

Allen R, Pereira L, Raes D, Smith M (1998) Crop evapotranspiration—Guidelines for computing crop water requirements. FAO irrigation and drainage papers 56, Rome

Biastoch A, Rühs S, Ivanciu I, Schwarzkopf FU, Veitch J, Reason C, Zorita E, Tim N, Hünicke B, Vafeidis A, Santamaria-Aguilar S, Kupfer S, Soltau F (2024) The Agulhas current system as an important driver for oceanic and terrestrial climate. In: von Maltitz GP, Midgley GF, Veitch J, Brümmer C, Rötter RP, Viehberg FA, Veste M (eds) Sustainability of Southern African

ecosystems under global change, ecological studies, Vol 248, Springer Cham, pp 191–220. https://doi.org/10.1007/978-3-031-10948-5_
Brandt P, Bordbar MH, Coelho P, Imbol Koungue RA, Körner M, Lamont T, Lübbecke JF, Mohrholz V, Prigent A, Roch M, Schmidt M, van der Plas AK, Veitch J (2024) Physical drivers of Southwest African coastal upwelling and its response to climate variability and change. In: von Maltitz GP, Midgley GF, Veitch J, Brümmer C, Rötter RP, Viehberg FA, Veste M (eds). Sustainability of Southern African ecosystems under global change, ecological studies, Vol 248, Springer Cham, p. 221–257. https://doi.org/10.1007/978-3-031-10948-5_9
Burls NJ, Blamey RC, Cash BA, Swenson ET, al Fahad A, Bopape, M-J M, Straus DM, Rason CJC (2019) The cape town "day zero" drought and Hadley cell expansion. NPJ Clim Atmos Sci 2:27. https://doi.org/10.1038/s41612-019-0084-6
Czernecki B, Głogowski A, Nowosad J (2020) Climate: an R package to access free in-situ meteorological and hydrological datasets for environmental assessment. Sustainability 12(1):394
Dubey A, Choubey AP (2017) A systematic review on k-means clustering techniques. Int J Sci Res Eng Technol (IJSRET, ISSN 2278–0882) 6(6)
Engelbrecht FA, Steinkopf J, Padavatan J, Midgley GF (2024) Projections of future climate change in southern Africa and the potential for regional tipping points. In: von Maltitz GP, Midgley GF, Veitch J, Brümmer C, Rötter RP, Viehberg FA, Veste M (eds). Sustainability of Southern African ecosystems under global change, ecological studies, Vol 248, Springer Cham, pp 169–190. https://doi.org/10.1007/978-3-031-10948-5_7
Engelbrecht CJ, Landman WA (2016) Interannual variability of seasonal rainfall over the Cape south coast of South Africa and synoptic type association. Clim Dyn 47:295–313. https://doi.org/10.1007/s00382-015-2836-2
Everitt BS, Landau S, Leese M, Stahl D (2011) Cluster Analysis. Wiley series in probability and statistics. Wiley, Hoboken
Geldenhuys H, Lötze E, Veste M (2023) Fruit quality and yield of mandarin (*Citrus reticulata*) in orchards with different windbreaks in the Western Cape, South Africa. Erwerbs-Obstbau 65:259–269. https://doi.org/10.1007/s10341-022-00725-3
Goliger AM, Retief JV (2002) Identification of zones of strong wind events in South Africa. J Wind Eng Ind Aerodyn 90(11):1227–1235
Haeseler S, Ziese M (2016) El Niño 2015/16 und seine klimatischen Folgen im Vergleich zu 1982/83 und 1997/98. (El Niño 2015/16 and its climatic consequences) Deutscher Wetterdienst. https://www.dwd.de/DE/leistungen/besondereereignisse/verschiedenes/20160615_el_nino_global_2015-2016.pdf
Hartigan JA, Wong MA (1979) Algorithm AS 136: a k-means clustering algorithm. J R Stat Soc Ser C (Appl Stat) 28(1):100–108
Hughes DA, Wright A (1988) Spatial variability of short-term rainfall amounts in a coastal mountain region of southern Africa. Water SA 14(3):131–138
Hurry L, van Heerden J (1982) Southern Africa' weather patterns. Goodwood, Cape Town
De Kock WM, Blamey RC, Reason CJC (2021) Large summer rainfall events and their importance in mitigating droughts over the South Western Cape, South Africa. J Hydrometerology 22(3) https://doi.org/10.1175/JHM-D-20-0123.1
Kruger AC, Goliger AM, Retief JV, Sekele S (2010) Strong wind climatic zones in South Africa. Wind Struct 13(1):19p
Kruger AC, Retief JV, Goliger AM (2013) Strong winds in South Africa: part 1 application of estimation methods. J S Afr Inst Civ Eng 55(2):29–45
Legendre P, Legendre L (2012) Numerical ecology. Elsevier, Amsterdam
Littmann T (1991) Rainfall, temperature, and dust storm anomalies in the African Sahel. Geographical J 157:136–160
Littmann T (2000) An empirical classification of weather types in the Mediterranean and their interrelation with rainfall. Theo Appl Climatol 66:161–171

Longley RW (1976) Weather and weather maps of South Africa. S Afr Weather Bureau Tech Pap 3, Pretoria

Mahlobo DD, Ndarana T, Grab SW, Engelbrecht FA (2019) Granger causality of the local Hadley cell and large-scale cloud cover over South Africa. S Afr J Sci 115(9/10):10. 1017159/sajs.2019/5724

Mason SJ, Jury MR (1997) Climatic variability and change over southern Africa: a reflection of underlying processes. Prog Phys Geogr 21(1):23–50

McInnes L, Healy J, Melville J (2018) Umap: uniform manifold approximation and projection for dimension reduction. arXiv:1802.03426

Reason CJ (2002) The wet winter of 2001 over the southwestern Cape, South Africa: potential large-scale influences. S Afr J Sci 98(5):307–310

Reason CJ, Jagadheesha D, Tadross M (2003) A model investigation of inter-annual winter rainfall variability over southwestern South Africa and associated ocean-atmosphere interaction. S Afr J Sci 99(1):75–80

RStudio Team (2020) RStudio: integrated development for R. RStudio, PBC, Boston, MA. http://www.rstudio.com/

Sousa PM, Blamey RC, Reason CJC, Ramos RM, Trigo RM (2018) The 'day zero' cape town drought and the poleward migration of moisture corridors, Environ Res Lett 13(12):124025. https://doi.org/10.1088/1748-9326/aaebc7

South African Weather Service (2024) Cape Town's record-breaking July 2024 rainfall. CS-CMS-LETT-003, pp 1–4

Taljaard J (1995) Atmospheric circulation systems, synoptic climatology, and weather phenomena of South Africa. S Afr Weather Bureau Tech Rep 30, Pretoria

Theron SN, Archer E, Midgley SJE, Walker S (2021) Agricultural perspectives on the 2015–2018 Western cape drought, South Africa: characteristics and spatial variability in the core wheat growing regions. Agric For Meteorol 304–305:108405. https://doi.org/10.1016/j.agrformet.2021.108405

Veste M, Littmann T, Kunneke A, Du Toit B, Seifert T (2020) Windbreaks as part of climate-smart landscapes reduce evapotranspiration in vineyards, Western Cape Province, South Africa. Plant Soil Environ 66:119–127. https://doi.org/10.17221/616/2019-pse

von Maltitz GP, Midgley GF, Veitch J, Brümmer C, Rötter RP, Viehberg FA, Veste M (eds) (2024a) Sustainability of Southern African ecosystems under global change, ecological studies 248, Springer Cham. https://doi.org/10.1007/978-3-031-10948-5

von Maltitz GP, Midgley GF, Veitch J, Brümmer C, Rötter RP, Viehberg FA, Veste M (2024b) Synthesis and outlook on future research and scientific education in Southern Africa. In: von Maltitz GP et al Sustainability of Southern African ecosystems under global change, ecological studies vol 248. Springer, Cham, pp 933–964. https://doi.org/10.1007/978-3-031-10948-5_32

Van der Walt AJ, Fitchett JM (2020) Statistical classification of South African seasonal divisions on the basis of daily temperature data. S Afr J Sci 116(9/10):15 Art. #7614. https://doi.org/10.17159/sajs.2020/7614

Weischet W, Endlicher W, (2000) Regionale Klimatologie Teil 2 Die Alte Welt. Teubner, Stuttgart-Leipzig (Regional climatology part 2 the old world)

Wickham H, Averick M, Bryan J, Chang W, McGowan LD, François R, Yutani H (2019) Welcome to the Tidyverse. J Open Source Softw 4(43):1686

Wickham H (2006) An introduction to ggplot: an implementation of the grammar of graphics in R. Statistics, pp 1–8

Wolski P (2018) How severe is the cape town´s "day zero" drought? Significance 15:24–27. https://doi.org/10.1111/j.1740-9713.2018.01127.x

Wolski P, Conradie S, Jack C, Tadross M (2021) Spatio-temporal patterns of rainfall trends and the 2015–2017 drought over the winter rainfall region of South Africa. Int J Climatol 41(Suppl. 1):E1303–E1319. https://doi.org/10.1002/joc.6768

Chapter 5
Simulation Modeling of Windbreak Types Affecting Wind Speed, Evapotranspiration, and Water Balance in the Cape Winelands

Thomas Littmann, Tobias Recke, Daniel Littmann, Anton Kunneke, Ben du Toit, and Maik Veste

Abstract Based on wind field modeling, the effects of various model hedge and technical windbreak types on wind speed and evapotranspiration in the cultivars were investigated. Computations were calibrated with microclimatic field measurements at citrus orchard and vineyard farm site in the Western Cape near Somerset West and Stellenbosch, respectively. The hedge types reduce evapotranspiration over the entire plot by 7.9% (Chinese Poplar), 11.1% (beefwood), and 12.8% (dense beefwood) at Babylonstoren and by 10.4%, 14.8%, and 17.4%, respectively, at Rust en Vrede farm. This implies considerable annual water consumption reductions of 318 to 533 m^3 ha^{-1} a^{-1} including estimated water consumption of the hedges at Rust en Vrede farm. Because of their limited height, technical windbreaks such as single vertical shading nets and solar fences show limited wind speed and evapotranspiration reductions (2.9% to 5.7%) Their effect on reducing water consumption is also quite low with 173 to 184 m^3 ha^{-1} a^{-1} at Rust en Vrede farm. Therefore, technical windbreaks only seem to be a feasible option when combining their moderate effects on irrigation demands with technical benefits. Although reducing wind speed and evapotranspiration significantly, first results on the effect of complete four-side netting of orchards on the water balance imply a significant increase in irrigation water demand because rainfall under the net is inhibited by 90%.

T. Littmann (✉) · T. Recke · D. Littmann
DLC—Dr. Littmann Consulting, Ennepetal, Germany
e-mail: littmann@littmann-consulting.com

A. Kunneke · B. du Toit
Department of Forestry and Wood Sciences, Stellenbosch University, Stellenbosch, South Africa

M. Veste
CEBra—Center for Energy Technology Brandenburg e.V., Cottbus, Germany

Department of Molecular Botany, University of Hohenheim, Institute of Biology, Hohenheim, Germany

M. Veste et al. (eds.), *Trees in South African Agriculture*, Innovations in Landscape Research, https://doi.org/10.1007/978-3-032-00993-7_5

Keywords Hedgerow · Technical windbreaks · Nets · Citrus orchard · Vineyard · Field modeling · Western Cape

5.1 Introduction

The efficient management of local water resources is a prerequisite to sustain agricultural production (Du Plessis 2017) and will become of more importance for the South African farmers under climate change (von Malitz et al. 2024). A major approach to reduce water demands in agriculture, especially in arid and semi-arid regions, is the introduction of smart eco-engineering measures directly influencing soil evaporation and crop transpiration. It is generally accepted that agroforestry as an integrated land-use system can provide various ecosystem services and can mitigate climate change effects (Hérnandez-Morcillo et al. 2018; Sheppard et al. 2020; Veste et al. 2024). This applies specially to perennial crops with their high-water demand, such as vineyards and fruit orchards (Veste et al. 2020). The data-based establishment of specially designed obstacles to airflow will significantly influence the near-ground wind field. For centuries it is a well-known fact that the establishment of tree shelterbelts is an appropriate method to reduce wind speeds to minimize soil erosion and to enhance soil quality (Burel 1996; Cui et al. 2012). The overall effect of shelterbelts on microclimate and ecological processes is common knowledge; however, it is still not clear what are possible positive effects on crop production systems. Understanding the direct effects on microclimatic boundary conditions needs more detailed and process-oriented information about the interactions between trees shelterbelts and crops. The effects on evapotranspiration and water use efficiency need more detailed information since only a few studies were conducted in agroforestry systems in Europe and South Africa, respectively (Campi et al. 2009, 2012; Kanzler et al. 2019; Markwitz et al. 2020; Veste et al. 2020).

The physical processes of evapotranspiration are mainly driven by the saturation deficit of the air depending on temperatures and the radiation balance, and the near-ground wind speed (Littmann and Veste 2006, 2008). The evapotranspiration processes are also depending on turbulent exchange and increase exponentially with wind speed over the canopy of a crop stand. There is little chance for agro-engineering measures to influence processes based on the saturation deficit of the air, but they can drastically reduce wind speeds. Some explicit investigations have shown the reduction of wind speed on the downwind side of a shelterbelt as a function of distance, aerodynamic porosity, and height (Cleugh 1998; Cleugh and Hughes 2002; Foereid et al. 2002; Frank and Ruck 2005; Campi et al. 2009). Effective reductions of wind speed were measured within the distance of four to six times the height of the shelterbelt, and minor reductions are effective up to 35 times the height (Wang and Tackle 1995, 1997). While distance to a windbreak is a parameter dependent on height and porosity, it has great significance for the planning of field width between two rows of windbreaks. For leeside distances of five times the height of a low porosity windbreak, wind speed reductions of 90% were reported and within ten times the height of up to 50% (Vigiak et al. 2003). Turbulence, however, may be increased in case of larger windbreak and crop canopy spacing when stationary eddies lead to higher momentum, heat, and vapor flux while densely spaced canopies like grapevine show only low turbulent flow (Bailey and Stoll 2013).

Older empirical approaches applying such parameterization focused on the reduction of wind erosion only (Fryrear et al. 1998). The effect of windbreaks on field evapotranspiration was investigated in more recent studies. De Vries et al. (2010) showed that leeside shading of a coniferous windbreak (height 10 m, porosity 0.4) in New Zealand reduced evapotranspiraton only over a few meters while the simultaneous temperature increase did not have any influence. Over the entire distance of 20 times the height of the windbreak, 16% water savings over the irrigation period were computed. Ohana-Levi et al. (2022) modeled evapotranspiration and trickle-irrigation water demands in a vineyard in Israel and derived nonlinear model functions to determine crop-specific evapotranspiration. The results showed that there is no need to apply kc-values in the Penman–Monteith approach to compute crop-specific evapotranspiration when sufficient meteorological measurement data are available. In any case, the proper determination of evapotranspiration is compulsory for the irrigation scheme as grape yield, and quality is closely dependent on the correct water balance.

Recently, the replacement of windbreak hedgerows by nets (Rolshausen et al. 2019), mainly on all four sides of an orchard, is a relatively recent development in the Western Cape. Alone in the Western Cape, the area covered by shade net increased from 559 ha in 2013 to over 3000 ha in 2020 (Kriel 2020) and with steadily further increase. Apart from the benefit of fruit sunburn damage protection, the role of shading nets in decreasing wind damage and evapotranspiration was investigated in a few recent publications. Tanny et al. (2009) found that horizontal netting (transparency 70%) reduced radiation, momentum, and mass fluxes between the boundary and canopy layers of citrus and apple orchards in northern Israel, thus weakening vertical air movements and evapotranspiration by lower temperatures and a lower vapor pressure deficit. Wind speeds were reported to be 9% lower as compared to the non-netted part of the orchards. Lulane et al. (2022) investigated the water saving benefits of a white horizontal shading net (transparency 80%) over an apple orchard over two growing seasons. Wind speed reductions were reported as 36% within the canopy. However, absolute wind speeds may be considered too low (0.3 m s^{-1}) to have considerable effect on evapotranspiration computations. Without a vertical net enclosure on the sides of the orchard, temperatures and relative humidity did not show significant difference to the outside control measurements. The computed water savings were 4% and originated mainly from a reduction of plant transpiration.

Therefore, our simulation approach compares the effects of different types of tree-based sheltbelts and furthermore includes artificial windbreaks to support climate-resilient development planning on the farm and landscape level. We present results from wind field modeling, evapotranspiration, and water balance computations based on and calibrated with field measurements at Rust en Vrede and Babylonstoren farm sites near Stellenbosch (Chap. 3). The aim is to compare the effect of various biotic and technical model obstacles to near-ground airflow on wind speed which in turn effects the evapotranspiration process within the crop stands and, consequently, the local water balance and irrigation water demand.

5.2 Methods

5.2.1 Wind Field Modeling and Wind Speed

Wind field modeling of the study sites was conducted with the non-hydrostatic model WAsP 12 implemented in EMD WindPro 3.5.587. WAsP contains several physical models (Fig. 5.1) to describe the wind flow over different terrain characterized by elevation and roughness changes and close to sheltering obstacles including a stability model for average heat flux conditions. A detailed summary of the physical background was given by Astrup and Larsen (1999). The digital recording of elevation data in a square of 20 km side length around the site is carried out using the SRTM elevation dataset (2014) of the U.S. Geological Survey, EROS Data Center, Sioux Falls, SD. with a horizontal resolution of 1 arcsec which can be transferred to the map scale used independently of scale and thus enables a high-resolution representation of the relief. The terrain roughness is digitized in the same radius as the terrain model with CORINE Landcover 2018 data (100 m grid). From the resulting roughness map, the roughness changes relevant for the calculation of the internal boundary layer (IBL) are calculated in the simulation model.

The wind data base applied was the reanalysis dataset EMD-WRF South Africa (grid node S33.812 E18.91 for Babylonstoren Farm and S34.001 E18.802 for Rust en Vrede Farm). These data (horizontal wind speed and wind direction) were available for a representative 25-year period (1994 to 2019) as hourly values and computed for 10 m above ground. As we consider the growth period during the summer half year for wind and evapotranspiration analyses explicitly, the reanalysis data were filtered for October to March.

In a first model run, mean wind data were calibrated by iterative adjustment of the wind energy level to the measured mean wind speeds at both farms (Table 5.1). For the calibration runs, we applied the actual roughness and wind flow obstacles as given by the windbreak hedgerows present. As a result, the mean model wind speeds for the control points show overestimates of 2.0–2.9% which is still within the expected range of measurement uncertainties. The results of the porosity measurements at the tree shelterbelts at Babylonstoren and Rust en Vrede farms were inserted into the flow obstacle definitions and could be confirmed within the model calibration procedure. In general, the derived porosity values for the different hedge types are consistent with windbreak classes introduced by Vigiak et al. (2003).

Further modeling was done for defined increments along a transect perpendicular to the existing windbreak hedges and also perpendicular to the prevailing wind directions over the summer half year (Fig. 5.2). The Babylonstoren transect covers

(a)

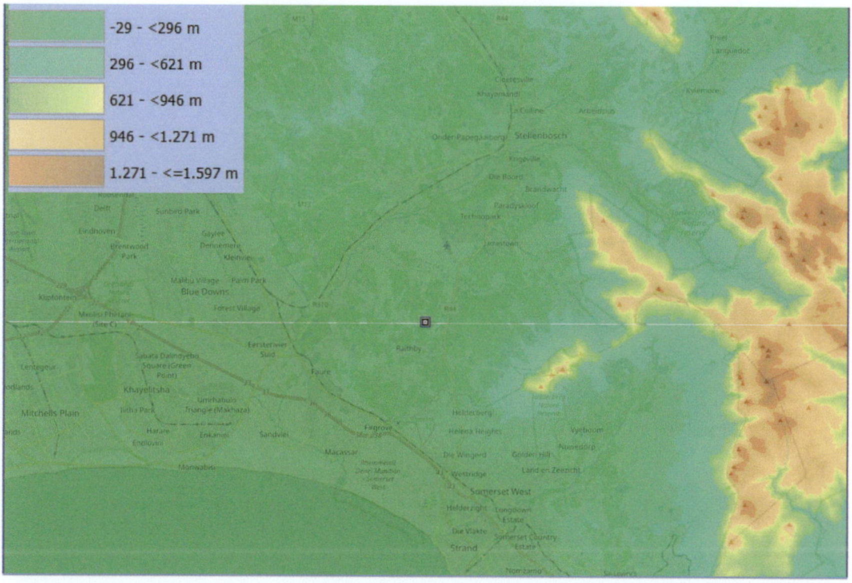

Fig. 5.1 Digital terrain (**a**) and roughness (**b**) models for Rust en Vrede farm, Stellenbosch

the vineyard of a distance of 260 m downwind of the windbreak in 10 m intervals (starting 30 m upwind) while the Rust en Vrede transect covering the citrus orchard has a length of 120 m downwind.

5.2.2 Evapotranspiration and Water Balance

There is a wide range of physical and empirical approaches to estimate actual evapotranspiration and crop-specific evapotranspiration (Novák 2012). Some are based on the radiative energy balance and the saturation deficit only, such as the original Penman and Penman–Monteith equations; others include water vapor and wind gradients above ground or a stand of crops such as the Thornthwaite–Holzman equation or focus on turbulent water vapor exchange above the crop canopy such as eddy covariance approaches (Littmann et al. 2004; Littmann and Veste 2008). In an extensive overview, Allen et al. (1998) recommended the FAO combined Penman–Monteith approach as the sole method for determining reference evapotranspiration. In the FAO approach, the bulk surface and aerodynamic resistances r_s and r_a of the original Penman–Monteith equation

$$\lambda ET = \frac{\Delta (R_n - G) + \rho_a c_p \frac{(e_s - e_a)}{r_a}}{\Delta + \gamma \left(1 + \frac{r_s}{r_a}\right)}$$

(b)

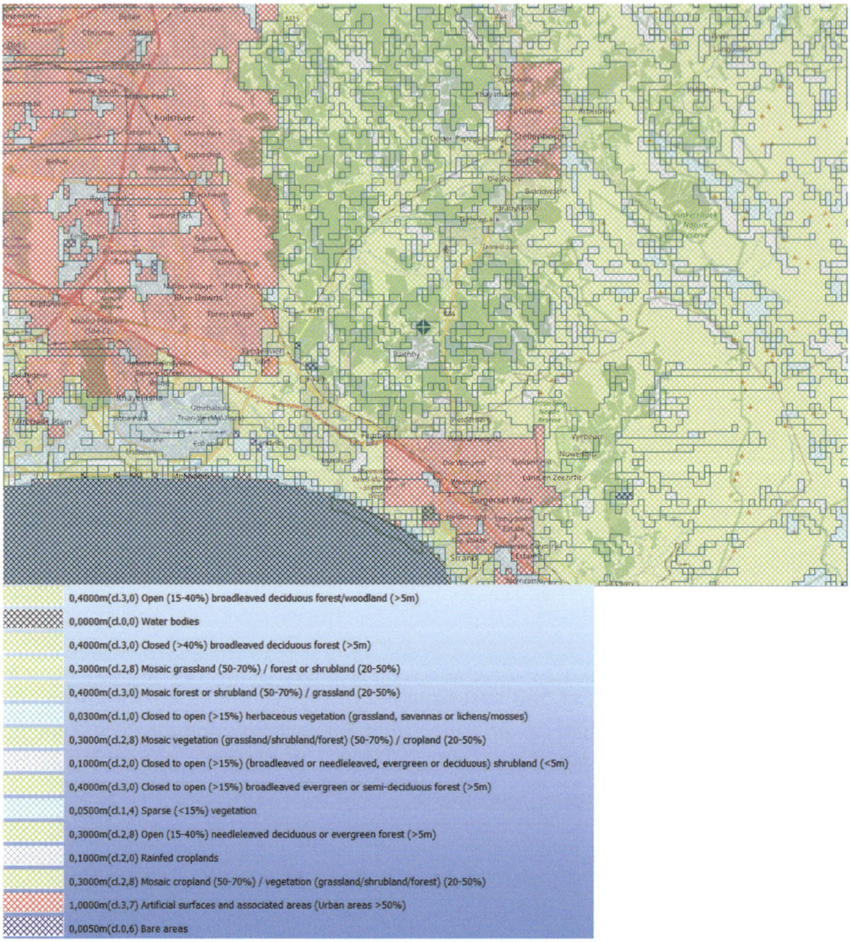

Fig. 5.1 (continued)

Table 5.1 Wind model calibration for different microclimate stations at Rust en Vrede and Babyslonstroren (for location see Chap. 3)

Farm/control	Measured [m s^{-1}]	Model [m s^{-1}]	Deviation [%]
Rust en Vrede 2020–2021			
F3 (3 m)	2.00	2.05	2.50
F4 (3 m)	1.25	1.28	2.40
Reference (3 m)	2.50	2.55	2.00
Babylonstoren 2015–2020			
US 1 (2 m)	2.21	2.26	2.26
US 2 (2 m)	1.71	1.76	2.92

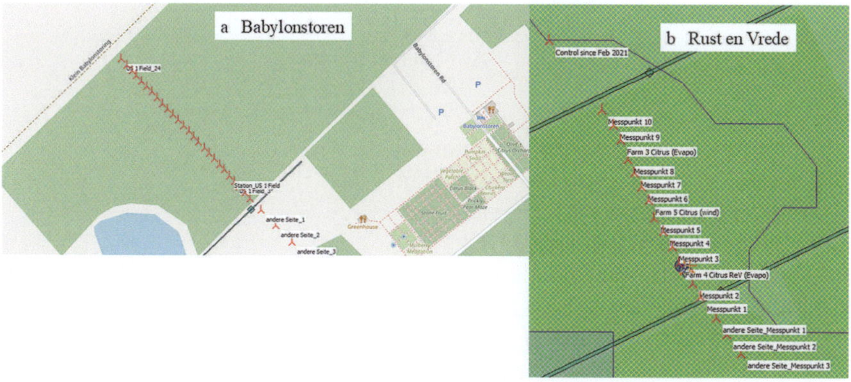

Fig. 5.2 Model transect at Babylonstoren (**a**) and Rust en Vrede (**b**). Windbreaks cover the entire width of the fields in southeast–northwest direction (Babylonstoren 260 m; Rust en Vrede 241 m)

are transformed into explicit terms of momentum (wind) and heat transfer

$$r_a = \frac{\ln\left[\frac{Z_m - d}{Z_{om}}\right] \ln\left[\frac{Z_h - d}{Z_{oh}}\right]}{K^2 u_z}$$

where

r_a aerodynamic resistance [s m^{-1}],
z_m height of wind measurements [m],
z_h height of humidity measurements [m],
d zero plane displacement height [m],
z_{om} roughness length governing momentum transfer [m],
z_{oh} roughness length governing transfer of heat and vapor [m],
k von Karman's constant, 0.41 [-],
u_z wind speed at height z [m s^{-1}], normally 2 m above ground.

$$r_s = \frac{r_l}{LAI_{active}}$$

where

r_s (bulk) surface resistance [s m^{-1}],
r_l bulk stomatal resistance of the well-illuminated leaf [s m^{-1}],
LAI_{active} active (sunlit) leaf area index [m^2 (leaf area) m^{-2} (soil surface)].

Reference values for irrigated crops are $r_s = 70$ s m^{-1} and $r_a = 208/u_2$ s m^{-1} (Allen et al. 1998). The resulting combined FAO Penman–Monteith equation (Allen et al. 1998) provides an explicit tool to compute reference evapotranspiration values on different time scales (hour, day, and month), depending on the measured input data resolution:

$$ET_o = \frac{0.408\Delta(R_n - G) + \gamma\frac{900}{T+273}u_2(e_s - e_a)}{\Delta + \gamma(1 + 0.34u_2)}$$

where

ET_oreference evapotranspiration [mm day^{-1}],
$R_n$net radiation at the crop surface [MJ m^{-2} day^{-1}],
Gsoil heat flux density [MJ m^{-2} day^{-1}],
Tair temperature at 2 m height [°C],
$u_2$wind speed at 2 m height [m s^{-1}],
$e_s$saturation vapor pressure [kPa],
$e_a$actual vapor pressure [kPa],
e_s–$e_a$saturation vapor pressure deficit [kPa],
Δslope vapor pressure curve [kPa $°C^{-1}$],
γpsychrometric constant [kPa $°C^{-1}$].

Estimates of crop-specific evapotranspiration over the crop growth period introduce the crop coefficient K_c which is used as multiplication factor for reference ET_0 of the FAO Penman-Monteith equation. For a better comparison and transfer to other sites of the results obtained here, we do not apply any further correction of the computed reference evapotranspiration.

The weather station data from 3 microclimate station (Farm 1 in Babylonstoren, Farm 3 and Farm 4 in Rust en Vrede, Fig. 5.3) serve as the input for ET calculations. To calculate ET as a function of the different windbreaks, wind speed data from weather stations (Chap. 3) were replaced with modeled transect wind speeds (both with and without windbreaks). ET was then calculated for the transects in the summer period 2020/2021. The 10-min raw meteorological data such as: air temperature, relative humidity, wind speed and heat flux, as well as incident and reflected radiation, were initially processed through an error analysis followed by linear interpolation of data gaps and data aggregation to daily measurements. Farm 3 experienced a total data loss in the period from December 2020 to March 2021, and thus, the analysis of the influence of windbreaks on evapotranspiration was accomplished using the daily measured values from Farm 4.

5.2.3 Windbreak Types

The tree-based windbreak types defined for the wind field model runs and further computations were derived from the existing hedgerows at Babylonstoren and Rust en Vrede farms (Chap. 2) as well as from another experimental site at Welgevallen near Stellenbosch (Chap. 6). Those types should be clearly understood as fully developed model types with the same reference height and porosity values as derived from Lidar measurements (see 5.2.4) and model calibrations. They do not represent the actual situation at the respective farm sites. For the wind simulation, we defined three widely used tree-based windbreak types (Chap. 2).

Fig. 5.3 Microclimate stations at Babylonstoren (**a**, Station Farm 1) and Rust en Vrede (**b**, Station Farm 4). (Photographs Maik Veste)

(I) Chinese Poplar (*Populus simonii*), (II) beefwood (*Casuarina cunninghamiana*), and (III) beefwood with dense canopy at the ground (Fig. 5.4).

5.2.4 Structural Analysis of Windbreaks Using LiDAR

For spatial structure analysis of the windbreaks, a Riegl VZ-2000i terrestrial LiDAR (Light Detection and Ranging) scanner was used to obtain height and porosity of the tree shelterbelts. The data were collected in both partially leaf-on (9 Sep 2020) and leaf-on conditions (17 Nov 2020) of poplar trees at Babylonstoren farm, while beefwood trees were investigated at Rust en Vrede farm. Scan positions were registered to a common coordinate system, and a single point cloud created from the combined scans. This point cloud was then filtered to remove outliers and ambiguous points. Single trees were segmented from the point cloud and breast height diameter (dbh), height and other crown parameters extracted. The resulting point cloud was analyzed

Fig. 5.4 Windbreak types. **a** Chinese poplar (type A), **b, c** Beefwood (type B), **d** Beefwood with dense cover at the ground (type C) (Photographs Maik Veste)

for biomass and density of the biomass. The point cloud, which consist of single points (x, y, z), was first transformed to voxels of 2 × 2 × 2 cm. The reason for this size is to reconstruct the volume of the biomass as realistically as possible because the scan pattern registers a point roughly every 2 cm. Voxels thus represent biomass (returned laser pulse) and not air. This is then evaluated against the total volume of the rectangle representing the windbreak (air and biomass).

To estimate the wind permeability, voxels were counted in a transect across the windbreak in the direction of the southeasterly wind which is the dominant wind direction in the area. The number of voxels and volume represented by the voxels were then expressed as a ratio of the total volume of the 1 m wide transect over the vegetation (Fig. 5.5). In the figure the voxels are coded by real color value (RGB). This index indicates the volume of the barrier to wind flow. The higher the index the denser the windbreak and the less wind will penetrate trough the barrier.

The windbreak of *Populus simonii* on the southeastern side of the orchard is represented by transect profiles with no. 1 on the western side on bottom left in and profile number 181 on the eastern side (Fig. 5.6a). The profile indicates gaps in the canopy and not a very dense structure as indicated by low index values.

The citrus orchard at Rust en Vrede is surrounded by *Casuarina cunninghamii* windbreaks (Fig. 5.4b,c). The dominant wind is from the southeast (bottom right to top left in Fig. 5.6b). Figure 5.6b shows the results of the height and density analysis where values from 1–163 indicate cross sections from SW to NE.

Fig. 5.5 Voxel cross section of windbreak in the dominant wind direction. The yellow box in this figure is the volume used for calculating the index shown in Fig. 5.6

The index is almost 10 times that found at Babylonstoren and can be explained by the denser structure of evergreen beefwood. Only the northern windbreak was scanned. It is assumed that values will be similar in the windbreak to the south. To provide an estimate of wind reduction in the downwind position of the windbreak, a *gross windbreak density index* was calculated, based on the volume of biomass occupying a volume represented by the total theoretic maximum volume of the windbreak (Table 5.2). For example, a windbreak of 100 m length with width 2 m at the base and height 10 m will occupy a maximum volume of 2000 m^3 in the landscape if it was a solid rectangle. However, if the biomass volume elements (voxels) occupy only 1000 m^3, then the *gross windbreak density index* will be 50%. The index cannot be directly transformed into "residual" porosity values as the relative porosity values applied to the calibrated model computations are smaller.

5.2.5 Simulation of Technical Windbreaks

Apart from the hedgerow types, we also considered technical windbreak types as there is a recent trend to either replace existing hedges by nets for fruit trees or to

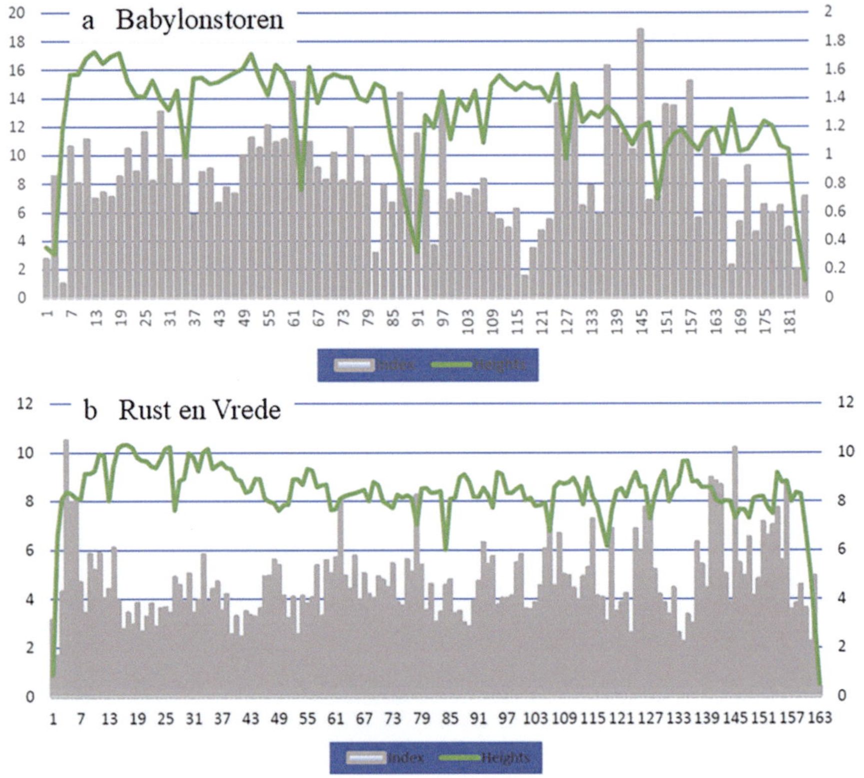

Fig. 5.6 Profile of windbreak height and density for the orchards at Babylonstoren (**a**) and Rust en Vrede (**b**) farms. Height is shown on the left axis (in m) and index on the right. The shorter the bar the less dense the windbreak is, and the more wind will penetrate

Table 5.2 Density index and height of actual windbreaks derived from LiDAR measurements

Windbreak	Gross windbreak density index (%)	Height (m)
Babylonstoren Citrus	60	13
Babylonstoren Shiraz (vineyard)	42	6
Rust en Vrede Citrus South	61	8
Rust en Vrede Citrus North	63	8

use such nets as a combined protection measure to prevent fruit damage by solar radiation (horizontal netting) and wind (vertical netting). The nets are developed to be used to influence various biotic and abiotic stress treating the fruit production, including sunburn by excess sunlight, hail damages, diseases, insect pests, and wind damages (Manja and Aoun 2019).

Following product information of several suppliers, shading nets show a typical light transparency of 30% to 50% which is equal to airflow porosity. For modeling purposes, we applied the netting layout of the Rust en Vrede citrus orchard with 5 m high vertical nets showing a porosity of 0.5. Horizontal netting could not be considered in the wind field model as it is no flow obstacle. On the other hand, it may reduce canopy surface roughness and thus slightly increase horizontal wind speed above the canopy while turbulence within the orchard should be reduced.

Nowadays, bi-facial solar module technology allows the establishment of vertical solar fences (Fig. 5.7). These rather new technology allows the integration into agri-photovoltaics systems in combination with crops (Reher et al 2024). Due to their structure, they serve simultaneously as a windbreak when mounted perpendicular to the main wind directions and producing solar energy. Water savings of up to 1410 m^3 ha^{-1} were found in the study with the vertical AV installation in Chile mainly due to the reduced irradiation combined with windbreak effects (Bruhwyle et al. 2023).For our model approach, we used with design solar fence a system with vertical alignment of special solar modules with solar cells on both sides (Fig. 5.7, Next2Sun Technology GmbH, Dillingen, Germany,) with a total height of 2.5 m, two modules mounted in landscape format and a free space above ground of 0.5 m. Integrated over the entire exposed cross section this is equal to a porosity of 0.2.

Fig. 5.7 Vertical Agricultural photovoltaic system as solar fence allows to combine agricultural production between energy production (Picture with courtesy of Next2Sun Technology GmbH, Dillingen, Germany)

5.3 Results and Discussion

5.3.1 *Modeled Wind Conditions at Canopy Level*

Long-term modeled wind conditions at canopy level (Babylonstoren vineyard 2 m; Rust en Vrede citrus orchard 3 m) for the summer half year show some differences at the two farm sites (Fig. 5.8). At Babylonstoren, the dominant wind directions are east-southeast and south-southeast (22% each) while south-southwest and west-southwest show frequencies around 10% only. This is an effect of orographic shielding by the Simonsberg mountain southwest of Babylonstoren site. The average wind speed is 2.31 m s^{-1} while highest sectoral wind speeds are from southeasterly directions. Rust en Vrede site shows a broader wind direction distribution with frequencies between 12 and 13% from the east-southeast to south-southwest sectors. The average wind speed is higher (3.01 m s^{-1}), and sectoral means are highest in the east-southeast and south-southwest sectors. This leads to significant differences in the Weibull distributions of wind speed (Fig. 5.9): while hours with mean wind speeds > 5 m s^{-1} are practically absent at Babylonstoren they show a frequency of 12.5% at Rust en Vrede. These findings should be kept in mind when discussing the relative wind speed reduction efficiency of the model windbreak types exposed to weak (Babylonstoren) and strong (Rust en Vrede) summerly wind fields.

Mean wind speeds at an elevation of 2 m above ground were computed with and without the digital terrain model at Babylonstoren as elevations show some differences between 3 and 10 m along the model transect (Fig. 5.10). Although the difference between models with and without terrain increases with the porosity of

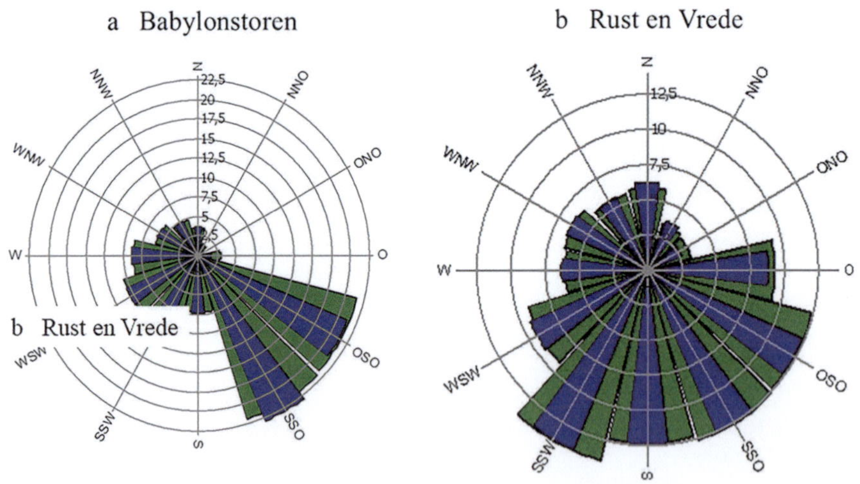

Fig. 5.8 ERA5-wind directions (green) over the summer half year (October to March 1994–2019) and modeled wind direction distribution (blue) for Babylonstoren (**a**) and Rust en Vrede (**b**) in %

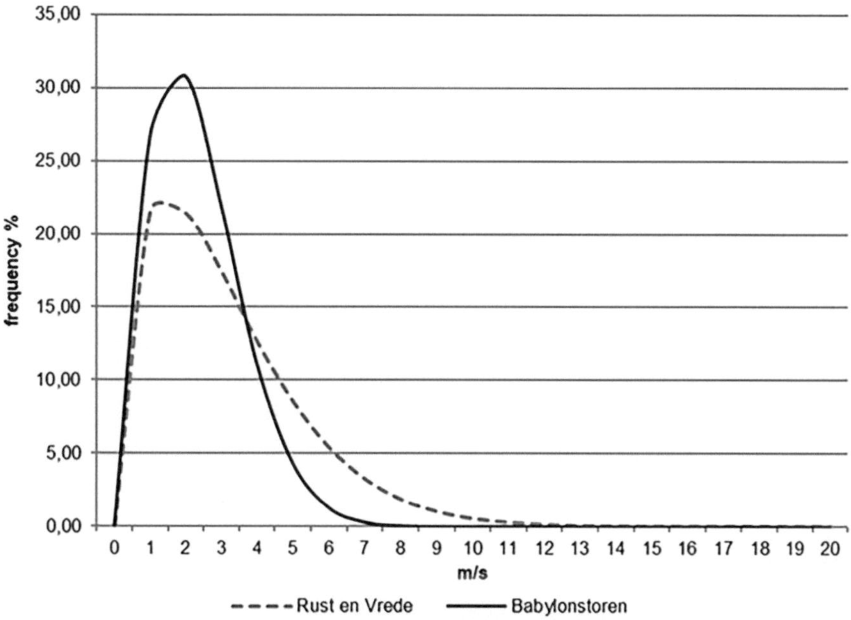

Fig. 5.9 Summary Weibull distributions of wind speed at canopy level

the hedge type (the influence of the windbreak on the wind speed transects increases) and the average difference is 5.66%, the overall interrelation of elevation and near-ground wind speed is linear and shows about 91.5% of wind speed variation along the transect to be explained by elevation (Fig. 5.11). Therefore, the terrain influence on wind speed transects may be considerable in case of complex terrain and should be considered, respectively, in further applications which also implies that direct comparisons of the impact of different windbreak types on the local wind field are confined to wind field models without terrain sub-models (Table 5.3).

As height and porosity of the windbreak are the dominant factors controlling the efficiency of wind speed reduction, the technical windbreaks show the least reductions over the entire transect although leeside reductions are considerable (around 20%) within the first 50 m behind the windbreak (Fig. 5.12). This implies distances between two windbreaks of 40 m to 50 m to obtain a minimum protection of > 10% wind speed reduction. The hedge types show a considerably higher reduction of 15% to 22% over the entire transect and up to 50% within the first 50 m behind the windbreak (Table 5.4). Therefore, the minimum distances required for a significant wind protection are large (160 m to 200 m). At Rust en Vrede farm site, the model transect does not show differences in elevation of more than 2 m. Therefore, terrain influence can be neglected. Because of the stronger wind field and the additional frequencies from southwesterly directions, relative wind speed reductions over the entire transect are higher (Table 5.5) as compared to Babylonstoren (factor 1.5 and especially for the non-biotic windbreaks with a factor of 2,0). However, leeside reductions near

Fig. 5.10 Modeled wind speed at 2 m a.g.l. along the Babylonstoren transect

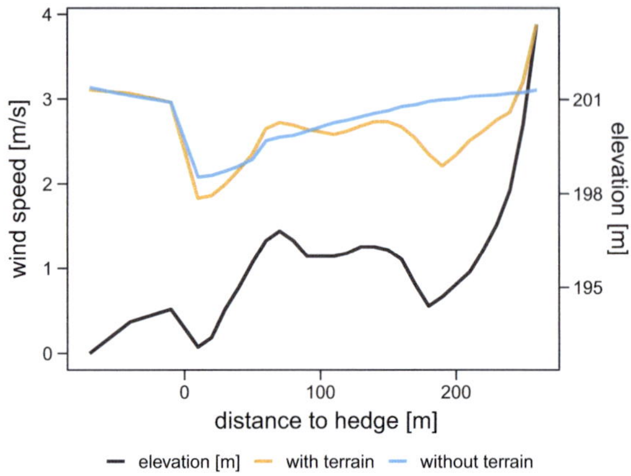

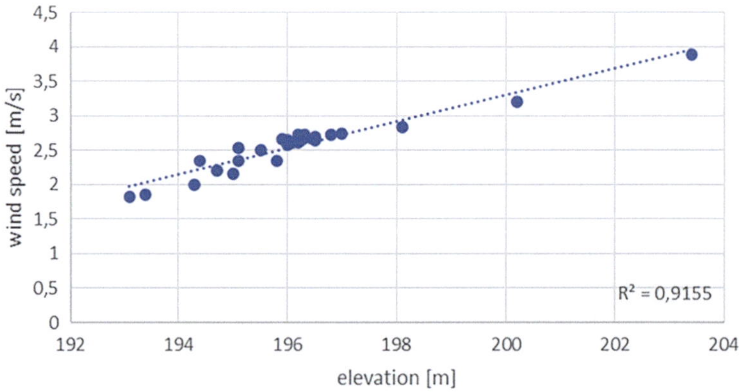

Fig. 5.11 Interrelation of wind speed and terrain elevation along the transect Babylonstoren

Table 5.3 Terrain influence on wind speed for different windbreak types at Babylonstoren

Difference of wind speed	With terrain [m s^{-1}]	Without terrain [m s^{-1}]	Difference[%]
Hedge type A	2.62	2.76	−5.2
Hedge type B	2.45	2.60	−5.8
Hedge type C	2.37	2.52	−6.0
Average	2.48	2.63	−5.7

the windbreak are similar and lower in a distance of 60 m to 100 m downwind of the windbreak. The higher reduction values near the end of the transect (100 m to 120 m) may be an effect of the vicinity to a following hedgerow on the northern site of the citrus orchard.

The horizontal wind profiles show different progressions mainly due to the porosity and height of the windbreaks, with the V-shape being much more pronounced

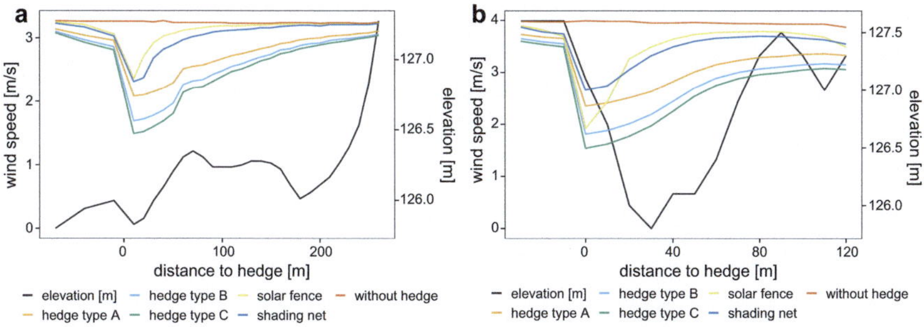

Fig. 5.12 Modeled wind speeds at 2 m along the transect at Babylonstoren (**a**) and Rust en Vrede (**b**)

Table 5.4 Results of modeled wind speed reduction of the different windbreak types (without terrain influence) at Babylonstoren

Windbreak type	Height [m]	Porosity [-]	Mean speed without windbreak [m s^{-1}]	Mean wind speed with windbreak [m s^{-1}]	Difference total transect [%]
Shading net	5	0.5	3.24	3.06	−5.7
Solar fence	2.5	0.2	3.24	3.07	−5.4
Hedge type A	15	0.4	3.24	2.76	−14.9
Hedge type B	15	0.2	3.24	2.60	−19.8
Hedge type C	15	0.1	3.24	2.52	−22.3
	Distance from windbreak 0 m-50 m [%]	Distance from windbreak 60 m-100 m [%]	Distance from windbreak > 100 m [%]	Distance > 10% reduction [m]	
Shading net	−19.7	−5.9	−1.6	50	
Solar fence	−21.5	−4.1	−0.9	40	
Hedge type A	−33.5	−20.3	−8.9	160	
Hedge type B	−44.6	−27.0	−11.9	190	
Hedge type C	−50.2	−30.4	−13.4	200	

Table 5.5 Results of modeled wind speed reduction of the different windbreak types (without terrain influence) at Rust en Vrede

Windbreak type	Height [m]	Porosity [-]	Mean speed without windbreak [m/s]	Mean wind speed with windbreak [m s^{-1}]	Difference total transect [%]
Shading net	5	0.5	3.95	3.49	−11.7
Solar fence	2.5	0.2	3.95	3.50	−11.6
Hedge type A	15	0.4	3.95	3.11	−21.2
Hedge type B	15	0.2	3.95	2.84	−28.3
Hedge type C	15	0.1	3.95	2.69	−31.8
	Distance from windbreak 0 m-50 m [%]	Distance from windbreak 60 m-100 m [%]	Distance from windbreak > 100 m [%]	Distance > 10% reduction [m]	
Shading net	−20.8	−6.6	−3.5	40	
Solar fence	−22.3	−4.2	−3.4	30	
Hedge type A	−33.9	−17.1	−14.2	120	
Hedge type B	−45.3	−22.7	−19.0	120	
Hedge type C	−50.9	−25.6	−21.3	120	

for the non-biotic windbreaks. To increase the quality of curve fittings, the wind profiles were decomposed into sections and fitted with different model functions (see Fig. 5.13). The derived model functions (Tables 5.6, 5.7) can be used as a basis for general wind protection models.

5.3.2 *Water Balance*

The weather station data from Farm 1 in Babylonstoren and Farm 3 and Farm 4 in Rust en Vrede (Chap. 3), respectively, serve as the input for evapotranspiration calculation after Allen et al. (1998). To calculate evapotranspiration as a function of different windbreaks types, wind speed data from weather stations were replaced with modeled transect wind speeds (both with and without windbreaks). Evapotranspiration was then calculated for the transects in the summer period 2020/2021. The 10-min raw meteorological data (air temperature, relative humidity, wind speed and heat flux,

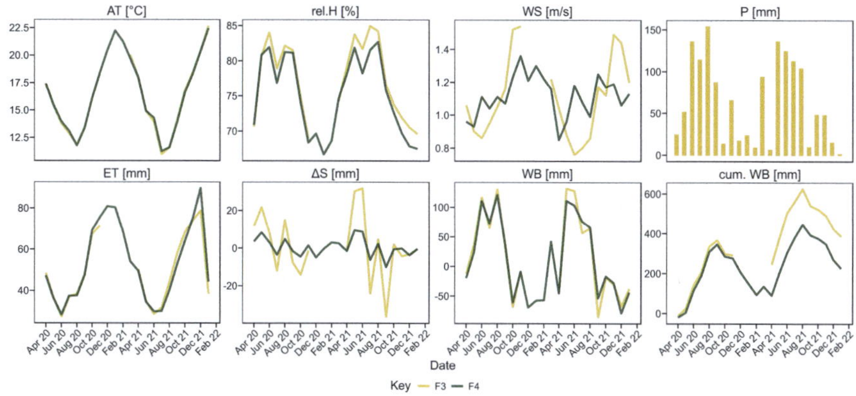

Fig. 5.13 Calculated water balance at Rust and Vrede (Farm 3 and Farm 4), during the study period April 2020 to January 2022

incident, and reflected radiation) were initially processed by error analysis followed by linear interpolation of data gaps and data aggregation to daily measurements.

Farm 3 experienced a total data loss in the period from December 2020 to March 2021. Therefore, the analysis of the influence of windbreaks on evapotranspiration was accomplished using the daily measured values from Farm 4. At Rust en Vrede, the logging of precipitation and soil moisture enabled the calculation of a simple water balance:

$$\mathrm{WB} = \mathrm{P} - \mathrm{ET} + \Delta\mathrm{S}$$

where.

WB = water balance.

P = precipitation.

ET = evapotranspiration.

ΔS = soil water storage.

In addition, data from the water balance were used to estimate the water consumption of the citrus orchard assuming the different windbreak types. Evapotranspiration was computed with the combined FAO Penman–Monteith equation (Sect. 5.2.2) applying the respective wind speed reduction of the windbreak types.

The influence of windbreaks on evapotranspiration reduction (Table 5.7) varies in the range of −2.9% (vertical shading net) to −12.8% (hedge type C) in Babylonstoren and −5.4% to −17.4% in Rust en Vrede, respectively, with the reduction of evapotranspiration by technical windbreaks being the lowest at both sites. In addition, this results in a deviation in the water balance of 33.8 to 85.7 mm for Rust en Vrede when comparing technical and biotic windbreak types.

Detailed data from the measurement stations at Rust en Vrede (F4 located directly northwest of the southeasterly hedge and F3 located directly southeast of the northwesterly hedge) show higher summerly wind speeds but not necessarily higher ET

Table 5.6 Curve fittings (without terrain influence) at vineyard Babylonstoren

Windbreak type	Distance to hedge [m]	Model	R^2	p-value
Shading net	−70 to −40	$y = -0.05*x-4.06$	0.99	<0.01
	−40 to −10	$y = -0.29*x-13.47$	0.97	<0.05
	−10 to 10	$y = -1.5*x-26.98$	1.00	<0.05
	10 to 30	$y = +1.26*x-54.03$	1.00	<0.05
	30 to 80	$y = +15.05*\log(x)-69.42$	0.93	<0.001
	> 80	$y = +0.02*x-4.05$	0.82	<0.001
Solar fence	−70 to −40	$y = -0.03*x-2.92$	0.97	<0.01
	−40 to −10	$y = -0.17*x-8.11$	0.97	<0.05
	−10 to 10	$y = -0.95*x-16.68$	1.00	<0.05
	10 to 30	$y = +0.78*x-33.63$	1.00	<0.05
	30 to 80	$y = +9.48*\log(x)-43.66$	0.93	<0.001
	> 80	$y = +0.01*x-2.55$	0.80	<0.001
Hedge type A	−70 to −10	$y = -0.084*x-10.39$	0.98	<0.001
	−10 to 80	$y = -1.43*x + 0.045*x^2-4.4e-04*x^3 + 1.4e-06*x^4-23.53$	0.98	<0.001
	80 to 240	$y = +15.84*\log(x)-91.75$	0.98	<0.001
Hedge type B	−70 to −10	$y = -0.12*x-13.72$	0.99	<0.001
	−10 to 80	$y = -1.56*x + 0.047*x^2-3.7*x^3 + 5e-07*x^4-32.83$	0.97	<0.001
	80 to 240	$y = +18.79*\log(x)-109.43$	0.98	<0.001
Hedge type C	−70 to −10	$y = -0.12*x-13.8$	0.98	<0.001
	−10 to 80	$y = -1.57*x + 0.04*x^2-2.9e-04*x^3 + 2e-07*x^4-32.66$	0.98	<0.001
	80 to 240	$y = +21.5*\log(x)-123.9$	0.98	<0.001

at the less shielded station F3 as temperatures were equal and relative humidity was even higher than at the F4 measurement better shielded from dry southeasterly winds (Fig. 5.13). However, rainfall events always led to higher fluctuations of soil water storage at F3, probably an effect of different infiltration properties. With two relatively good rainy seasons and in the study period, the cumulated water balance was mainly positive over the summer half years. In the model computations applying the higher long−term model wind speeds, the water balance results are negative but clearly show.

A simple water balance model (Table 5.8) can be used to analyze the water saving potential in relation to the different windbreak types for the citrus cultivar in Rust en Vrede using the meteorological and hydrological data from the FarmImpact weather stations, and the evapotranspiration model calculations described above.

Table 5.7 Curve fittings (without terrain influence), citrus orchard, Rust en Vrede

Windbreak type	Distance to hedge [m]	Model	R^2	p− value
Shading net	−30 to −10	y = −0.16*x−5.96	0.95	0.139
	−10 to 0	y = −2.8*x−32.58	1.00	–
	0 to 100	y = −40.9 + ((−2.81 + 40.9)/ (1 + exp((+11.2−x)/ + 8.5)))	0.99	<0.01
Solar fence	−30 to −10	y = −0.25*x−9.3	0.97	0.109
	−10 to 0	y = −4.48*x−51.8	1.00	–
	0 to 100	y = −63.61 + ((−4.6–63.61)/ (1 + exp((+11.76−x)/ + 8.17	0.99	<0.01
Hedge type A	−30 to −10	y = −0.09*x−8.99	0.99	0.055
	−10 to 0	y = −3.3*x−41.1	1.00	–
	0 to 100	y = −42.9 + ((−14.6 + 42.9)/ (1 + exp((+39.78−x)/ + 14.89)))	1.00	<0.001
Hedge type B	−30 to −10	y = −0.11*x−11.77	1.00	<0.05
	−10 to 0	y = −4.41*x−54.64	1.00	–
	0 to 100	y = −56.81 + ((−19.68 + 56.81)/ (1 + exp((+39.96−x)/ + 14.25)))	1.00	<0.001
Hedge type C	−30 to −10	y = −0.13*x−13.37	1.00	<0.01
	−10 to 0	y = −4.93*x−61.4	1.00	–
	0 to 100	y = −63.88 + ((−22.19–63.88)/(1 + exp((+39.97−x)/ + 14.37)))	1.00	<0.001

Table 5.7 Modeled evapotranspiration for Babylonstoren

Windbreak type	Wind speed [m s^{-1}]	Difference wind speed [%]	ET[mm]	Difference ET [%]
Without obstacle	3.2	0.0	735.8	0.0
Shading net	3.1	−5.9	714.4	2.9
Solar fence	3.1	−5.5	714.6	2.9
Hedge type A	2.7	−15.9	677.7	7.9
Hedge type B	2.6	−21.1	654.3	11.1
Hedge type C	2.5	−23.8	641.6	12.8

With a cultivated area of about 2.58 ha, an estimated water consumption of the hedge types of about 5% and the further assumption based on drone data from Seifert et al. that about 65% of the area is not covered with vegetation, the cumulative water balance for the summer half year 2020/2021 was calculated to determine the water saving potential compared to the reference model without windbreak. For the

Table 5.8 Cumulated water balance and modeled evapotranspiration with long–term wind data at Rust en Vrede farm

Windbreak type	Wind speed [m s^{-1}]	Difference wind speed [%]	Et [mm]	Difference et [%]	Cum. Wb [mm]	Difference cum. Wb [%]
Without obstacle	3.9	0	713.7	0	−609.1	0
Shading net	3.4	−13.4	675.5	−5.4	−570.9	−6.3
Solar fence	3.4	−13.2	673.1	−5.7	−568.5	−6.7
Hedge type A	3.0	−24.5	639.8	−10.4	−535.1	−12.1
Hedge type B	2.7	−32.6	607.8	−14.8	−503.2	−17.4
Hedge type C	2.5	−36.7	589.8	−17.4	−485.2	−20.3

comparison of the effects of different windbreaks types, the modeling approach is subject to rough estimates, and parameters such as crop coefficient, rooting depth, and soil texture variation were not considered. Compared to the reference model without windbreak, the water saving potential for the technical windbreaks is in the range of 248 m^3 ha^{-1} a^{-1} for the shading net to 263 m^3 ha^{-1} a^{-1} for the agri–PV solar fence (Table 5.9). On the other hand, the water saving potential for the biotic hedges increases significantly and varies from 457 to 765 m^3 ha^{-1} a^{-1} (Table 5.9) In relation to the total area cultivated, this corresponds to an annual saving potential of 641 to 681 m^3 for the technical and 1179 to 1974 m^3 for the tree–based windbreak types.

Table 5.9 Modeling results of the annual water saving potential with the use of different windbreaks at Rust en Vrede

Windbreak type	Abs mean difference cum wb [mm m^{-2}]	Abs mean difference cum. Wb with 5% hedge water uptake [mm m^{-2}]	Water consumption reduction [m^3 ha^{-1}]	Total water consumption reduction [m^3]
Shading net	38.2	38.2	248.3	640.6
Solar fence	40.6	40.6	263.9	680.9
Hedge A	74.0	70.3	457.0	1178.9
Hedge B	105.9	100.6	653.9	1687.1
Hedge C	123.9	117.7	765.1	1973.9

5.4 Shading Nets and Their Influence on Microclimate in a Citrus Orchard

Based on different positive reports, the existing beefwood hedges at Rust en Vrede farm were replaced by horizontal and vertical nets in August 2022 (Fig. 5.5a). This new setting allowed us to the test the influence of the technical nets on the microclimate in the citrus orchard (Fig. 5.5b, see also Chap. 3). For the measurement period from August 2022 to January 2023, we analyzed the microclimatic conditions inside the net and adjacent open field (Fig. 5.14). The results indicate that complete four-side netting modifies the microclimatic conditions over the entire season. Obviously, mean wind speed and wind gust reduction is significantly higher with 50% and 40.7%, respectively, than modeled for the single vertical shading net, and solar radiation was decreased by 16.4% (Table 5.10). This leads to a massive reduction of computed evapotranspiration (−32.4%; about six times higher than computed for the single vertical net) and corresponds to the findings of Tanny et al. (2009) that horizontal nets largely suppress turbulent flow and decrease evapotranspiraton. However, temperatures were not significantly lower than in the open field except of a slightly higher mean minimum temperature, and relative humidity was increased by 4.8% (Table 5.10). Horizontal netting seems to inhibit rainfall to a very large extent (−90.5% as compared to the open field rainfall). Over the test period, it was only one strong summer rainfall event in December that led to 22 mm precipitation within the citrus orchard. Although effectively reducing evapotranspiration, complete netting led to 30% more water deficit than in the open field (precipitation—evapotranspiration: −290.5 mm versus −223.2 mm).

5.5 Conclusion

It was shown that different windbreak types implemented quasi-perpendicular to the main wind directions (southwest-northeast orientation) reduce wind speed, evapotranspiration, and water balance deficits to different extent, depending on their height and aerodynamic porosity, i.e., their ability to reduce horizontal wind speed. The reduction of wind speed and evapotranspiration shows characteristic curves over a horizontal transect on the leeside of a windbreak that are approximated by curve fittings for several intervals behind the windbreak. The derived model functions can be used as a basis for general wind protection models.

Because of their limited height, technical windbreaks such as single vertical shading nets and solar fences show limited wind speed reductions (about 5% in a weak wind field environment like the Simonsberg area and about 11% in a stronger wind field environment like the Cape Flats). Minimum spacing of those windbreaks to obtain a significant wind speed reduction of > 10% ranges between 30 and 50 m. The contribution of technical windbreaks to the reduction of evapotranspiration is also quite limited (reductions of 2.9% in the weak wind field environment and of 5.4%

Fig. 5.14 Shading nets covering a citrus orchard at the Rust en Vrede farm in September 2022 (**a**) and microclimate measurements inside the orchard (**b**) (Photographs Maik Veste)

to 5.7% in the strong wind field environment) as well as their effect on improving the water balance (6.3% to 6.7% and water consumption reductions of 173 to 184 m^3 ha^{-1} at Rust en Vrede farm). Therefore, technical windbreaks only seem to be a feasible option when combining their moderate effects on irrigation demands with technical benefits, such as the generation of photovoltaic power of the solar fence at farm level.

On the other hand, first results on the effect of complete four-side netting of orchards on the water balance imply a significant increase in irrigation water demand as compared to a single vertical shading net windbreak. Additional horizontal netting

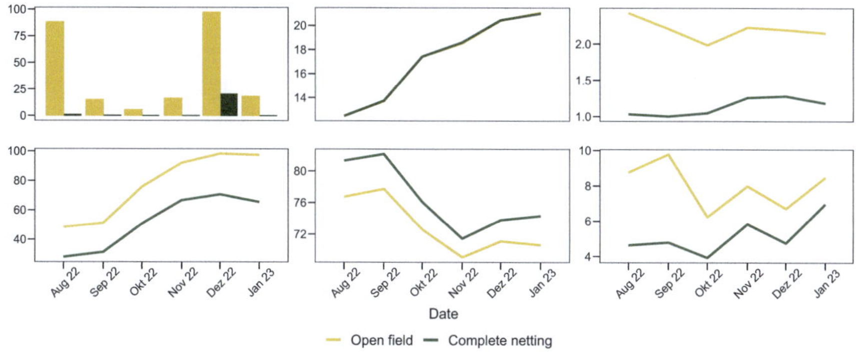

Fig. 5.15 Comparison open field and complete netting (see Fig. 5.14) at Rust en Vrede farm

Table 5.10 Comparison of measured microclimate in the open field and within the netted orchard at Rust en Vrede farm. Measurement period: August 6, 2022 to January 31, 2023. P: precipitation, ET_0: evapotranspiration

	Air temperature (min–max) [°C]	Rel, humidity [%]	Solar radiation [MJ m^{-2}]	Wind speed [m s^{-1}]	Wind speed gust [m s^{-1}]	Sum P [mm]	Sum ET_0 [mm]
Open field	17.5 (4.6–34.9)	73.0	3866	2.21	4.2	239.9	463.1
Under shading net	17.3 (4.9–34.6)	76.5	3233	1.11	2.49	22.6	313.1
Differences			−16.4%	−50%	−40.7%	−90.5%	−32.4%

indeed reduces wind speed and turbulent flow within the orchard, resulting in reductions of computed evapotranspiration of 32% but simultaneously inhibits 90% of rainfall which may pass the horizontal net in case of strong rainfall events only. Therefore, the resulting climatic water balance deficit is about 30% higher than without netting and results in a, respectively, higher irrigation water demand. This finding has not been considered yet in existing investigations of microclimatic effects of horizontal and complete four-side netting.

Tree windbreaks were considered as fully developed model hedgerows with the same height (15 m) but different porosity (*Populus* hedge type A 0.4, *Casuarina* hedge with open stem area type B 0.2, and *Casuarina* hedge with dense stem area type C 0.1) as derived from LiDAR measurements and wind field model calibrations at the test farms. Modeled wind speed reductions range between 15% (type A) and 22% (type C) in the weak wind field Simonsberg area and between 21 and 32% in the strong wind field environment of the Cape Flats. Minimum spacing of those windbreaks to obtain a significant wind speed reduction of >10% ranges between 160 and 200 m at Babylonstoren and is 120 m at Rust en Vrede, i.e., covers the entire plot considered. The hedge types reduce evapotranspiration over the entire plot by 7.9% (type A), 11.1% (type B), and 12.8% (type C) at Babylonstoren and by 10.4% (type A), 14.8% (type B), and 17.4% (type C) at Rust en Vrede farm. These are considerable mean reductions over the entire model plots, and the effects on improving the water balance (12% to 20% and water consumption reductions of 318 to 533 m $ha^{-1}a^{-1}$ including estimated water consumption of the hedges at Rust en Vrede farm) correspond to that.

To conclude, dense hedgerows such as fully developed *Casuarina* hedges with dense cover of the stem area (model type C) show the largest reductions of wind speed and evapotranspiration (including water uptake of the hedge) and have the highest potential of improving the water balance and effectively reducing the irrigation water demand of all windbreak types considered. However, other aspects like the time it takes to establish a fully-grown hedge, probable shading impact on crops, and root water competition will have to be considered in the individual farm context.

Acknowledgements We thank Johan and Gideon Roos from Paul Roos Wine, Rust en Vrede, Stellenbosch, and Ian de Villiers at Babylonstoren, Paarl, for their support and permission for the field experiments.

Thanks for the financial support of the Federal Ministry of Education and Research (BMBF) and management by PT-DLR for FarmImpact project (grant no. FKZ 01LZ1711A–F) as part of "CLIENT II—International Partnerships for Sustainable Innovations" funding initiative and ASAP—Agroforestry in Southern Africa—new pathways of innovative land-use systems under a changing climate (grant no. FKZ 01LL1803C) as part of SPACES II—Science Partnership for the Adaptation to Complex Ecosystem Processes in Southern Africa.

References

Allen R, Pereira L, Raes D, Smith M (1998) Crop evapotranspiration—guidelines for computing crop water requirements. FAO irrigation and drainage papers 56, Rome

Astrup P, Larsen SE (1999) WAsP Engineering Flow Model for Wind over Land and Sea. Risø National Laboratory, Roskilde, Denmark

Bailey BN, Stoll R (2013) Turbulence in sparse, organized vegetative canopies: a large–eddy simulation study. Bound Layer Met 147:369–400. https://doi.org/10.1007/s10546-013-9796-4

Bruhwyler R, Sánchez H, Meza C, Lebeau F, Brunet P, Dabadie G, Dittmann S, Gottschal R, Negron JJ (2023) Vertical agrivoltaics and its potential for electricity production and agricultural water demand: a case study in the area of Chanco Chile. Sustain Energy Technol Assess 60:103425

Burel F (1996) Hedgerows and their role in agricultural landscapes. Crit Rev Plant Sci 15(2):169–190

Campi P, Palumbo AD, Mastrorilli M (2009) Effects of tree windbreak on microclimate and wheat productivity in a Mediterranean environment. Eur J Agron 30:220–227

Campi P, Palumbo AD, Mastrorill M (2012) Evapotranspiration estimation of crops protected by windbreak in a Mediterranean region. Agricult Water Manag 104:153–162

Cleugh HA (1998) Effects of windbreaks on airflow, microclimates and crop yields. Agrofor Syst 41:55–84

Cleugh HA, Hughes DE (2002) Impact of shelter on crop microclimates: a synthesis of result from wind tunnel and filed experiments. Aust J Exp Agric 42:679–701

Cui Q, Feng Z, Pfiz M, Veste M, Küppers M, He K, Gao J (2012) Trade–off between shrub plantation and wind–breaking in the arid sandy lands of Ningxia, China. Pak J Bot 44:1639–1649

De Vries TT, Cochrane TA, Galtier A (2010) Saving irrigation water by accounting for windbreaks. WIT Trans Ecol Envir 134:121–131. https://doi.org/10.2495/SI100111

du Plessis A (2017) South Africa's water availability and use. In: Freshwater challenges of South Africa and its upper vaal river. Springer Water. Springer, Cham. https://doi.org/10.1007/978-3-319-49502-6_4

Foereid B, Bro R, Overgaard Mogensen V, Porter JR (2002) Effects of windbreak strips of willow coppice—modelling and field experiment on barley in Denmark. Agr Ecosyst Environ 93:25–32

Frank C, Ruck B (2005) Double—arranged mound—mounted shelterbelts: influence of porosity on wind reduction between the shelters. Environ Fluid Mech 5:267–292

Fryear D, Saleh A, Bilbro J, Schomberg M (1998) Revised wind erosion equation. Tech Bull No 1, Agric Res Serv Lubbock, USA

Geldenhuys H, Lötze E, Veste M (2023) Fruit quality and yield of mandarin (*Citrus reticulata*) in orchards with different windbreaks in the Western Cape, South Africa. Erwerbs—Obstbau 65:959–969. https://doi.org/10.1007/s10341-022-00725-3

Hernández—Morcillo M, Burgess P, Mirck J, Pantera A, Plieninger T (2018) Scanning agroforestry—based solutions for climate change mitigation and adaptation in Europe. Environ Sci Policy 80:44–52

Kanzler M, Böhm C, Mirck J, Schmitt D, Veste M (2019) Microclimate effects on evapotranspiration and winter wheat (*Triticum aestivum* L.) yield within a temperate agroforestry system. Agrofor Syst 93:1821–2184. https://doi.org/10.1007/s10457-018-0289-4

Kriel G (2020) The costs and pitfalls of buying nets. Farmers Weekly 8 October 2020. https://www.farmersweekly.co.za/agri—business/agribusinesses/the—costs—and—pitfalls—of—buying—nets/. Last accessed 21 Jun 2023

Littmann T, Veste M (2008) Evapotranspiration, transpiration and dewfall. In: Breckle S—W, Yair A, Veste M (eds.), Arid dune ecosystems—the Nizzana sands in the Negev desert. Ecological studies 200, Springer, Heidelberg, pp 183–200. https://doi.org/10.1007/978-3-540-75498-5_13

Littmann T, Steinrücke J, Bürger M (2004) Elemente des Klimas. Reihe Geographie Kompakt, Klett—Perthes, Gotha

Littmann T, Veste M (2006) Determination of actual evapotranspiration and transpiration in desert sand dunes of the Negev using different microclimatic approaches. For Stud China 8(1):1—9. https://doi.org/10.1007/s11632-006-0001-z

Lulane EB, Dzikiti S, Volschenk T, Loetze E, Midgley SJ (2022) Quantifying water saving benefits of fixed white protective netting in irrigated apple orchards under Mediterranean—type climate in South Africa. Sci Hort 305. https://doi.org/10.1016/j.scientia.2022.111439

Manja K, Aoun M (2019) The use of nets for tree fruit crops and their impact on the production: a review. Sci Hortic 246:110–122. https://doi.org/10.1016/j.scienta.2018.10.050

Markwitz C, Knohl A, Siebicke L (2020) Evapotranspiration over agroforestry sites in Germany. Biogeosciences 17(20):5183–5208. https://doi.org/10.5194/bg-17-5183-2020

Novák V (2012) Evapotranspiration in the soil-plant-atmosphere system. In: Progress in soil science, Dordrecht, NL

Ohana—Levi N, Ben—Gal A, Munitz S, Netzer Y (2022) Grapevine crop evapotranspiration and crop coefficient forecasting using linear and non—linear multiple regression models. Agricult. Water Manag 262. https://doi.org/10.1016/j.agwat.2021.107317

Reher T, Lavaert C, Willockx B, Huyghe Y, Bisschop J, Martens JA, Diels J, Cappelle J, Van de Poel B (2024) Potential of sugar beet (*Beta vulgaris*) and wheat (*Triticum aestivum*) production in vertical bifacial, tracked, or elevated agrivoltaic systems in Belgium. Appl Energy 359:122679. https://doi.org/10.1016/j.apenergy.2024.122679

Rolshausen P, Barry G, Schumann A (2019) Growing citrus under net. Citrograph 10(1):40–43

Sheppard JP, Bohn Reckziegel R, Borrass L, Chirwa PW, Cuaranhua CJ, Hassler SK, Hoffmeister S, Kestel F, Maier R, Mälicke M, Morhart C, Ndlovu NP, Veste M, Funk R, Lang Seifert F, du Toit TB, Kahle HP (2020) Agroforestry: an appropriate and sustainable response to a changing climate in Southern Africa? Sustainability 12(17):6796. https://doi.org/10.3390/su12176796

Tanny J, Cohen S, Grava A, Naor A, Lukyakov V (2009) The effect of shading screens on microclimate of apple orchards. Acta Hortic 807:103–108. https://doi.org/10.17660/ActaHortic.2009.807.11

Veste M, Littmann T, Kunneke A, Du Toit B, Seifert T (2020) Windbreaks as part of climate−smart landscapes reduce evapotranspiration in vineyards, Western Cape Province, South Africa. Plant Soil Environ 66:119–127. https://doi.org/10.17221/616/2019-pse

Veste M, Sheppard JP, Abdulai I, Ayisi KK, Borrass L, Chirwa PW, Funk R, Kapinga K, Morhart K, Mwale SE, Ndlovu NP, Nyamadzaw G, Nyoka BI, Sebola P, Seifert T, Senyolo MP, Sileshi GW, Syampungani S, Kahle H−P (2024) The need for sustainable agricultural land−use systems: benefits from integrated agroforestry systems. In: von Maltitz GP, Midgley GF, Veitch J, Brümmer C, Rötter RP, Viehberg, FA, Veste M (eds) Sustainability of southern African ecosystems under global change: Science for management and policy interventions. Ecological studies 248, Springer Nature, Heidelberg. https://doi.org/10.1007/978-3-031-10948-5_21

Vigiak O, Sterk G, Warren A, Hagen L (2003) Spatial modeling of wind speed around windbreaks. CATENA 52:273–288

Von Maltitz GP, Midgley GF, Veitch J, Brümmer C, Rötter RP, Rixen T, Brandt P, Veste M (2024) Synthesis and outlook on future research and scientific education in Southern Africa. In: von Maltitz GP, Midgley GF, Veitch J, Brümmer C, Rötter RP, Viehberg, FA, Veste M (eds) Sustainability of southern African ecosystems under global change: science for management and policy interventions. Ecological studies 248, Springer Nature, Heidelberg. 933–964. https://doi.org/10.1007/978-3-031-10948-5_32

Wang H, Takle E (1995) A numerical simulation of boundary layer flows near shelterbelts. Bound Layer Met 75:141–173

Wang H, Takle E (1997) Momentum budget and shelter mechanism of boundary layer flow near a shelterbelt. Bound Layer Met 82:417–435

Chapter 6
Benefits of Windbreaks for Fruit Production and Quality of Mandarins and Plums in the Cape Winelands

Heinrich Geldenhuys, Maik Veste, Thomas Littmann, Tobias Recke, and Elmi Lötze

Abstract Windbreaks and shade nets are used in South Africa to protect fruit orchards from wind scar damage, thereby avoiding major export quality losses. In the Western Cape, strong southeasterly winds (>3.6–4.5 m s^{-1}) during spring and summer are the main causes of significant external peel damage. In this study, wind scar damage on 'Tango' and 'Nadorcott' mandarins, as well as 'Ruby Star' plums, was quantified at various distances from tree-based windbreaks around Stellenbosch, Western Cape, South Africa. The wind speed was consistently higher at 16H compared to 3H after flowering (November–February). The highest pack out

The chapter is partly based on an earlier publication: Geldenhuys, H., Lötze, E. & Veste, M. Fruit Quality and Yield of Mandarin (*Citrus reticulata*) in Orchards with Different Windbreaks in the Western Cape, South Africa. Erwerbs-Obstbau 65, 959–969 (2023). https://doi.org/10.1007/s10341-022-00725-3.

H. Geldenhuys (✉) · E. Lötze
Department of Horticultural Science, Stellenbosch University, Stellenbosch, South Africa
e-mail: hg@sun.ac.za

E. Lötze
e-mail: elmi.lotze@agritechnovation.com

H. Geldenhuys
Citrus Research International, Stellenbosch, South Africa

M. Veste
CEBra–Centre for Energy Technology Brandenburg e.V., Cottbus, Germany

Department of Molecular Botany, Institute of Biology, University of Hohenheim, Hohenheim, Germany

M. Veste
e-mail: veste@cebra-cottbus.de

T. Littmann · T. Recke
DLC–Dr. Littmann Consulting, Ennepetal, Germany
e-mail: littmann@littmann-consulting.com

E. Lötze
Agri Technovation South Africa, Wellington, South Africa

M. Veste et al. (eds.), *Trees in South African Agriculture*, Innovations in Landscape Research, https://doi.org/10.1007/978-3-032-00993-7_6

(class 1 fruit) was observed near the windbreak (3H) and decreased as the distance increased towards 10H and 16H. There were no significant differences in the incidence of severe wind scar damage between the distance treatments. Most of severe wind scar damage (86%) occurred within the first 12 weeks after petal drop, during the highest wind incidence. Maximum wind speed reductions were observed between 2H and 4H from the established windbreaks. Our results confirmed the impact of wind on external fruit quality and highlighted the limitation of tree-based windbreaks as distance increases, which partly supports the current trend towards full enclosure of orchards with shade nets.

Keywords Shelterbelts · Wind analysis fruit orchard · 'Tango' and 'Nadorcott' mandarin · Japanese plum · Western Cape

6.1 Introduction

South Africa is the largest fruit producer in Africa and has gained a high reputation for the exceptional quality of its fruits, which are understandably in high demand, especially in Europe. In addition to biochemical properties, storage and transportability and visual appearance play an important role in assessing the quality of the fruit for export and determine the retail price at the end of the value chain (Chap. 12, Geldenhuys et al. 2023).

However, citrus and plums are wind-sensitive fruits and require special protection under high wind speeds with strong winds of 5–8 m s^{-1} at 10 m above ground in the cultivated areas and up to 20 m s^{-1} along the ridges of the Western Cape (Western Cape Department of Agriculture 2021). The severe southeasterly winds (Chap. 4) in the region during late spring and early summer are the main cause of significant external fruit quality reductions (Hamlet 2000; Pienaar 2005; Geldenhuys et al. 2023). Especially during times, the small immature fruits (Fig. 6.1a) are most susceptible to wind scarring, and wind protection is crucial. Rind appearance (blemishes), colour, and fruit size are the main external quality parameters (Lado et al., 2018). To avoid major financial losses in South African orchards (Chap. 12), protecting citrus fruit from extreme climatic conditions is critical (Brown 2018; Veste et al. 2024). Wind damage in citrus orchards may result in reduced yields because of poor fruit set (Tarara et al. 2005) and reduced fruit quality because of lesions and marks (Holmes and Farrel 1993, Fig. 6.1c). Plum production too in the Western Cape is very susceptible to wind scar damage, resulting in massive losses on the export market (Malik 2020). The consumers' perception and the main driver for the initial purchase are based on the rind coloration, as well as the absence of any blemishes on the rind (Botes 2018). In this context, wind scar damage of citrus could be one of the main causes of a loss in export quality, and the reduction in pack out varies between 10 and 70%, depending on the citrus type and production region (Holmes and Farrel 1993; Outspan 1997; Turner and Hardy 2006; Chap. 12).

Fig. 6.1 **a** ‘Tango’ mandarin fruit 12 weeks after petal drop. **b** Mechanical damage from moving leaves. **c** Severe wind scar damage (class 3) on ‘Tango’ mandarin fruit during stage two (cell enlargement) (Photos Heinrich Geldenhuys **a**, **c**, Maik Veste **b**)

There is currently limited research on the effect of shelterbelts on wind scar damage of soft citrus cultivars such as ‘Tango’ and ‘Nadorcott’ mandarin, and Japanese plums (‘Ruby Star’) under the specific climatic conditions of the Western Cape. The objectives of the study in the Cape Winelands (Geldenhuys et al. 2023) were to evaluate the influence of established beefwood (*Casuarina cunninghamiana*), Chinese Poplar (*Populus simonii*), and Eucalyptus species windbreaks. The wind scar damage was quantified according to three classes on commercial orchards in the Cape Winelands in Stellenbosch and Paarl.

6.2 Fruit Wind Sensitivity and Fruit Quality

Fruit damage depends on threshold wind speeds that impact both the vegetative and reproductive development of the tree. In citrus, wind speeds exceeding 4.5 m s^{-1} for 1 h, which is a gentle breeze with 3 Bt. By definition of the Beaufort scale at 3 Bt leaves, small twigs are in constant motion. This leads to mechanical friction between the hard leaves and the sensitive surface of the fruit (Fig. 6.1b). Citrus fruits are especially sensitive within the first twelve weeks after petal drop and may result in 90% of fruit being damaged (Freeman 1976; Outspan 1997; Turner and Hardy 2006; Cataldo et al. 2013). In plums, the fruit is most susceptible to wind scarring during the first three weeks after full bloom, if wind speed exceeds 5.8 m s^{-1} (4 Bt) (Michailides and Morgan 1992). For classification of fruit quality, the wind scar damage of citrus and plum can be classified into three groups: class 1 (no damage), class 2 (light wind damage), and class 3 (severe wind damage). The incidence of wind damage (classes 2 and 3) in citrus orchards has severe economic implications (Chap. 12), as only class 1 fruit can be exported.

6.3 Experimental Approach

6.3.1 Experimental Sites

The studies with mandarins and Japanese plums (Table 6.1) were conducted on three commercial orchards located in the Cape Winelands of Western Cape province, South Africa. (Meadows 2015, Chap. 1). The climate in the Western Cape of South Africa is regarded as Mediterranean, with wet winters and dry summers (Chap. 4). The average annual rainfall is 742 mm in Stellenbosch and 770 mm in Paarl, mm, respectively (Meadows 2015). To protect the citrus and plums fruit orchards with different windbreak types were established (Chap. 1).

Table 6.1 Farm locations, fruit trees, spacing, irrigation systems, planting year and soils (after Soil Classification Working Group 1991; Bargmann 2003) of investigated trials

Trial	Fruit trees	Tree spacing	Plantation year	Irrigation system	Soil
Rust en Vrede, (34°00′42″ S, 18°48′35″ E)					
Trial 1	'Tango' mandarin (*Citrus reticulata*) grafted on 'Carrizo' citrange rootstock (*Citrus sinensis* × *Poncirus trifoliata*)	5 m between rows and 2.5 m within rows	2015	Double line drip irrigation	Tukulu soil type
Babylonstoren (33°49′34″ S, 18°55′18″ E)					
Trial 2	'Tango' mandarin trees grafted on 'Carrizo' citrange rootstock	5 m between rows and 2.5 m within rows	2015	Micro-sprinklers	The Simonsberg vineyard area is characterised by granites, conglomerates, and loamy soils
Experimental farm Welgevallen (33°56′57″ S, 18°52′20″ E)					
Trial 3	'Nadorcott' mandarin (*Citrus reticulata* Blanco) grafted on a 'Carrizo' citrange rootstock	5.5 m between rows and 3.0 m within the row,	2015	Micro-sprinklers	Oakleaf soil type
Trial 4	Japanese plum 'Ruby Star' trees were planted on a Marianna rootstock with 'Songold' as a cross-pollinator	4 m between rows and 1.5 m within the row	2017	Micro-sprinklers	Oakleaf soil type

- **Rust en Vrede, Paul Roos Wine Stellenbosch**. The orchard was surrounded by a beefwood windbreak (Fig. 6.2a) (*C. cunninghamiana*) planted perpendicular to the orchard with a height of 9 m at the beginning of the investigation. The windbreak trees were uniform, with no discontinuities (gaps or missing trees) in the windbreak row.
- **Babylonstoren, Paarl**. The orchard was protected by two Chinese Poplar windbreaks (Fig. 6.2b) (*P. simonii*) planted parallel to the orchard at heights of 14 m (southern side) and 8 m (northern side) at the beginning of the experiment.
- **Experimental farm of Stellenbosch University Welgevallen, Stellenbosch**. A multi-row windbreak (MRW, Fig. 6.3a) was planted on 6 June 2019, parallel to the orchards, in the prevailing southeast wind direction. The windbreak height at the beginning of the trials (14/01/2020) was 2 m for the *Populus simonii* (Chinese Poplar) trees (trial 3) and 1 m for the *Eucalyptus* species (trial 4). Trees had thus not filled the space sufficiently (porosity) to serve as a windbreak, but measurements

Fig. 6.2 **a** Established Beefwood windbreak at Rust en Vrede, Stellenbosch (trial 1), and **b** Chinese Poplar at Babylonstoren, Paarl (trial 2). Windbreaks on the southern side of the orchards

indicated the initial stages of establishment for future projects. In addition, adjacent to the Chinese Poplar windbreak (trial 3), existing natural woody vegetation (NV) protected one section of the citrus orchard (Fig. 6.3b).

Fig. 6.3 Welgevallen Experimental Farm, Stellenbosch: **a** Multi-row windbreak, consisting of a parallel row of rose gum/red mahogany hybrid (*Eucalyptus grandis x resinifera*) and sugar gum hybrid (*Eucalyptus grandis x urophylla*). **b** Natural woody vegetation next to the 'Nadorcott' citrus orchard at the Welgevallen Experimental Farm, for trial 3

6.3.2 *Experimental Layout and Treatments*

Trial 1 Rust en Vrede: In the citrus orchard (trial 1), two microclimate stations (Chap. 3) were installed for wind analysis at different distances, 3H (Farm 4) and 16H (Farm 3) from the windbreak (Fig. 6.4a). 3H is the most wind-protected during summer with southeasterly winds, but the most wind-exposed during winter, with northerly winds. For wind scar damage analysis of 'Tango' mandarin trees, 10 trees were randomly selected for measurements at 3H (30 m), 10H (90 m), and 16H (140 m) (Fig. 6.4a). The experiment was conducted over two consecutive seasons, 2019/20 (season 1) and 2020/21 (season 2).

Trial 2 Babylonstoren: In the citrus orchard at Babylonstoren, wind analysis was carried out at 2H (Farm 1) and 4H (Farm 6) (Fig. 6.4b). Also, for this study, the windbreak on the southern side of the orchard was chosen as the reference point for distance treatments, and fruit of 10 trees were collected at 2H (25 m) and 4H (60 m). The experiment was conducted over two consecutive seasons, 2019/20 (season 1) and 2020/21 (season 2).

Trial 3 Welgevallen: The effects of two different windbreak types (natural, woody vegetation, and a multi-row windbreak) on citrus were evaluated at the experimental farm Welgevallen (Fig. 6.5a). Individual citrus trees served as experimental units, and 10 trees per treatment were randomly selected for data collection. Furthermore, wind and leeward of the trees were analysed.

Trial 4 Welgevallen: Fifteen plum trees per treatment were randomly selected at distances 10 and 50 m from the windbreak for data collection (Fig. 6.5b). This trial was only performed during one season (2020/21) and performed on bearing plum trees.

6.3.3 *Harvest and Damage Analysis*

Wind Scar Damage at Harvest

The visual wind damage analysis for soft citrus (trials 1, 2, and 3) follows the approaches of the Citrus Research International (Figs. 6.6 and 6.7), while for plums (trial 4) of the Unifruco wind damage score (Fig. 6.8) was used. The scales range from 1 (no blemish) to 8 (very severe wind scar damage). The wind damage scale for citrus (Figs. 6.6 and 6.7) and plums ranged from 1 (no blemish) to 8 (very severe wind scar damage). For export only fruits classified in class 1 with blemishes scored 1, 2, or 3 (no wind damage/very light). However, there is zero tolerance for major fruit damage (>12 mm^2), and fruit classified as 3 can still be downgraded to class 2. Fruit graded 4 or 5 was classified as class 2 (light wind damage), being still marketable as fresh produce, but not for export. Fruit graded 6, 7, or 8 were classified as class 3 (severe wind damage) and were only suitable for juicing or the local market. To

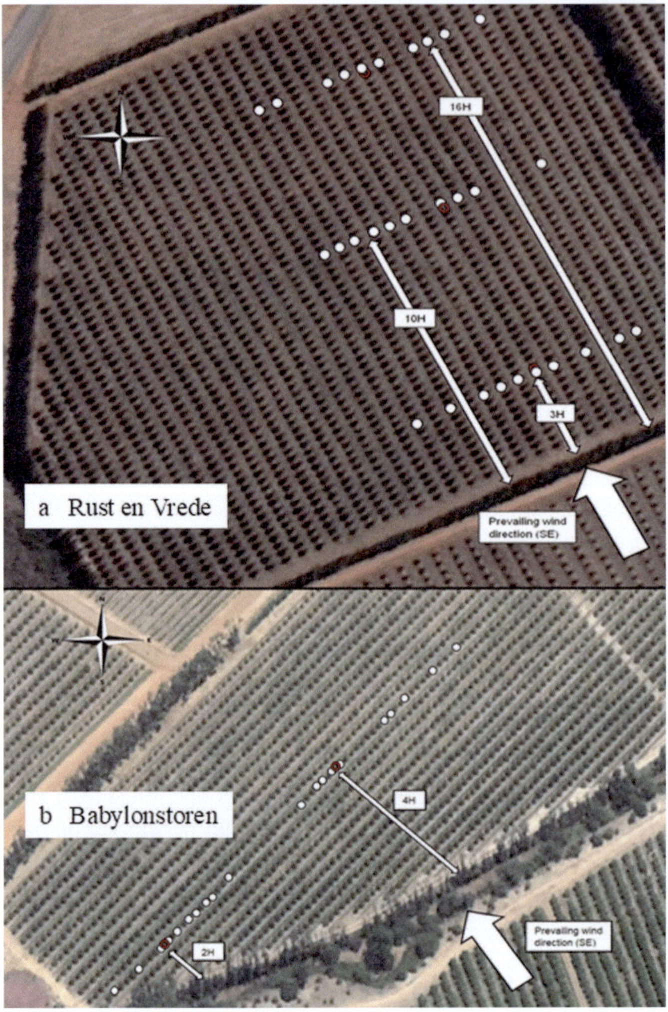

Fig. 6.4 Experimental layout at **a** Rust en Vrede, Stellenbosch (trial 1) and **b** Babylonstoren, Paarl (trial 2). The location of the microclimate station of each distance treatment is indicated with a red circle (own design, Photo Google Maps)

determine wind damage intensity, the wind mark was covered with an index finger as a size measurement. When the mark was covered completely, it was still classified as class 1. If the mark was bigger than the finger, the fruit was classified as class 2, and when there were marks on both sides of the fruit, it was classified as class 3. Trees in trials 1, 3 (citrus) and 4 (plums) were strip-picked for the evaluation of external quality. For trial 2 (citrus), a representative sample of ±25 kg fruit per tree was collected for wind scar damage evaluation. The percentage of fruit per class distribution was calculated for each treatment.

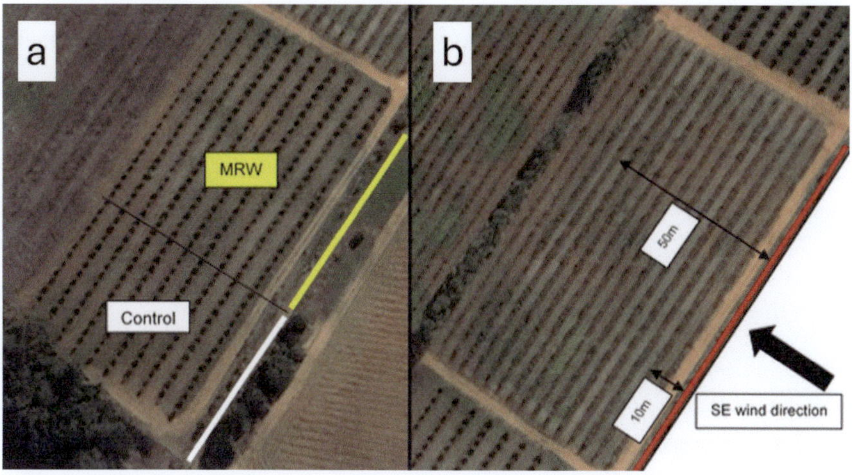

Fig. 6.5 The experimental layout at Welgevallen Experimental Farm, Stellenbosch for trial 3 (**a**) and trial 4 (**b**). Trial 3, a newly established multi-row windbreak (MRW)—*P. simonii* (yellow) and existing, natural woody vegetation (NV) (white). Trial 4, a parallel row of *Eucalyptus. grandis x resinifera* and *Eucalyptus. grandis x urophylla* (red) with different distances from windbreak (own design, photo Google Maps)

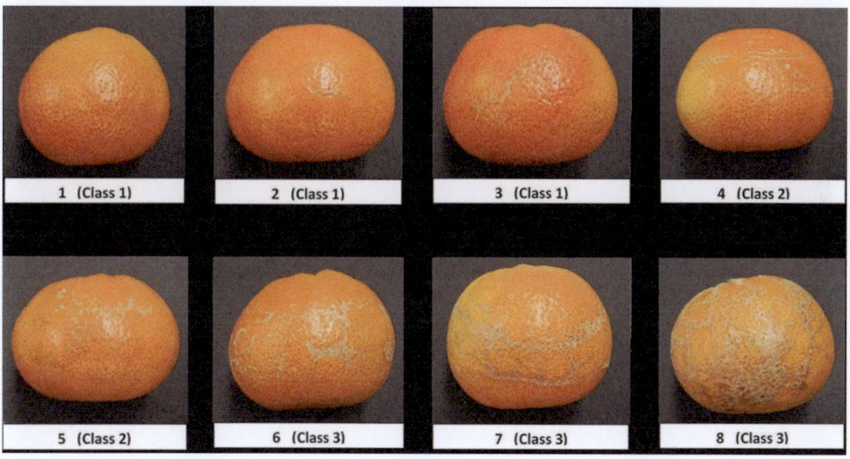

Fig. 6.6 Wind scar damage scale for soft citrus ('Tango' and 'Nadorcott') and fruit class distribution for mature fruit

Wind Scar Damage After Fruit Set

To test the wind sensitivity of citrus after fruit set, unripe fruits were also examined during the mid-summer period the season 2020/21. Ten trees were selected at random, per windbreak type treatment and were strip-picked and scored according to the above-described procedure. Furthermore, an additional monitoring of citrus were

Fig. 6.7 Wind scar damage scale for soft citrus ('Tango' and 'Nadorcott') and fruit class distribution for immature fruit

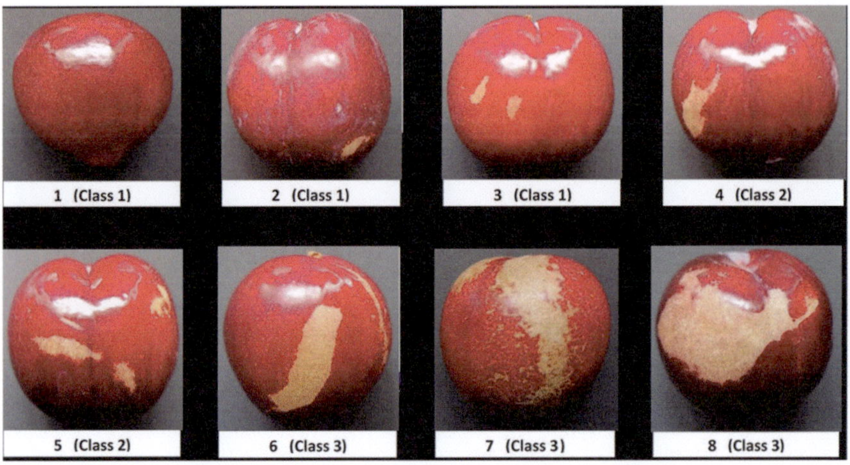

Fig. 6.8 Wind scar damage scale for plums ('Ruby Star') and fruit class distribution

carried out and evaluated wind damage for the same fruit on four occasions between December 2021 and January 2022. 10 fruit per tree, with 5 trees per treatment were selected for the investigation.

6.3.4 Statistical Analysis

Statistical analysis software (SAS) Enterprise Guide 7.1 (SAS Institute, Cary, NC, USA) and the GLM (general linear models) procedure were used for analyses. XLSTAT (2024.1.1) was used for data analyses for the early wind scar evaluation during the 2021/22 season. Data were subjected to a one-way analysis of variance (ANOVA) to compare means. Interpretation of the results should thus reflect the layout of the trials. A logit function was used to analyse the percentage of wind damage, as fruit numbers varied between trees. For post hoc testing, means were separated at a 5% significance level (p-value < 0.05) using the Fisher least significant difference (LSD) test.

6.4 Wind Analysis in the Citrus Orchard

In citrus, regular wind speeds of >3 Beaufort (>3.6–4.5 m s^{-1}) is crucial threshold for the development of wind scar damages on the citrus fruits. As illustration, for the Cape Winelands, a detailed analysis of the local wind field of Rust en Vrede and Babylonstoren conducted (Chap. 3). It was clear that the established windbreaks reduced the average wind speed (Chap. 3) inside the citrus orchards and minimized the incidences of wind speeds above the threshold for fruit damage. The average wind speed at distance 3H (Farm 4) was 2.4 m s^{-1}, with the highest wind speed of 4–6 m s^{-1} from NNW to NNE. At 16H (Farm 3), the average wind speed was 2.4 m s^{-1}, with the highest wind speed of 4–6 m s^{-1} from SSE, S, N (Fig. 6.9a). During the investigation period, the average wind speed outside the orchard was 3.6 m s^{-1} with the highest wind speed of 6–12 m s^{-1} from NNW to NEE. A clear daily pattern of highest wind speed could be observed at the different measurement positions (Fig. 6.9b).

- 3H (Farm 4) highest wind speed in the midday and evening hours from NW
- 16H (Farm 3) highest wind speed in the midday and evening hours from SE, S and SW
- Reference (highest wind speed in the midday and evening hours from NW).

This implication of the windbreaks is confirmed by previous studies (Nuberg 1998; Hamlet 2000; Peri and Bloomberg 2002). For a semipermeable tree windbreak (45–55% porosity), the distance at which maximum wind speed reduction occurs is between 2 and 4H.

However, for fruit damage, the seasonal variation of critical wind speeds has a higher relevance than the average and depends on the severity as well as wind directions. In general, outside the wind-protected orchard, the limit value is exceeded almost daily, and peak gust speeds are sometimes over 10 m s^{-1} in the vegetation periods (Fig. 6.10). Inside the orchard, the wind speed is drastically reduced, and wind speeds above 3.6 m s^{-1} are less frequent at Farm 3 and Farm 4 compared to the

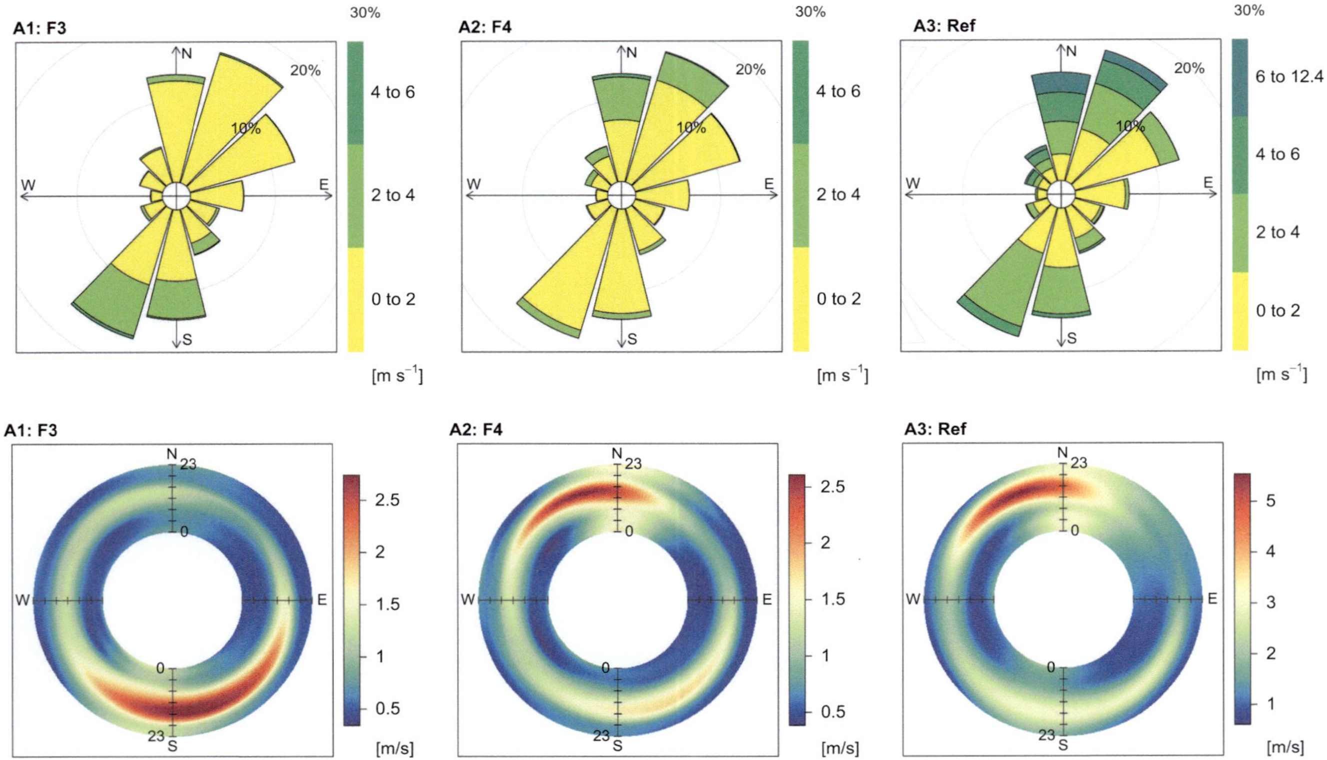

Fig. 6.9 **a** Wind direction and wind speed analysis. **b** Diurnal fluctuation of wind speed. At 3H (Farm 4), 16H (Farm 3) and the reference station at Rust en Vrede, Stellenbosch (22 April 2021–15 January 2022)

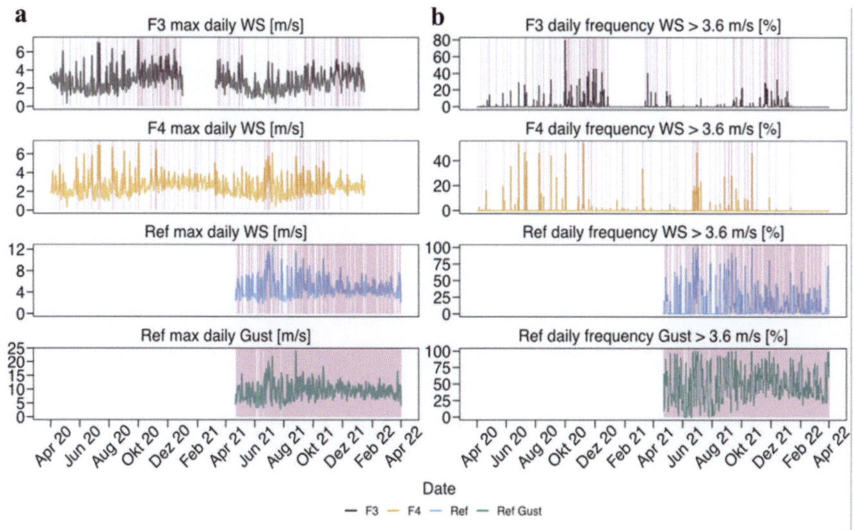

Fig. 6.10 **a** Maximum wind speed. **b** Daily frequency of wind speed. At 3H (Farm 4), 16H (Farm 3) and the reference station at Rust en Vrede, Stellenbosch (22 April 2021–15 January 2022)

reference station (Fig. 6.10). Even with the same average wind speed at 3H and 16H, the frequency and intensity of critical wind speeds differed between the different locations in the orchard due to the seasonal changes of the wind direction (Chaps. 3 and 4).

6.5 Influence of Windbreaks on Fruit Quality

6.5.1 Wind Sensitivity of Citrus at Rust En Vrede and Babylonstoren

In, trial 1, the influence of the distance from the shelterbelt was illustrated clearly by the distribution of different degrees of wind damage to the outer peel (skin), which differed between seasons (Fig. 6.11). In particular, the high valued class 1 fruit for export was associated with the closest distance to the protective windbreak, with the proportion decreasing with distance from the trees. Significantly more class 1 fruit was found at 3H than 10H or 16H. As the distance from the windbreak increased, the proportion of class 2 fruit (slightly damaged) increased significantly and was associate with the majority of the crop harvested at distances 10 H and 16 H. There was no significant spatial distribution in class 3 fruit in the orchard. This comprised between 22 and 25% of the total harvest. Variation in fruit damage due to exposure to wind was also observed at tree level.

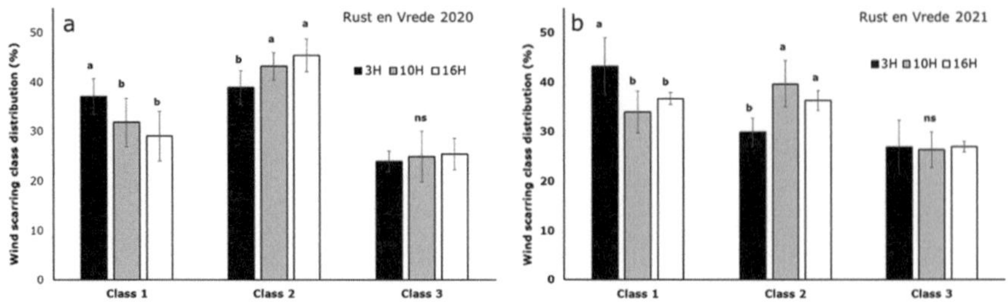

Fig. 6.11 Wind scar damage analysis of 'Tango' mandarin trees at Rust en Vrede, Stellenbosch. **a** At harvest for the 2019/2020 season (06/08/2020). **b** At harvest for the 2020/2021 season (25/08/2021). For damage classification see Fig. 6.6

There were significantly more class 1 and less class 3 fruit on the leeward than wind side (Table 6.2). No significant differences between treatments were recorded for class 2 fruit. Significantly more class 1 fruit was recorded at the bottom and significantly more class 2 and 3 fruit at the bottom. Our results confirmed previous reports that the highest pack out is expected closest to the windbreak or sheltered area of the orchard (McAneney et al. 1984) due to a reduction of turbulent airflow. Similarly, reports by Gravina et al. (2011) were confirmed with significantly more class 2 fruit at the most wind-exposed treatment (16H), as was results from Tarara et al. (2005), with the leeward side recording significantly more class 1 fruit. No significant differences were observed between treatments in classes 2 and 3. Our significantly more class 1 fruit from the bottom (<2 m) and class 3 fruit from the top (>2 m up to 3 m), confirmed results from Jacobs and Van Boxel (1991). A possible explanation for this observation is the increase in wind speed with an increase in height of the canopy also reported by Bosco et al. (2017). During first season of trial 1, the percentage class 1 fruit recorded at 10H and 16H confirmed findings by Freeman (1976) for 'Valencia' oranges grown without wind protection (30%). No significant differences were observed for the percentage class 3 fruit between the three distance treatments, despite the higher wind speeds (up to 17%) recorded at 16H compared to 3H. Even in Babylonstoren (trial 2), with an efficient windbreak using polar trees, treatments were not significantly different (Fig. 6.12). Observations by Hamlet (2000) and Peri and Bloomberg (2002) reported similar results for distances from 2 to 4H.

6.5.2 *Influence of Windbreak Types at Welgevallen*

The effect of different windbreak types on citrus fruit quality was quantified in trial 3. In season 1, the natural woody vegetation (NV) had a significantly higher percentage of class 1 fruit compared to the multi-row windbreak (MRW), with no significant differences between treatments for fruit in classes 2 and 3 (light and severe damage, Fig. 6.13). The percentage of severe damage confirms findings in

Table 6.2 Wind scar damage analysis of tree position of 'Tango' mandarin trees at Rust en Vrede, Stellenbosch, at harvest for the growth season 2019/20 and 2020/21. Harvest: 6 August 2020 and 25 August 2021

	2019/20			2020/21		
	% Class 1	% Class 2	% Class 3	% Class 1	% Class 2	% Class 3
Wind	30.6 b*	43.6^{ns}	25.8^{ns}	36.5 b	35.6^{ns}	27.9 a
Leeward	35.8 a	41.6	22.6	43.6 a	31.8	24.6 b
Top	22.8 b	43.8^{ns}	33.4 a	28.0	41.3 a	30.7 a
Bottom	42.6 a	41.6	15.8 b	47.4	29.0 b	23.6 b

ns non-significant differences within a column
* Means with a different letter within a column differ significantly at the 5% level (least significant difference test)

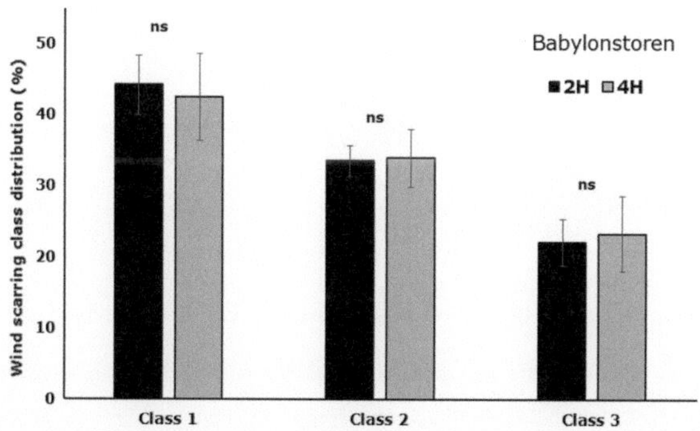

Fig. 6.12 Wind scar damage analysis of 'Tango' mandarin trees at Babylonstoren, Paarl, at harvest for the 2020/2021 season (05/08/2021)

lemon trees (Cataldo et al. 2013). Percentage class 1 fruit in both NV (43.7%) and MRW (37%) treatments were similar to those reported for 'Ellendale' tangor (44–31%) in wind-protected zones next to an artificial windbreak of 6.5 m (Gravina et al. 2011). In contrast, no significant differences were observed between the treatments during the following season, with the class distribution similar to that observed in the first season. Although wind speeds and hours above the threshold of 4.5 m s^{-1} varied between seasons, this did not affect wind scar damage.

Wind scar damage in 'Nadorcott' mandarins were related to fruit position in the tree (Table 6.3). A significantly higher class 1 fruit was recorded in the leeward side in the NV and MRW treatments, confirming previous results (Pienaar 2005) on significantly lower wind speeds on the leeward side as well.

The leeward side also had significantly more class 1 fruit compared to the wind side, which also had significantly more class 2 fruit (Table 6.3). No significant differences were observed in class 3. During the second season, the leeward side had significantly more class 1 fruit than the wind side, which had significantly more class 3 fruit (Table 6.3). No significant treatment effects were recorded for class 2 fruit.

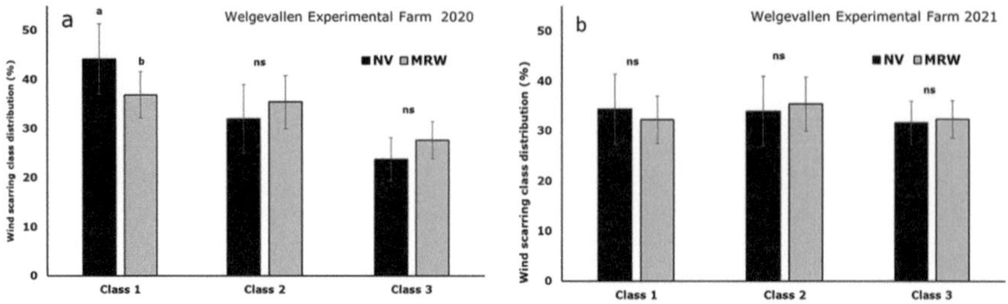

Fig. 6.13 Wind scar damage analysis of 'Nadorcott' mandarin trees at Welgevallen Experimental Farm, Stellenbosch. **a** At harvest for the 2019/2020 season (28/07/2020). **b** At harvest for the 2020/2021 season (18/06/2021)

Table 6.3 Wind scar damage analysis of tree position of 'Nadorcott' mandarin trees at Welgevallen Experimental Farm, Stellenbosch, the growth season 2019/20 and 2020/21. Harvest: 28 July 2020 and 18 June 2021

	2019/20			2020/21		
	% Class 1	% Class 2	% Class 3	% Class 1	% Class 2	% Class 3
Wind	36.7 b*	36.3 a	27.0^{ns}	28.8 b	34.8^{ns}	36.4 a
Leeward	44.4 a	31.1 b	24.5	37.8 a	34.6	27.6 b

ns non-significant differences within a column
* Means with a different letter within a column differ significantly at the 5% level (least significant difference test)

In early sampled fruitlets (2020/21, 12 weeks after petal drop), no significant differences were observed between windbreak types (Fig. 6.14). However, there was still a 12% decrease in class 1 fruits and 14% increase in class 3 between the early strip (25/01/2021) and harvest (18/06/2021) time. The proportion of class 2 fruit did not differ between the sampling dates. Thus, the summer winds from 12 weeks after petal drop were responsible for approximately 86% of severe wind scar damage previously reported (Outspan 1997; Turner and Hardy 2006; Cataldo et al. 2013). These findings were supported by the highest wind speeds recorded during late spring/early summer (from November until the end of January). The remaining 14% damage was associated with winter winds during the last 24 weeks of the growing season (Fig. 6.10).

6.5.3 *Wind Sensitivity of Plum at Welgevallen*

For Japanese plums at Welgevallen (trial 4), less wind damage was observed at harvest (January 2021) compared to citrus fruit at the same time. The incidence of severe wind damage (class 3) in plums (Fig. 6.15) was less than 10% with a considerably higher packout percentage (74.5%) compared to the citrus (33.9%).

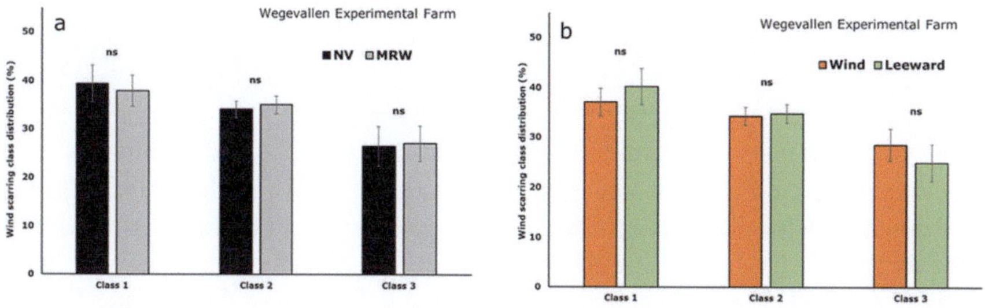

Fig. 6.14 Wind scar damage analysis of 'Nadorcott' mandarin trees at Welgevallen Experimental Farm, Stellenbosch during summer (12 weeks after petal drop) for the 2020/2021 season (25/01/2021). **a** Different wind breaks, (NV) natural vegetation and (MRW) multi-row. **b** Tree side, wind vs leeward

Further studies by Malik (2020) in the Worcester region, Western Cape, reported wind damage values of 22.4% and 12.4% for 'Larry Ann' and 'Midnight Gold' plums grown without wind protection. This can be attributed to the fact that for wind scar damage to develop on plums, a much higher threshold value must be exceeded (5.8 m s^{-1}) to cause damage (Michailides and Morgan 1992). During the 2020/21 season, these values were very rarely reached during the first three weeks after full bloom (fruit most susceptible). As the young Chinese Poplar and Eucalyptus windbreaks are still growing, the percentage of class 1 fruit is expected to increase along with the incidence of severe damage to decrease as the windbreaks become more effective in wind reduction in the coming seasons.

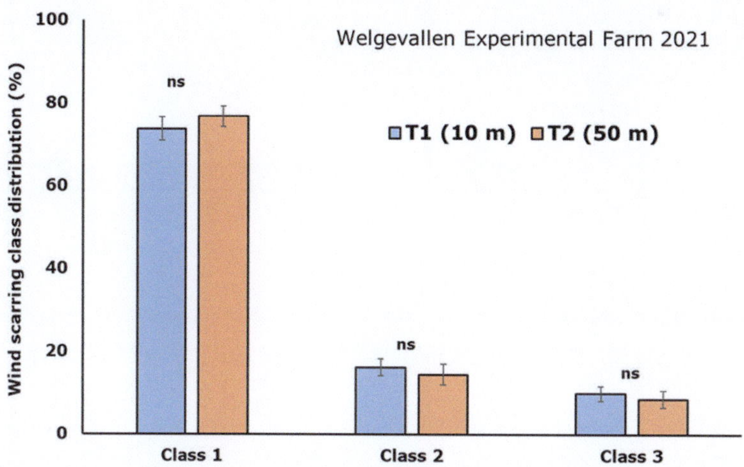

Fig. 6.15 Wind scar damage analysis of 'Ruby Star' plum trees at Welgevallen Experimental Farm, Stellenbosch, at harvest for the 2020/2021 season (24/02/2021). For damage classification see Fig. 6.8

6.6 Seasonal Wind Effects on Fruit Quality

Due to the bi-directional wind system in the Western Cape (Chap. 4), the seasonality of the winds and their effects on the phenological development of the fruit trees must also be considered. As an example for the Western Cape, we analysed the different wind conditions for the farm Rust en Vrede and related them to the development of citrus (Fig. 6.16). During the flowering period and early fruit development (November 2020–February 2021), the average wind speed at 3H (Farm 4) was 1.3 m s^{-1} with 7 days wind speed exceeding 4 m s^{-1}, and, respectively, at 16H (Farm 3-without January, February) average wind speed was 1.6 m s^{-1} and 2 days with wind speed > 4 m s^{-1}. During further fruit development, the average wind speed was 1.0 m s^{-1} for 10 days with wind speeds exceeding 4 m s^{-1} at both locations (3H and 16H). During the harvest period (July–August 2021), the average wind speed at 3H (Farm 4) was 1.0 m s^{-1} with 7 days with wind speed > 4 m s^{-1}, and at 16H (Farm 3) average wind speed was 1.6 m s^{-1} and 2 days with wind speed > 4 m s^{-1}.

Particularly during the summer period from November to February southern high wind speeds are characteristic of the Cape Region (Chap. 3). For the 2020/21 season the highest mean wind speeds were recorded from October until February and the lowest, from May until August. The prevailing wind direction during early spring/

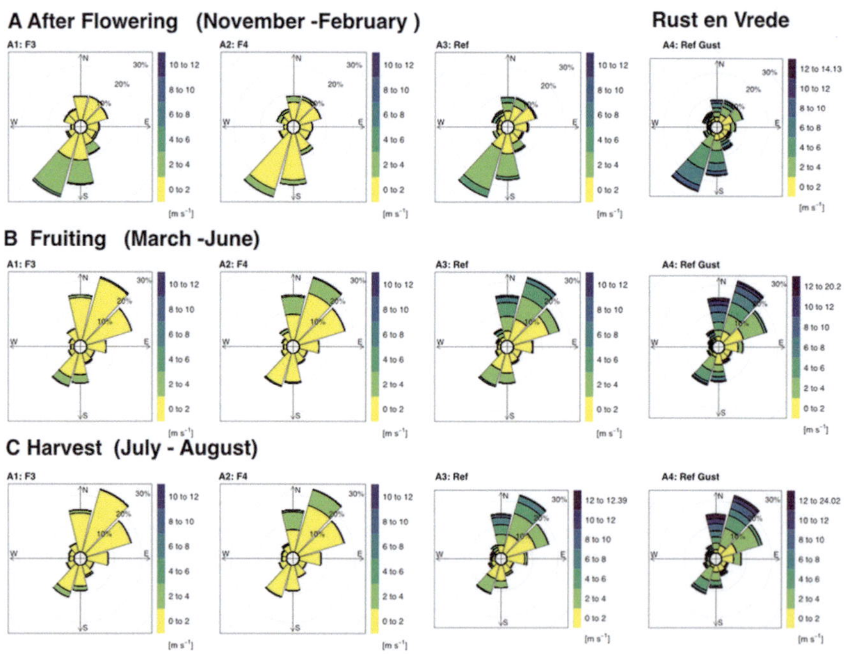

Fig. 6.16 Wind direction and wind speed analysis at 3H (Farm 4), 16H (Farm 3) and the reference station at Rust en Vrede, Stellenbosch. **a** after flowering (November–February), **b** fruiting (March–June) and **c** harvest (July–August)

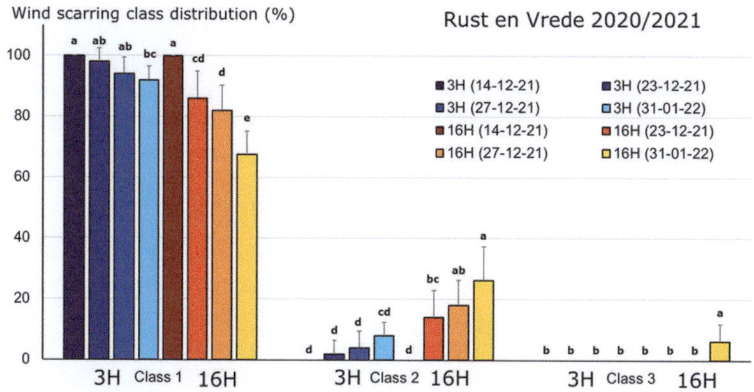

Fig. 6.17 Early wind scar damage analysis of 'Tango' mandarin trees at Rust en Vrede, Stellenbosch, after petal drop for the 2021/2022 season (14/12/2021–31/01/2022). For damage classification see Fig. 6.7

summer (1 October 2020–31 December 2020) was southeast, with higher wind speeds at 16H than at 3H. It is precisely then that citrus fruits are sensitive to wind scar damage, and these increased significantly on immature fruits with the distance from the wind-protecting shelterbelt. As a consequence of the strong southeasterly winds, mandarins in particular show an increase in damage in the sensitive phase of fruit development (Fig. 6.17).

6.7 Conclusions

Our study analyses the effects of tree-based windbreaks on the fruit quality of 'Tango' and 'Nadorcott' mandarins, and 'Ruby Star' plums in the Cape Winelands. For the establishment of the windbreaks in the Western Cape, the different seasonal wind regimes, with southeastern (summer) and northern (winter), need to be considered. The results confirmed that maximum wind protection occurred between 2H and 4H from the windbreak. As the distance from the windbreak increased (10–16H), a considerable increase in wind speed was observed, which resulted in non-uniform wind protection in the orchards. The highest percentage of class 1 quality fruit was observed closest to the beefwood windbreak and decreased with the distance. Especially when growing mandarins, the seasonality of the winds that occur must also be considered, as they are particularly sensitive during the twelve weeks after fruit set (October–December). This is also when the typical south-southeasterly winds occur, which reach the highest speeds. Fruit at the top is also more vulnerable to wind and show a higher proportion of severely damaged fruit than at the bottom. The increase of pack out by maintaining the tree canopy at 2 m instead of 3 m, is not a possible option for the producer, because that result in an unsustainable decrease in yield. The marginal wind protection provided by the tree-based windbreaks in our study was reflected in the low pack out (less than 45%) and requires further attention. A denser planting of the windbreaks would increase wind protection but would inevitably lead to a reduction in the area available for growing fruit trees. Especially with beefwood windbreaks, this increases the risk of competition with the lateral roots of the trees, which penetrate the irrigated areas of the fruit trees. Citrus trees in the sheltered area (0.5–2H) can directly compete for water, nutrients, and solar radiation (shading), resulting in potentially lower yields (Brandle et al. 2009).

However, the integration of poplars within the orchard could also be an alternative here, as they are less competitive than beefwood. While these, as evergreens, protect the border with their dense foliage.

The latest trend towards establishment of nets provides a higher degree of wind protection as well as immediate protection after establishment, but tree-based shelterbelts have a lower establishment cost compared to the nets.

Acknowledgements We thank Johan and Gideon Roos from Paul Roos Wine, Rust en Vrede, Stellenbosch, and Ian de Villiers at Babylonstoren, Paarl, for their support and permission for the field experiments. The FarmImpact project was supported by the Federal Ministry of Education and Research (BMBF) and managed by PT-DLR (grant no. FKZ 01LZ1711A-F) as part of the 'CLIENT II—International Partnerships for Sustainable Innovations' funding initiative.

References

Bargmann CJ (2003) Geology and Wine 7. Geology and wine production in the coastal region, Western Cape province, South Africa. GeoSci Can 30:161–182. https://id.erudit.org/iderudit/geocan30_4ser02

Bosco L, Bergamaschi H, Cardoso L, Paula V, Marodin G, Brauner P (2017) Microclimate alterations caused by agricultural hail net coverage and effects on apple tree yield in subtropical climate of Southern Brazil. Bragantia 77(1):181–192. https://doi.org/10.1590/1678-4499.2016459

Botes J (2018) Impact of shade netting on internal and external quality of 'Nadorcott' mandarin fruit. Master's Thesis, Stellenbosch University, Department of Horticultural Science. https://scholar.sun.ac.za/handle/10019.1/105141?show=full

Brandle JR, Hodges L, Tyndall J, Sudmeyer RA (2009) Windbreak practices. In: Garrett HE (ed) North American agroforestry: an integrated science and practice, vol 2. American Society of Agronomy, WI, USA, pp 75–97. https://doi.org/10.2134/2009.northamericanagroforestry.2ed.c5

Brown R (2018) Effect of permanent shade netting on 'Nadorcott' mandarin tree phenology and productivity. Master's Thesis, Stellenbosch University, Department of Horticultural Science. https://scholar.sun.ac.za/items/f7429214-30ff-41c3-b545-5c159fe2c4e5

Cataldo J, Durañona V, Pienika R, Pais P, Gravina A (2013) Wind damage on citrus fruit study: Wind tunnel tests. J Wind Eng Ind Aerodyn 116:1–6. https://doi.org/10.1016/j.jweia.2013.01.008

Freeman B (1976) Artificial windbreaks and the reduction of wind scar of Citrus. Proc Fla State Hortic Soc 89:52–54

Geldenhuys H, Lötze E, Veste M (2023) Fruit quality and yield of mandarin (*Citrus reticulata*) in orchards with different windbreaks in the Western Cape, South Africa. Erwerbs-Obstbau 65:959–969. https://doi.org/10.1007/s10341-022-00725-3

Gravina A, Cataldo J, Gambetta G, Pardo E, Fornero C, Galiger S, Pienika R (2011) Relation of peel damage in citrus fruit to wind climate in orchard and its control. Sci Hortic 129(1):46–51. https://doi.org/10.1016/j.scienta.2011.03.002

Hamlet GM (2000) The effect of tree windbreaks on the microclimate and crop yields in the Western Cape region of South Africa. Master's Thesis, Stellenbosch University, Department of Forestry Sciences. https://scholar.sun.ac.za/handle/10019.1/51971

Holmes M, Farrel D (1993) Orchard microclimate as modified by windbreaks: A preliminary Investigation. S Afr Avocado Grow Assoc Yearb 16:59–64

Jacobs A, Van Boxel J (1991) Horizontal and vertical distribution of wind speed in a vegetation canopy. Neth J Agric Sci 39(3):165–178. https://doi.org/10.18174/njas.v39i3.16539

Lado J, Gambetta G, Zacarias L (2018) Key determinants of citrus fruit quality: Metabolites and main changes during maturation. Sci Hortic 233:238–248. https://doi.org/10.1016/j.scienta.2018.01.055

Malik N (2020) Shade net studies on two Japanese plum cultivars *(Prunus salicina* Lindl.). Master's Thesis, Stellenbosch University, Department of Horticultural Science. https://scholar.sun.ac.za/items/e52b3176-54f1-4940-9bff-a0257534d7fe

Meadows ME (2015) The cape winelands. In: Grab S, Knight J (eds) Landscapes and landforms of South Africa, World geomorphological landscapes. Springer, Heidelberg, pp 103–109

McAneney K, Judd MJ, Trought MCT (1984) Wind damage to kiwifruit (*Actinidia chinensis* Planch.) in relation to windbreak performance. N Z J Agric Res 27:255–263. https://doi.org/10.1080/00288233.1984.10430427

Michailides TJ, Morgan DP (1992) Development of wind scab and predisposition of French prune fruits to preharvest and postharvest fungal decay by wind scab and russet scab. California Dried Plum Board, pp 149–156. https://ucanr.edu/repository/a/?get=78912

Nuberg I (1998) Effect of shelter on temperate crops: a review to define research for Australian conditions. Agrofor Syst 41:3–34. https://doi.org/10.1023/A:1006071821948

Outspan (1997) Sitrus produksie riglyne—Deel 1, Windbeskerming, pp 1–16

Peri PL, Bloomberg M (2002) Windbreaks in southern Patagonia, Argentina: A review of research on growth models, wind speed reduction, and effects on crops. Agrofor Syst 56:129–144. https://doi.org/10.1023/A:1021314927209

Pienaar JW (2005) The effect of wind on the performance of the grapevine. Master's Thesis, Stellenbosch University, Department of Forestry Sciences. https://scholar.sun.ac.za/handle/10019.1/50486

Soil Classification Working Group (1991) Soil classification: A taxonomic system for South Africa. Department of Agricultural Development, Pretoria, Republic of South Africa (2), pp 1–257

Tarara JM, Ferguson JC, Hoheisel GA, Perez Pena JE (2005) Asymmetrical canopy architecture due to prevailing wind direction and row orientation creates an imbalance in irradiance at the fruiting zone of grapevines. Agric for Meteorol 135:144–155. https://doi.org/10.1016/j.agrformet.2005.11.011

Turner J, Hardy S (2006) Windbreaks for citrus. CITT groups Australia, pp 1–13.

Veste M, Sheppard JP, Abdulai I, Ayisi KK, Borrass L, Chirwa PW, Funk R, Kapinga K, Morhart K, Mwale SE, Ndlovu NP, Nyamadzaw G, Nyoka BI, Sebola P, Seifert T, Senyolo MP, Sileshi GW, Syampungani S, Kahle H-P (2024) The need for sustainable agricultural land-use systems: benefits from integrated agroforestry systems. In von Maltitz GP, Midgley GF, Veitch J, Brümmer C, Rötter RP, Viehberg FA, Veste M (eds) Sustainability of Southern African ecosystems under global change, Ecological Studies 248. Springer Cham. pp 587–623. https://doi.org/10.1007/978-3-031-10948-5_21

Western Cape Department of Agriculture (2021) Cape farm mapper 2.6.6. https://gis.elsenburg.com/apps/cfm/

Chapter 7
Spatial Analysis of Windbreak Effects in Citrus Orchards Using Drones—A Sensor Fusion Approach

Stefan Seifert, Erich Seifert, Anton Kunneke, and Thomas Seifert

Abstract Variable rate application in precision agriculture requires an exact quantification of the variability within the orchard. A sensor fusion framework integrating visible-light optical imagery with infrared thermal data acquired by an uncrewed aerial vehicle was employed to demonstrate the potential of drone-based data for quantifying the spatial influence of windbreaks in a citrus orchard in the Western Cape of South Africa. Sensor fusion employed a 3D reconstruction derived from structure-from-motion techniques to delineate individual citrus tree crowns. Following georectification of both datasets, these crown delineations were integrated with the thermal imagery. Modelling techniques such as Voronoi growing space analyses and generalised additive regression models were then applied to quantify the spatial effect of the windbreak structure on the surface temperature and the crown volume of the citrus trees. The main results show the dominant impact of the distance to the windbreak, which led to increased crown surface temperatures in proximal trees and smaller crowns as a consequence. Also, the height difference of the windbreak to the citrus trees had an effect, however, a much weaker one. All in all, the newly developed methodology showed a promising way to analyse the variability within the orchard as a base for a more precise management.

Keywords Agroforestry · IR thermography · Temperature variation · Structure from motion · Sensor fusion · Precision agriculture · Individual tree analysis · UAV

S. Seifert (✉) · E. Seifert
Scientes Mondium UG, Altomünster, Germany
e-mail: s.seifert@scimond.com

E. Seifert · A. Kunneke · T. Seifert
Department of Forest and Wood Science, Stellenbosch University, Matieland, South Africa

T. Seifert
Chair of Forest Growth and Dendroecology, Albert-Ludwigs-University Freiburg, Freiburg, Germany

M. Veste et al. (eds.), *Trees in South African Agriculture*, Innovations in Landscape Research, https://doi.org/10.1007/978-3-032-00993-7_7

7.1 Introduction

Precision agriculture is an agricultural management paradigm based on the utilisation of best available data to improve the management of agricultural production. Precision agriculture has matured over the past decades and has been widely adapted in annual and perennial production systems (see e.g. Nowak 2021 and Misara et al. 2022 for reviews). One cornerstone of precision agriculture is the concept of variable rate application (Misara et al. 2022). This means to acknowledge the natural variability in a field and adapt soil cultivation, seeding, fertilisation, and irrigation as well as pesticide application accordingly. A necessary precondition for production gains through variable rate application is always high-resolution spatial data with regards to soil and crop conditions. Remote sensing techniques have the potential to deliver the necessary high-resolution data and paved the way to promote further efficiency gains in production (Zhang and Kovacs 2012; Sishodia et al. 2020). Investments in high-tech equipment pay off especially for production systems that require a long-term investment in the establishment such as fruit tree orchards and vineyards. In the following an example is given where modern remote sensing technology is used to obtain more information on the variability of a citrus orchard, to create an information base for more efficient management.

7.2 Drone Application in Orchards

With the advancing development of drone technology, data from uncrewed aerial vehicles (UAVs) are increasingly used in precision agriculture to characterise the variability of growing conditions and crop growth (refer to Zhang and Kovacs 2012; Hunt and Daughtry 2018; Velusamy et al. 2022 for reviews). On the sensor side, classic RGB cameras are being applied on drone platforms, but also increasingly multispectral, and, more recently, thermal sensors are also used for perennial crops such as fruit trees (Gonzalez-Dugo et al. 2013, Csillik et al. 2018; Ampatzidis and Partel 2019). Here, special challenges arise in the analysis of drone data. For plantation crops such as grapevines, citrus, or other fruit trees, results are usually sought per individual tree (Ampatzidis and Partel 2019), whereas traditional sensing and analysis of annual field crops typically involve areal analyses in a map format without looking at individual plants (e.g. Warren and Metternicht 2005; Nguy-Robertson et al. 2012; Maresma et al. 2016). Another special feature of perennial crop sensing is that the canopies of the target plants are usually taller than a typical annual field crop. As a result, verticality plays a more important role, and pure 2D approaches, as typically used for annual crops, cannot be easily transferred.

7.3 Windbreaks in Orchards

Windbreaks have been introduced long time ago as agroforestry systems to protect crops and fruit trees. Their positive effects on micro-climate have been studied in abundance (Bates 1911); Wang and Takle 1997; Cleugh 1998). In particular in wind-prone areas, windbreaks are implemented as biologic infrastructure to dissipate wind and to protect commercial fruit orchards and vineyards (Chaps. 1, 2 and 12). Freeman (1976) found that artificial windbreaks with a 44% permeability and about 5 m height reduced wind speeds between 25 and 70%. Veste et al. (2020) reported a similar wind speed reduction of about 20% in leeward direction of vineyards shielded by Chinese poplar poplar windbreaks in the vineyards of Babylonstoren, Western Cape. There is good evidence that the porosity of a windbreak is decisive for its ability to reduce wind speed (Brandle and Finch 1991); Wang and Takle 1997; Cleugh 1998; Chap. 5). An outcome of these wind tunnel studies and laminar boundary layer simulations was that in principle denser windbreaks reduced the wind speed leewars more but if density was too high there was a risk of detrimental eddies leeward and a high air pressure windwards of the windbreak.

Windbreaks protect fruit from mechanical damage by reducing wind speed and have a substantial influence on fruit quality (Freeman 1976; Geldenhuys et al. 2023, Chaps. 6 and 12). Freeman (1976) reported substantially higher production and quality in citrus orchards. This led to a higher per canopy area yield of 17 and 21% in fruit numbers and weight, respectively, and amounted to 53% higher total fruit packouts and 30% higher top grade packouts. In the unprotected control blemish problems by wind scarring were frequent. Freeman (1976) further mentioned that wind exposed trees "were characterised by hardened leaves which were often rolled with signs of obvious mechanical injury. Trees behind the windbreak had larger and more flattened leaves and the trees appeared to be healthier and more vigorous." The hardened and rolled leaves might be an effect of wind-induced drought stress. Wind is causing drought stress in fruit trees by disturbing the thin air layer around the leaves (Livingston 1938; Bidwell 1979). The thin air layer forms close to the leave surface, and when the plant transpires water from the leaf stomata, it increases humidity locally around the leaf, decelerating the transpirational losses from the leaf. If this laminar layer is disturbed by wind, the leaf is exposed to the higher water vapour pressure deficit of the ambient air, which can lead to stomatal closure or cessation of gas exchange, in particular in hot and dry areas (Tranquillini 1969; Caldwell 1970; Monteith 1995). Dixon and Grace (1984) reported a reduction of transpiration according to the Penman–Monteith equation (Penman 1948; Monteith 1965) for three tree species with increasing wind speed in a wind tunnel experiment.

In effect this means a temperature change of leaves and canopies since transpiration has an overall cooling effect through the radiative and diffusive coupling between a leaf and the micro-climate (Monteith 1981; Chaves et al. 2016). An overheating of leaves leads to a reduction in photosynthesis rate. Recent findings imply that there is also a coupling between the rate of photosynthesis and stomatal closure rates (Duursma et al. 2014), which means that exceeding a certain temperature will not only tune down photosynthesis rates but also close the stomata based on this coupling additional to water supply in the leaves.

The previously described transpiration-related cooling effect of surface temperature is picked up by modern methods such as IR thermography (Stokes et al. 2006; Johnson et al. 2013; Vieira and Ferrarezi 2021), which can be used to either model canopy heat transfer, transpiration or be related to questions of tree and fruit production.

In this chapter we describe a method to analyse drone-based data from different sensor types in a sensor fusion approach in order to show directional effects of windbreaks on canopy surface temperature in a citrus plantation. These spatial windbreak effects are related with temperature and individual plant canopy size in models. The latter is used as a proxy for growth and fruit production. This information may be used to inform variable rate application concepts.

7.4 Material and Methods

7.4.1 Study Site

The test site was a citrus orchard is located on the commercial farm Babylonstoren (Western Cape, South Africa; 33°49′27.69″ S, 18°55′19.38″ E, altitude 194 m a.s.l., Chap. 2). The climate in the area is Mediterranean with wet winters and hot and dry summers with predominantly south-easterly winds and summer temperatures frequently exceeding 30 °C (Chap. 3, Veste et al. 2020). In case of the chosen site, a windbreak from Chinese poplar (*Populus simonii* (Carrière) Wesm.) was planted in multiple rows (Sheppard et al. 2024) to shield the citrus plantation from the predominantly south-easterly winds (Chap. 5). A shading effect of the windbreak can be ruled out since the windbreak is in southerly direction and the main sun direction is from the North on the southern hemisphere. A significant root competition influence could also be ruled out since the windbreak is separated from the orchard by a dirt road of about 4 m width. A single row windbreak on the northerly side might nonetheless influence the citrus growth of adjacent rows.

7.4.2 Flight Mission

A drone survey of the site was conducted in November at 2 pm. The special feature of the drone survey was the multi-sensor configuration, where RGB, red edge, NIR, and thermal data were acquired in the same flight to analyse different aspects of the variance on the study plots simultaneously. The aerial survey took place at an altitude at 67 m, covering an area of 19 ha with 5412 images in one flight. A ground resolution of 2.84 cm per pixel was achieved for the RGB data, with a back-projection error of 0.479 pixels. For the thermal data, the ground resolutions were correspondingly lower (22 cm per pixel, respectively). An overlap of pixel values from 9 pictures per pixel was ensured, and a good compromise for a focal length was found (7.84 mm) since the two factors were identified as crucial for canopy reconstruction with SfM in previous studies (Seifert et al. 2019; Denter et al. 2022). A MicaSense Altum1 sensor[1] was used that integrates different spectra channels with RGB, red edge, near infrared, and integrated thermal imager in one rig. A downwelling light sensor mounted on the top of the drone to determine ambient light and sun angle as control parameters was part of the configuration.

7.4.3 Image Processing to Obtain Geo-Rectified Information for all Spectral Channels

Two image processing pipelines were developed to retrieve the necessary information from the drone data: Pipeline 1 was used for the georectification of the RGB and thermal pixel data and a 3D reconstruction of the RGB data to obtain point clouds. Pipeline 2 was then developed for the fusion of thermal pixel data with height information, obtained from the RGB-based SfM process.

On the image data, a structure from motion (SfM) process was applied as shown in the flow chart in Fig. 7.1. The SfM Software package Metashape Pro 1.64[2] was used. The second step was the reconstruction of the exact camera positions. This was done by matching key points in each image to all other images, where the key points were identified. During the matching, the camera model parameters and the key points are used to set up a mathematical model of the scene. This model is solved to get the exact camera positions. With the now-known camera position, a so-called dense point cloud is computed which tries to assign a 3D position for every image pixel. This dense point cloud is triangulated to a 3D surface model.

The result of this first processing pipeline was the geo-rectified assignment of the pixel values of the IR thermal and the RGB sensor data) in different layers in the geographic information system (GIS). This was the basic requirement for merging the data in the analysis.

A second processing pipeline was created based on the results of the first step on analyses (Fig. 7.2). The main objective was to transfer the pixel values to individual plant canopies. This involved a spatial reconstruction of the trees in the windbreak hedge as well as the citrus trees in the plantation. Tree heights of both groups were determined by subtracting the SfM-derived absolute canopy height from the DTM. This allowed the creation of a mask separating tree crowns and soil based on height values. A Voronoi algorithm was used to determine the theoretical growing space of each citrus tree from the planting geometry. This was followed by delineation of

---

[1] https://support.micasense.com/hc/en-us/articles/360010025413-Altum-Integration-Guide (2023–08–07).

[2] https://www.agisoft.com/ (2023–07–16).

Fig. 7.1 Processing pipeline to obtain geo-rectified pixel information and 3D pointclouds

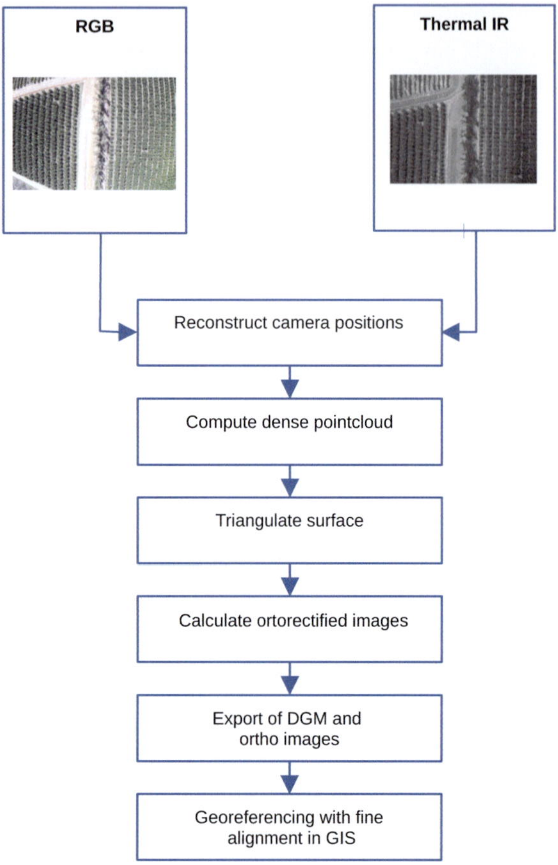

each citrus tree using the combination of the mask and the growing space, so that the crowns of each tree were separated.

Based on this, ortho-rectified pixel values from IR thermal sensor could be accurately assigned to individual tree crowns. In addition, the crown projection area (crown extent) of the citrus trees was also calculated.

To describe the windbreak structure, the hedge was segmented in overlapping sections. The sections were 8 m wide and each segment was assigned to the corresponding citrus row in wind direction. For each segment, the median plant height, and the maximum plant height was calculated. The ratio of median height to maximum height was used to describe the windbreak structure in terms of its homogeneity.

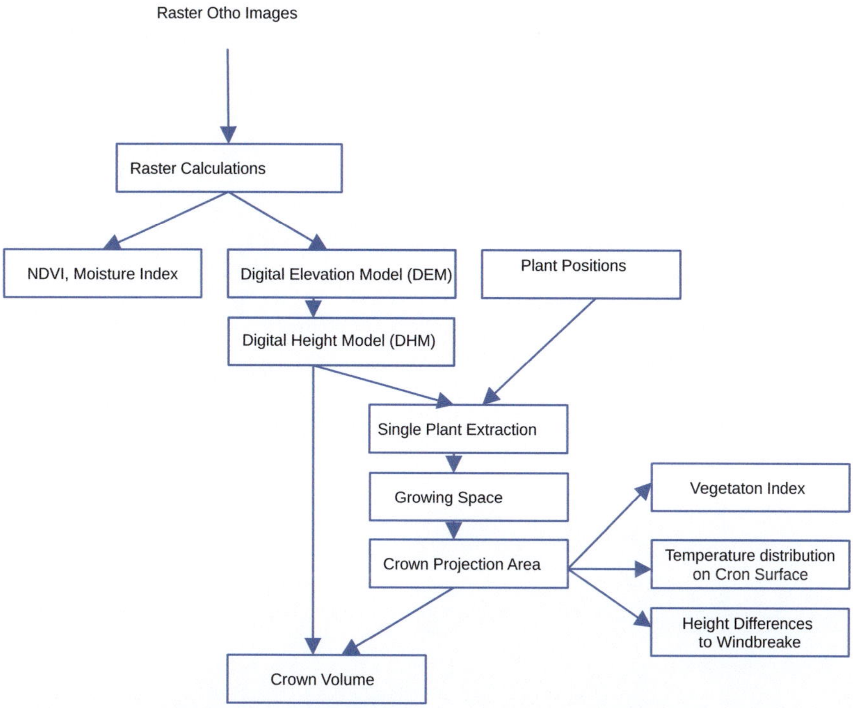

Fig. 7.2 Processing pipeline to assign the sensor information to individual tree crowns

7.4.4 Modelling of the Spatial Windbreak Influence

In a third analysis step, the spatial windbreak influence on the citrus trees was quantified with models. For this the spatial effect of maximum crown temperature and crown volume were modelled with a generalised additive regression models (GAM) and spline smoothers (mgcv package, Wood 2011) in R (R-Core Team Members 2022).

The input variable in the model were the distance of each citrus plant to the windbreak (distance) and the height difference of the citrus plant maximum height to the windbreak maximum height in direct wind direction (ΔH) (Fig. 7.3). The input variables in the GAM were interpolated by a penalised cubic regression spline.

7.5 Results

The results of the SfM reconstruction of the RGB values led to a high-resolution three-dimensional representation of the orchard that could also be used as a high-resolution ortho-photo (Fig. 7.4).

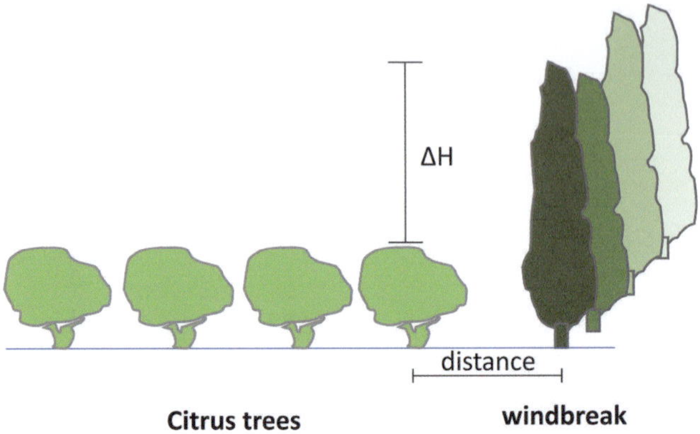

Fig. 7.3 Illustration of windbreak structural features used to describe influence on crown volume and crown temperature in the models

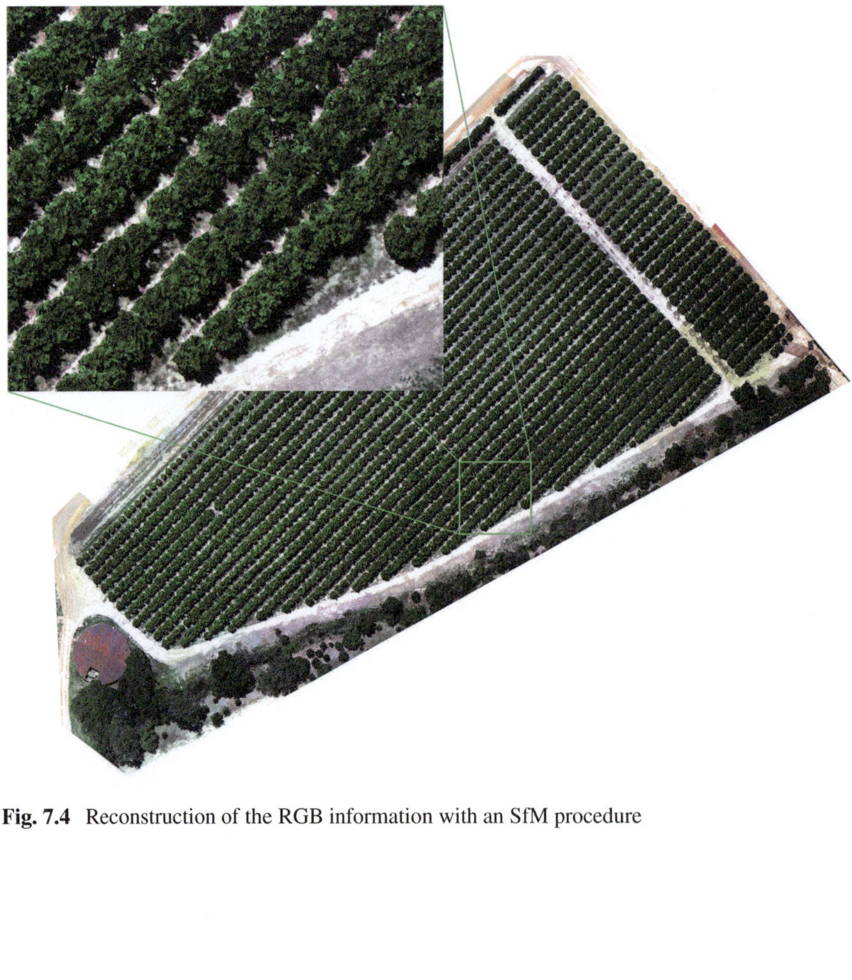

Fig. 7.4 Reconstruction of the RGB information with an SfM procedure

The Voronoi algorithm segmented the growing space of each citrus plant (Fig. 7.5a) and formed the base for further analyses of individual tree crown values. The SfM-based height reconstruction as a next important step shows a substantial variability in the orchard (Fig. 7.5b), with citrus trees being higher in more wind sheltered areas towards the windbreaks. The grey height values of the windbreak show also an influence of windbreak height. The resulting visible height differences of the citrus trees are about 1 m and more. This resembles the pattern of the crown volume of the trees with clearly larger crown volumes closer to the windbreak, where tree crowns had up to two times the volume compared to proximal citrus trees. The temperature distribution shown in a thousandth part of a degree in Fig. 7.5c completes the pictures and also shows a high degree of spatial correspondence with the height and crown volume pattern of the citrus trees. Trees in the vicinity of the windbreak were clearly cooler than trees further away from it.

The GAM regression models were able to depict the influence of the individual structural and spatial factors quantitatively. Both factors, the distance of the individual citrus tree to the windbreak (distance) and the height difference of the citrus plant maximum height to the windbreak maximum height in direct wind direction (ΔH), were found to modify maximum crown temperature and crown volume highly significantly (Table 7.1). The model for the maximum temperature variance explained

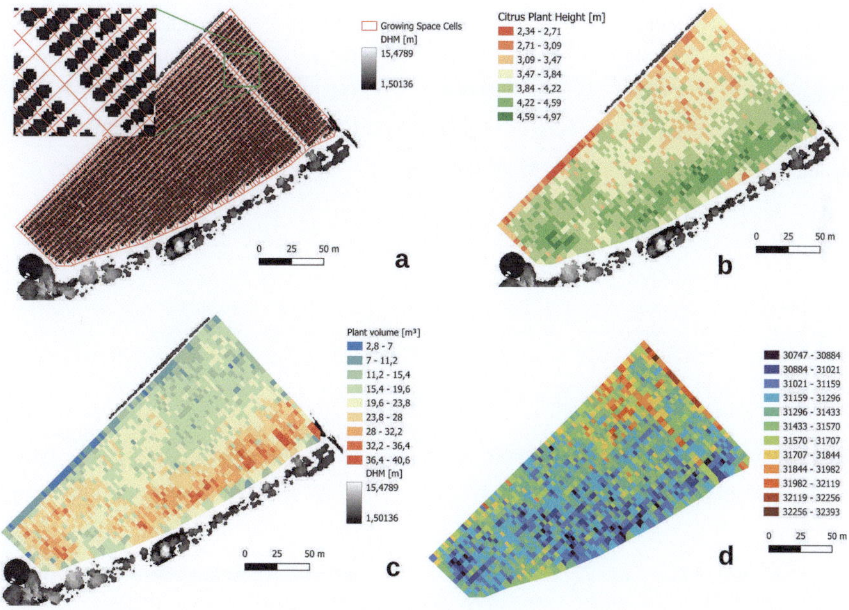

Fig. 7.5 Mapping of the spatial variation individual citrus tree-based values. Displayed are the growing space occupation of the citrus trees (**a**), the individual tree height in metres (**b**), the crown volume in m^3 (**c**), and the temperature distribution at 2 pm, displayed in thousandth parts of a degree (**d**). The grey values of the windbreak indicate height differences of the poplar trees with brighter values representing higher windbreak heights

28.3% of the variance and the crown volume model 41.6% according to the adjusted R^2 values.

The two cubic regression splines show the relative influence of the input variables on the estimated maximum canopy temperature (Fig. 7.6). It is a feature of GAMs that the individual splines for the factors can be interpreted as the partial influence of that factor that is centred around zero. Thus, positive values depict an increase in the independent variable compared to the average value around zero, values below zero a negative influence.

It is quite obvious that the distance to the windbreak has the major influence with increasing maximum crown surface temperatures with increasing distance to the windbreak. The influence of ΔH on temperature was more complex: ΔH values

Table 7.1 Regression statistics of the GAM models

Max. crown temperature model N = 1973, $R^2 adj. = 0.28$				
Coefficient	Value	Ref.df	F-value	Significance value
Intercept	31,365.586 8/	4.891	6413	< 2e-16 ***
Distance	3.282	3.714	133.64	< 2e-16 ***
ΔH	3.467	3.800	15.84	< 2e-16 ***
Crown volume model N = 1973, $R^2 adj. = 0.414$				
Coefficient	Value	Ref.df	F-value	Significance value
Intercept	21.292	0.101	210.9	< 2e-16 ***
Distance	3.963	3.999	338.14	< 2e-16 ***
ΔH	2.178	2.600	10.66	5.66e-06 ***

Significance codes used: *** 0.001, ** 0.01, * 0.05

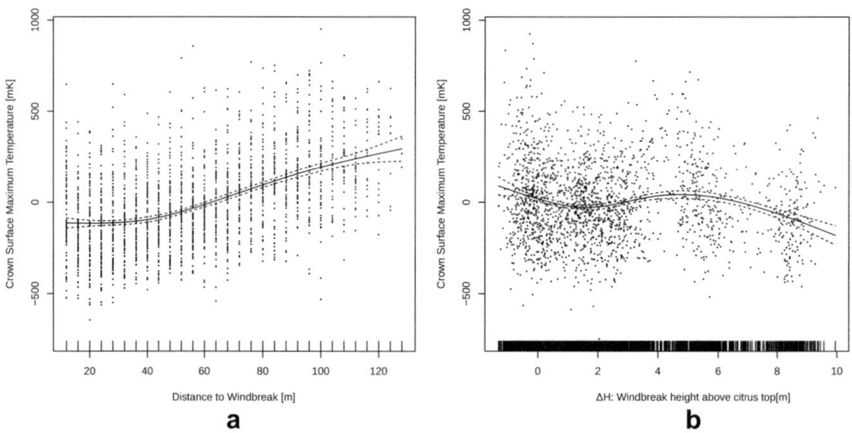

Fig. 7.6 Estimation of maximum canopy temperature in distance from the windbreak (**a**) and the windbreak height above citrus (**b**). For detailed explanation see text

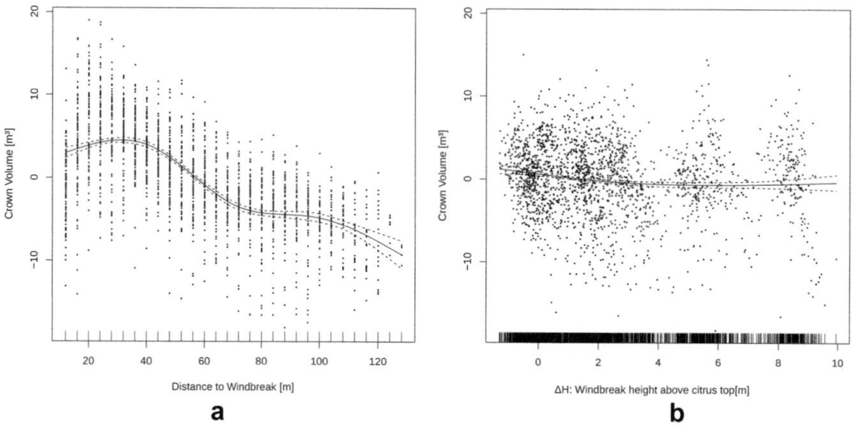

Fig. 7.7 Modelled crown volume in distance from the windbreak (**a**) and the windbreak height above citrus (**b**). For detailed explanation see text

smaller than the citrus trees were increasing maximum surface temperatures as well as ΔH values that indicated that the windbreak was 4–6 m metres higher than the citrus trees.

The same input variables were used to model the crown volume (Fig. 7.7). The intent was to show how crown volume of the citrus trees as a long-term integrator of climatic effects relates spatially to the windbreak structure. The crown model showed also that distance to the windbreak is a main variable to consider. With increasing distance, crown volume decreased almost linearly after a slight plateau of positive impact from 0 m to about 40 m distance. The influence of ΔH was much weaker and showed a positive effect on citrus tree crown volume when windbreak height was not more than 3 m higher than the canopy height of the citrus trees.

7.6 Discussion

A novel method was proposed to fuse 3D point cloud data with IR thermal data in order to analyse the spatial variability of a citrus plantation. The SfM-derived point cloud data were essential in tree crown delineation which was a necessary precondition to separate temperature effects on tree canopies and the soil effects. The fusion of point cloud data and thermal imagery brought further insights into the spatial effect of the windbreak height variation on the citrus crop. The GAMs developed to quantify these relationships were statistically significant, with the crown-volume model exhibiting a satisfactory level of explained variance. The R^2 of the temperature model was not excessively high but this was not to be expected since only a few factors derived from remote sensing data were considered. Further factors such as denser grid of ambient temperature and wind field measurements and repeated flights at many times of the day could have certainly improved the model outcome further. However, a third of the variance was explained just from remote sensing data.

Our results showed that crown temperatures increased in the leeward area of the hedge as the distance of the citrus trees from the hedge increased. Citrus tree height and crown volume decreased accordingly with increasing distance from the hedge. Wind as a stressor, especially in dry conditions of the study site, is an obvious explanation and well aligned with findings of Freeman (1976). The effect is probably because the citrus trees in the lee side of the hedge suffer less wind-induced disturbance of the laminar boundary layers around the leaves and therefore keep the stomata open longer during the day. This sustained transpiration cools the leaf, and the trees appear cooler in the thermal image. Thus, trees that can keep their stomata open longer can be expected to transpire and hence grow more, and show higher water use efficiencies than trees that suffer from frequent laminar layer disturbances by wind. Our findings are in accordance with the results of Freeman (1976) who recorded tree canopy area increases of 12% and 8% in the leeward zones, during a 12-month observation period in a windbreak experiment.

Translating the observed temperature differences to evaporation according to the Penman–Monteith equation the determined median difference of one degree Celsius between citrus trees near and far from the hedge (assumption: temperature increase from 30 °C to 31 °C) corresponds to an increase in evapotranspiration of about 6%. If we apply the observed spread of the median from 30 °C to 33 °C in the stand, this increases evapotranspiration by 20%, which, accumulated over time, explains the larger crown dimensions of the trees close to the hedge, which also amounted to up to 20%. When the well-known positive relationship between canopy size and fruit yield (Zaman et al. 2006) is included in the consideration, it becomes clear that windbreak hedges provide substantial positive effects on production and water use in citrus trees. Another positive effect can also be expected from a more complete utilisation of the growing space by the crowns of the trees close to the southerly windbreak, which was observed, as this also reduces the heating of the soil by shading, which in turn slows down water losses by soil evaporation.

We cannot completely rule out that the mono-row windbreak at the northern edge of the orchard did impact as well on tree crown volume to a certain degree by casting a shadow on the adjacent trees. However, the time of the drone flight was chosen at a time when the sun angle was close to the zenith so that the shading was minimised in order not to produce confounding effects for the citrus crown surface temperature measurement. Of course, data in the present study are limited, and repeated flights over many times of the day and over the years would have been preferable. However, the fact that even a once-off flight mission is able to provide meaningful information about the variability in a citrus orchard speaks much in favour of the potential of the use of drone-based sensor fusion data to obtain information that is relevant for variable rate management and also for the decision to establish and maintain windbreaks as a biologic infrastructure despite the obvious loss in productive area.

Acknowledgements We thank Ian de Villiers at Babylonstoren, Paarl for support and permission for the field experiments and Ben du Toit, Stellenbosch University for his advice. FarmImpact project was supported by the Federal Ministry of Education and Research (BMBF) and managed by PT-DLR (grant no. FKZ 01LZ1711D) as part of "CLIENT II—International Partnerships for Sustainable Innovations" funding initiative.

References

Ampatzidis Y, Partel V (2019) UAV-based high throughput phenotyping in citrus utilizing multispectral imaging and artificial intelligence. Remote Sensing 11(4):410. https://doi.org/10.3390/rs11040410

Bates CG (1911) Windbreaks: Their influence and value. Bulletin (United States. Forest Service), United States Forest Service 82–89, p 100

Bidwell RG (1979) Plant physiology, 2nd edn. Macmillan, New York, p 440

Brandle JR, Finch S (1991) EC 91-1763-B How Windbreaks Work (1991) Historical materials from university of Nebraska-Lincoln extension 4892. https://digitalcommons.unl.edu/extensionhist/4892

Caldwell MM (1970) Plant gas exchange at high wind speeds. Plant Physiol 46:535–537

Chaves MM, Costa JM, Zarrouk O, Pinheiro C, Lopes CM, Pereira JS (2016) Controlling stomatal aperture in semi-arid regions—the dilemma of saving water or being cool? Plant Science 251:54–64. https://doi.org/10.1016/j.plantsci.2016.06.015

Cleugh HA (1998) Effects of windbreaks on airflow, microclimates and crop yields. Agrofor Syst 41:55–84

Csillik O, Cherbini J, Johnson R, Lyons A, Kelly M (2018) Identification of citrus trees from unmanned aerial vehicle imagery using convolutional neural networks. Drones 2(4):39. https://doi.org/10.3390/drones2040039

Denter M, Frey J, Kattenborn T, Weinacker H, Seifert T, Koch B (2022) Assessment of camera focal length influence on canopy reconstruction quality. ISPRS Open J Photogramm Remote Sens 6:2022. https://doi.org/10.1016/j.ophoto.2022.100025

Dixon M, Grace J (1984) Effect of wind on the transpiration of young trees. Ann Bot 53(6):811–819. https://doi.org/10.1093/oxfordjournals.aob.a086751

Duursma RA, Barton CVM, Lin Y-S, Medlyn BE, Eamus D, Tissue DT, Ellsworth DS, McMurtrie RE (2014) The peaked response of transpiration rate to vapour pressure deficit in field conditions can be explained by the temperature optimum of photosynthesis. Agric For Meteorol 189–190:2–10. https://doi.org/10.1016/j.agrformet.2013.12.007

Freeman B (1976) Artificial Windbreaks and the reduction of windscar of citrus. Fla Proc Fla State Hort Soc 89:52–54

Geldenhuys H, Lötze E, Veste M (2023) Fruit quality and yield of mandarin (Citrus reticulata) in orchards with different windbreaks in the Western Cape, South Africa. Erwerbs-Obstbau 65:959–969. https://doi.org/10.1007/s10341-022-00725-3

Gonzalez-Dugo V, Zarco-Tejada P, Nicolás E, Nortes PA, Alarcón JJ, Intrigliolo DS, Fereres E (2013) Using high resolution UAV thermal imagery to assess the variability in the water status of five fruit tree species within a commercial orchard. Precis Agric 14:660–678. https://doi.org/10.1007/s11119-013-9322-9

Hunt ER Jr, Daughtry CST (2018) What good are unmanned aircraft systems for agricultural remote sensing and precision agriculture? Int J Remote Sens 39(15–16):5345–5376. https://doi.org/10.1080/01431161.2017.1410300

Johnson K, Sankaran S, Ehsani R (2013) Identification of water stress in citrus leaves using sensing technologies. Agronomy 3:747–756

Livingston BE Influences that affect the transpiration from leaves. Sigma Xi Quaterly 26(2):88–101

Maresma Á, Ariza M, Martínez E. Lloveras J, Martínez-Casasnovas JA (2016) Analysis of vegetation indices to determine nitrogen application and yield prediction in maize (*Zea mays* L.) from a standard UAV service. Remote Sens 8:973

Misara R, Verma D, Mishra N, Kant Rai S, Mishra S (2022) Twenty-two years of precision agriculture: A bibliometric review. Precision Agric 23:2135–2158. https://doi.org/10.1007/s11119-022-09969-1

Monteith JL (1965) "Evaporation and environment". Symp Soc Exp Biology 19:205–234. PMID 5321565

Monteith JL (1981) Coupling of plants to the atmosphere. In: Grace J, Ford ED, Jarvis PH (eds) Plants and their atmospheric environment. Blackwell, Oxford, pp 1–29

Monteith JL (1995) A reinterpretation of stomatal responses to humidity. Plant Cell Environ 18:357–364

Nebiker S, Annen A, Scherrer M, Oesch D (2008) A light-weight multispectral sensor for micro UAV: Opportunities for very high resolution airborne remote sensing. Int Arch Photogramm, Remote Sens Spat Inf Sciences XXXVII (Part B1):1193–1200

Nguy-Robertson A, Gitelson A, Peng Y, Viña A, Arkebauer T, Rundquist D (2012) Green leaf area index estimation in maize and soybean: Combining vegetation indices to achieve maximal sensitivity. Agron J 104(5):1336–1347

Nowak B (2021) Precision Agriculture: Where do we stand? A review of the adoption of precision agriculture technologies on field crops farms in developed countries. Agric Res 10:515–522. https://doi.org/10.1007/s40003-021-00539-x

Penman HL (1948) Natural evaporation from open water, bare soil and grass. Proc Roy Soc A 193:120–146

Seifert E, Seifert S, Vogt, H, Drew D, van Aardt J, Kunneke A, Seifert T (2019) Influence of drone altitude, image overlap, and optical sensor resolution on multi-view reconstruction of forest images. Remote Sens 11:1252. https://doi.org/10.3390/rs11101252

Sheppard JP, Larysch E, Cuaranhua CJ, Schindler Z, du Toit B, Malherbe GF, Kunneke A, Morhart C, Bohn Reckziegel R, Seifert T, Kahle H-P (2024) Assessment of biomass and carbon storage of a *Populus simonii* windbreak located in the Western Cape Province, South Africa. Agrofor Syst 98:697–714. https://doi.org/10.1007/s10457-023-00940-1

Sishodia RP, Ray RL, Singh SK (2020) Applications of remote sensing in precision agriculture: A review. Remote Sens 12(19):3136. https://doi.org/10.3390/rs12193136

Stokes VJ, Morecroft MD, Morison JIL (2006) Boundary layer conductance for contrasting shapes in a deciduous broadleaved forest canopy. Agric for Meteorol 13:40–54

Tranquillini W (1969) Photosynthese und Transpiration einiger Holzarten bei verschieden starkem wind. Centralblatt Für Das Gesamte Forstwesen 81:35–48

Velusamy P, Rajendran S, Mahendran RK, Naseer S, Shafiq M, Choi J-G (2022) Unmanned Aerial Vehicles (UAV) in precision agriculture: Applications and challenges. Energies 15(1):217. https://doi.org/10.3390/en15010217

Vieira GHS, Ferrarezi RS (2021) Use of thermal imaging to assess water status in citrus plants in greenhouses. Horticulturae 7:249. https://doi.org/10.3390/horticulturae7080249

Veste M, Littmann T, Kunneke A, Du Toit B, Seifert T (2020) Windbreaks as part of climate-smart landscapes reduce evapotranspiration in vineyards, Western Cape Province, South Africa. Plant Soil Environ 66:119–127. https://doi.org/10.17221/616/2019-pse

Wang H, Takle ES (1997) Momentum budget and shelter mechanism of boundary layer flow near a shelterbelt. Boundary-Layer Meteorol 82:417–435

Warren G, Metternicht G (2005) Agricultural applications of high-resolution digital multispectral imagery: Evaluating within-field spatial variability of canola (*Brassica napus*) in Western Australia. Photogramm Eng Remote Sens 71:595–602

Wood SN (2011) Fast stable restricted maximum likelihood and marginal likelihood estimation of semiparametric generalized linear models. J Roy Stat Soc (B) 73(1):3–36

Zaman Q, Schumann A, Hostler H (2006) Estimation of citrus fruit yield using ultrasonically-sensed tree size. Appl Eng Agric 22:39–44

Zhang C, Kovacs JM (2012) The application of small unmanned aerial systems for precision agriculture: A review. Precis Agric 13(6):693–712. https://doi.org/10.1007/s11119-012-9274-5

Chapter 8
Expert-N Agroforestry Model: Simulating the Influence of Windbreaks on Crop Growth and Water Use in a Citrus Orchard in the Western Cape, South Africa

Florian Heinlein, Xiaohong Duan, and Eckart Priesack

Abstract Agroforestry systems are an alternative to established agricultural production systems and may help mitigating environmental pollution, biodiversity loss, and climate change. Since there are only a few simpler agroforestry models, the modular model system Expert-N has been extended to include an agroforestry sub-model. This model was applied to a South African citrus agroforestry system focusing on the effect of windbreaks on the crop water use. Additionally, a control simulation without the agroforestry sub-model was carried out. Depending on the distance from the windbreak, the modelled wind speeds were reduced between 11.6% and 38.8% in comparison with the control simulation, while the incoming radiation was reduced between 6% and 45.4% due to the shading caused by the tree row. The application of the agroforestry sub-model resulted in lower crop yields and higher wood growth, while the water consumption was also reduced by about 20% in a simulation with constant irrigation throughout the growing period. However, these effects strongly depend on the distance to the windbreak. A simulation without irrigation provides the total plant water demand at different positions in the field. The results show that up to 40% of crop irrigation water may be saved due to the adjacent tree strip. The study confirms the benefits of agroforestry systems and thus supports their establishment. Additionally, Expert-N may also be applied for temporal and spatial irrigation planning, which is highly relevant in agroforestry systems.

F. Heinlein (✉) · X. Duan · E. Priesack
Helmholtz Zentrum München GmbH–German Research Center for Environmental Health, Neuherberg, Germany
e-mail: dr.florian.andreas.heinlein@inbox.lt

E. Priesack
Institute of Soil Science, University of Hohenheim, Stuttgart, Germany

M. Veste et al. (eds.), *Trees in South African Agriculture*, Innovations in Landscape Research, https://doi.org/10.1007/978-3-032-00993-7_8

8.1 Introduction

Agroforestry systems (AFS) are more often considered as alternative production systems that are helpful mitigating environmental pollution, biodiversity loss, and climate change, thereby providing enough food and fibre to meet the increasing demand of people and society. Due to the relatively high complexity of AFS with above- and below-ground competition between crops and trees for light, water, and nutrients, agroforestry models are helpful in decision-making and prediction to regulate the intended ecosystem services of the AFS. Numerous agricultural and silvicultural models have been developed in the last 50 years (Pretzsch et al. 2008; Priesack and Gayler 2009). However, only a few agroforestry models have been made available in the last two decades (Dupraz 2002; Palma et al. 2007; Luedeling et al. 2016). As far as these models are bio-physically based, the involved process descriptions are rather simple and mostly taken from agricultural and silvicultural process-based stand-level models and are complemented by descriptions of processes that are specific to tree–crop interactions such as shading or competition for water and nutrient resources (cf. among others: Brisson et al. 1998; Van Noordwijk and Lusiana, 1999; Mobbs et al. 1999). The more recently developed agroforestry model Yield-SAFE (Van der Werf et al. 2007) aims to simulate agroforestry stands in an even more simplified way based on the calculation of only seven simple state variables: biomass and leaf area of both trees and crop, shoot number of trees, heat sum, and available soil water. Thus, a robust model is made available to estimate the primary biomass production of agroforestry systems by rather simple representations of soil–plant processes on a regional scale (Graves et al. 2007). In contrast to more complex functional structural, often individual plant-based agricultural crop and silvicultural stand growth models (Garca-Barrios et al. 2001, Gayler and Priesack 2007; Bohn et al. 2014), only a few such models exist for agroforestry systems (Kraft et al. 2021). Furthermore, soil processes are described in a very simple way in available agroforestry models: it is often assumed that the soil is homogeneous for the entire root zone. Different soil horizons and their different soil physical, hydraulic and bio-geochemical properties, as well as differences in soil organic matter pools and their turnover rates are not considered. Therefore, in order to better account for the complex interactions between soil biota, nutrient cycling, water uptake, and plant growth, a more mechanistic and dynamic description of water flow, nutrient transport, and resource allocation in soil–plant systems is required. This is in particular important if mixed crop systems are considered to assess their sustainability. The agroforestry version of Expert-N has therefore been specifically developed, tested, and applied to meet these requirements.

A windbreak system (WBS) can be conceived as a special agroforestry system (Sheppard et al. 2020; Veste et al. 2024, Chap. 1). However, most agroforestry models lack a complex representation of wind reduction due to windbreaks, although there are various approaches to simulate wind attenuation (cf. among others: Brandle et al. 2004; Rosenfeldet al. 2010; Wilson 1985, 1987), which may be integrated into existing agroforestry models. Since the wind speed is a crucial parameter to determine the crop water use, a wind attenuation model was implemented into the Expert-N-agroforestry model.

It was the aim to apply and test the agroforestry system version of Expert-N on a windbreak system of the Western Cape in South Africa (Chap. 1) in order to calculate the current state of the agricultural production system as well as the expected situation of the production system within a week based on the weather forecast. This includes the simulation of the current nutrition status of the orchard plants based on the calculation of the prevailing carbon and nitrogen fluxes within soil and plants. In most WBS, the roots of the trees are prevented from penetrating the soils of the crop or orchard system and hence cannot compete for water and nutrients available in these soils (Chap. 2). Therefore, in contrast to a full simulation of an AFS in the case of WBS, only soil water and nutrient balances of the orchard systems were simulated. Furthermore, simulations with and without windbreak trees were compared to study and discuss possible effects of the WBS on the orchard tree growth. Also the effects of system simulations with and without irrigation are compared to get information about the additional water needed to provide optimal conditions for plant growth.

8.2 Materials and Methods

8.2.1 *The Expert-N 5.0 Model Configuration*

Originally, the model system Expert-N has been designed as a building block system of single modules representing the nitrogen dynamics in soil and plants. The relevant processes for this simulation study of a citrus orchard at Babylonstoren, South Africa, (Chap. 1) and the applied modules are shown in Table 8.1

A major focus of this study lies on the modelling of water consumption within the citrus orchard. In Expert-N, water losses of the soil–plant system may occur due to percolation to deeper soil layers or due to water exchange with the atmosphere through direct evaporation or indirect transpiration through plants. The actual evapotranspiration is defined by the sum of actual evaporation and transpiration. According to Nimah and Hanks (1973) the actual evaporation is defined by the minimum of the potential evaporation and of the maximal evaporative water flux, which is calculated when the Richards equation for soil water flow (Šimůnek et al. 1998) is solved. The potential evaporation depends on the potential evapotranspiration multiplied by the non-shaded fraction of the soil, which is defined through the simulated leaf area index LAI [m^2 m^{-2}] as $1 - LAI/3$.

The potential evapotranspiration is computed with a Penman–Monteith approach as suggested by the ASCE (American Society of Civil Engineers, Walter et al. 2005). The potential transpiration is calculated as the difference between potential evapotranspiration and potential evaporation. Within the plant model, the actual transpiration is simulated as the minimum of the potential transpiration and the potential root water uptake, which means actual transpiration is limited by the atmospheric

Table 8.1 Model configuration of Expert-N 5.0

Expert-N module	Module name	References
Soil heat transport	Daisy	Hansen et al. (1990)
Soil nitrogen transport	LEACHM 3.0	Hutson and Wagenet (1992)
Nitrification, denitrification, mineralization	Century	Developed from Parton et al. (1994)
Pedotransfer functions	Campbell	Campbell (1987)
Hydraulic functions	Van Genuchten and Mualem	Van Genuchten (1980), Mualem (1976)
Soil water flow	HYDRUS 6.0	Šimůnek et al. (1998)
Potential evapotranspiration, latent heat flux	Penman–Monteith ASCE	Walter et al. (2005)
Tree growth and actual transpiration	Treedyn	Bossel (1996)
Irrigation	Schaaf	Schaaf (1998)

water demand, the plants' leaf area, and the soil water availability (cp. Heinlein et al. 2017). The difference between the potential and actual transpiration can be considered as water demand. If the water demand is high, plant growth is limited or in severe cases the plants may even die. Hence, it is crucial to know the plants' water demand in order to counter steer, e.g. with irrigation. The simulations have been executed with the actual model version Expert-N 5.0 (Klein et al. 2017) comprising all the major modules and single processes from the stable long-term model version Expert-N 3.0 (Priesack et al. 2006; Biernath et al. 2011). Expert-N 5.0 benefits from a grid structure and has initially been designed to be coupled with the weather model WRF (Skamarock et al. 2008) by substituting its relatively simple with a more elaborate land surface model. The two-dimensional structure of Expert-N 5.0 has also played a significant role in the creation of the newly developed agroforestry module which requires data exchange between grid points representing different plant species interacting with each other.

8.2.2 The Expert-N Agroforestry Model

In general, the Expert-N agroforestry model represents crops, trees, or pasture (in the following called crop) growing next to a higher tree strip (in the following just called tree strip). This setup can be defined in two modes: the tree strip can either be simulated explicitly or be defined through some key parameters. The explicit simulation of the tree strip is necessary when two-way exchange processes such as competition for root water or nutrients between tree strip and crop need to be addressed. In this case, one tree strip grid point has to be defined which can interact with several adjacent crop grid points. If the focus lies on the influence of the tree strip on the crops and a two-way influence is unlikely, e.g. if the distance between tree strip and crop is large, the definition of tree strip key parameters is sufficient. In this case, the application of the agroforestry model results in a modification of the input weather data at the crop grid points. Since the distance between tree strip and crop is large, the key parameter approach has been chosen for this study. This results in a modification of input radiation and wind speed, both of which are key parameters in the simulation of crop water use, while the influence of agroforestry systems on these parameters is also well understood (Jacobs et al. 2022). The attenuated radiation R_{af} [W m^{-2}] is simulated with an exponential function like the Beer–Lambert law:

$$R_{af} = \mathrm{R}e^{LAI\, f_{ext} d_{ts}} \tag{8.1}$$

where R [W m^{-2}] is the measured input radiation, LAI [m^2 m^{-2}] is the leaf area index of the tree strip, and f_{ext} [–] is the extinction coefficient, which is a model input parameter. d_{ts} [m] denotes the average length through the tree strip. This length is calculated using the solar zenith and azimuth angles, which change throughout the day, the height, width, and orientation of the tree strip and the distance of the crop from the tree strip. Additionally, the attenuated radiation can be either calculated for a specific distance from the tree strip or averaged over a certain area adjacent to the tree strip. The air flow across windbreaks has been addressed in three-dimensional (Rosenfeld et al. 2010) and two-dimensional (Wilson 1985, 1987) numerical simulations. These studies in combination with empirical investigations (Brandle et al. 2004; Wright and Stuhr 2002) helped develop the Expert-N wind attenuation model. The concept of the wind attenuation simulation (Fig. 8.1) is based on the calculation of a marginal distance b [m] from the tree strip, up to which wind is still attenuated:

$$b = \frac{2t}{f_{ws}}, \tag{8.2}$$

where t [m] is the height of the tree strip, and $f_{ws} = f_{crop}\ \mathrm{f_d}\ \mathrm{f_s}\ \mathrm{f_{dir}}$ is the product of four different factors, which all have an influence on the attenuated wind speed. f_{crop} accounts for the difference between tree height t and crop height c [m]:

$$f_{crop} = \min\left(1.0; \sqrt{1.0 - \frac{t-c}{t}}\right) \tag{8.3}$$

This means, there is no wind attenuation due to height difference when tree and crop have the same height. The attenuation due to the tree density is represented through f_d:

$$f_d = \min\left(0.8; 0.8e^{-0.05(p-p_{base})}\right) \tag{8.4}$$

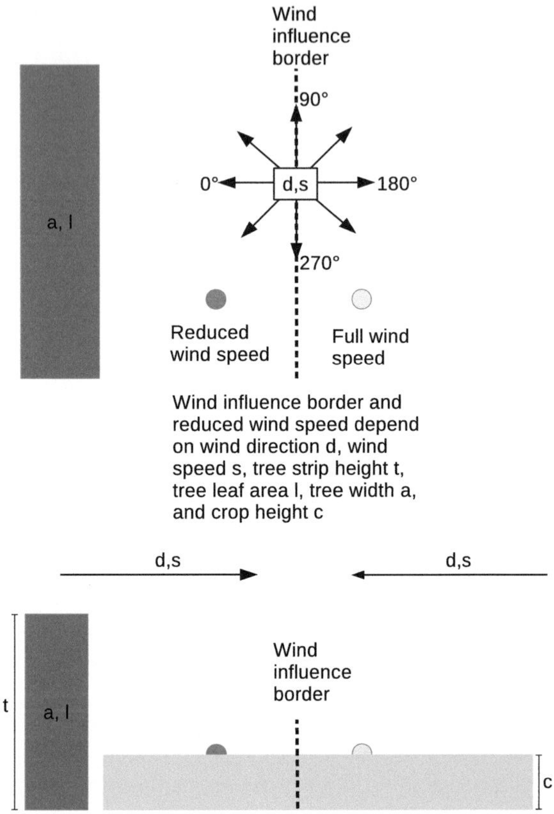

Fig. 8.1 Graphical representation of the wind model as implemented into the Expert-N agroforestry model. The wind influence border is the marginal distance from the tree strip, where the influence of the tree strip on the crop wind speed starts to become neglectable

with the porosity factor $p = \text{a} \cdot \text{LAI}$ [m] as the product of tree strip width a [m] and leaf area index LAI [$m^2\ m^{-2}$]. Thus, a wider tree strip or a tree strip with a higher leaf area index will lead to lower wind speeds at the crop grid points. The parameters a and p base [m] are specified in a model configuration file.

Additionally, higher (undisturbed) wind speeds s [$m\ s^{-1}$] are stronger attenuated than lower wind speeds:

$$f_s = \min\left(1.0; 0.7 + 0.3e^{-0.5(s - s_{base})}\right) \tag{8.5}$$

Wind attenuation due to this factor occurs only when s exceeds the marginal wind speed sbase [$m\ s^{-1}$], which is a model input parameter. The wind direction d [°] also has an influence on the wind attenuation:

$$f_{dir} = 1.0 + 0.25\cos(d) \tag{8.6}$$

This factor may be higher than 1 and thus even out the other factors. Thus, the wind attenuation may be reduced if the crop is situated on the windward side of the tree strip and higher if situated on the lee side of the tree strip. Depending on data availability, the wind direction may either be set as dominant wind direction or provided for each individual time stamp in coherence with the other weather input data.

Finally, the reduced wind speed sred [ms^{-1}] is calculated as

$$s_{red} = \min\left(1.0; s \log_b x\right), \tag{8.7}$$

where x [m] is the distance between tree strip and crop. The parameters t, c, and LAI can be either user-defined or calculated by other Expert-N modules, which allows for a site-specific model parameterization.

The impact of input parameter variation is shown in Fig. 8.2. With increasing distance from the tree strip, the simulated wind speed gets higher, while a higher porosity factor leads to lower wind speeds, e.g. there is no decrease in wind speed with a porosity factor of 25 m at distances higher than 75 m from the tree strip, while the porosity factors of 75 m and 100 m lead to wind attenuation for all represented distances and wind directions. The wind attenuation is lowest (highest) if the crop is on the windward (lee) side of the tree strip. The influence of the wind direction on the simulated wind speeds decreases with increasing porosity factor.

8.2.3 Site-Specific Model Input

The estimated soil hydraulic parameters, which are used to simulate the soil water flow, are shown in Table 8.2. These parameters were estimated using the ROSETTA program (Schaap et al. 2001) fed with measured soil textures. The soil in Expert-N is represented through 34 numerical soil layers ranging from 0 to 220 cm. The simulation period ranges from 12 February 2015 to 1 April 2021 using a model and output time step of 1 h. The weather input data (unshaded) global radiation [W m^{-2}], precipitation [mm d^{-1}], relative humidity [%], (undisturbed) wind speed [ms^{-1}], and temperature [°C] stem from measurements with 1 h resolution from the FarmImpact project (Chap. 3).

A typical model setup consists of a control simulation, i.e. without the agroforestry model, and 11 simulations with the agroforestry model. These eleven simulations represent the citrus rows, each at a different distance x from the tree strip as measured by project partners (Chap. 7). The distance between the rows amounts to 4 m. The nearest citrus row is located at 9 m distance from the tree strip. From this distance every second row is simulated until 81 m from the tree strip. A dominant wind direction, perpendicular to the tree strip with the citrus orchard on the lee side, was set in order to obtain maximum effects of the agroforestry model on the weather data and the water use. The tree strip in the model setup has a height t of 20 m and a width a of 8 m. The leaf area LAI of the tree strip was set to 3 m^2 m^{-2} and the extinction coefficient f_{ext} to 1.0. The base values for wind speed s_{base} and tree strip porosity p_{base} amount to 1.0 m s^{-1} and 10.0 m. Since knowledge about water use is

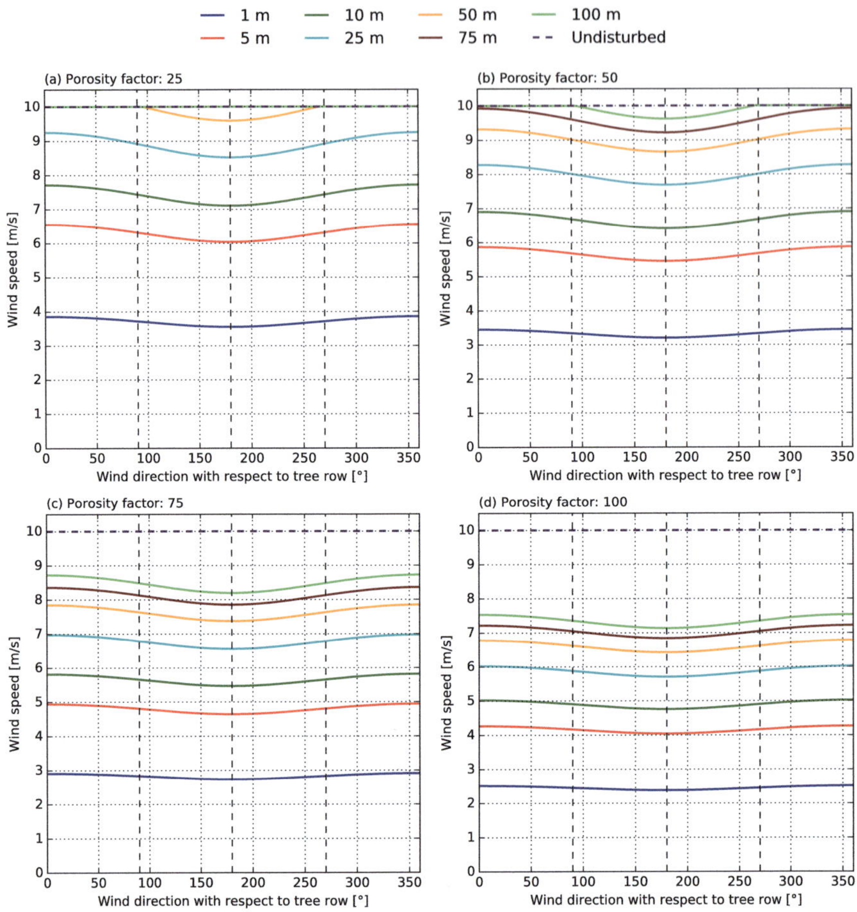

Fig. 8.2 Simulation of wind attenuation depending on the different input parameters. The colours show wind speeds at different distances from the tree strip x [m] ranging from 1 to 100 m as well as the undisturbed wind speed, which has been set to 10 m s^{-1}. The x-axes show the wind directions d with respect to the tree row, where 180° represents the lee side with respect to the tree strip. The subfigures show variations of the porosity factor p [m]: **a** 25, **b** 50, **c** 75, **d** 100

Table 8.2 Estimated soil hydraulic parameters of the citrus orchard in Babylonstoren, South Africa

Layer	Layer depth [cm]	θ_{sat} [cm^3 cm^{-3}]	θ_{sa} [cm^3 cm^{-3}]	K_{sat} [cm^3 d^{-1}]	α [cm^{-1}]	n
1	0–20	0.372	0.059	32.0	0.028	1.446
2	20–50	0.372	0.060	27.0	0.028	1.409
3	50–220	0.372	0.060	28.0	0.028	1.420

θ_{sat} denotes the water content at saturation, θ_{res} the residual water content, K_{sat} the saturated hydraulic conductivity, and α and n the van Genuchten parameters crucial, two model setups were created: one without irrigation and one with a daily irrigation amount of 6 mm from the beginning of August to the end of April. The stop date for irrigation is coherent with the first fruit harvest date, which happens at the end of April. A second harvest takes place at the end of May. Once a year at the beginning of August, the trees are pruned: the tree height is cut to 3 m, and 20% of the biomass is removed.

8.3 Effects of the Expert-N Agroforestry Model

Depending on the distance from the tree strip, wind and radiation are attenuated in a different way (Table 8.3). The simulated attenuated wind speed ranges between 61.2% (closest to the tree strip) and 88.4% (furthest away from the tree strip) in comparison with the undisturbed simulation without the agroforestry model (NoAf). The reduction in wind speed at 17 m distance amounts to 31.2% averaged over the whole simulation period. This is higher than the annual average of wind speed reduction measured by Veste et al. (2024) at the same study site. They find a wind speed attenuation of 27.6% at 18 m distance from the tree strip. However, over the growing season in summer the measured wind speed is reduced by 39.2%. The underlying modelled reductions of wind speed lie in a similar range than those found by Van Ramshorst et al. (2022), who simulated average wind speed reductions between 17 and 67% for a study site in Brandenburg, Germany. For radiation, these shares with respect to the simulation without the agroforestry module lie between 54.6% (closest to the tree strip) and 94.0% (furthest away from the tree strip). Both weather variables decrease with decreasing distance from the tree strip. This decrease intensifies when getting closer to the tree strip. However, this reduction is stronger for radiation than for wind speed. For wind speed, there is only one effect related to distance x as value of the logarithmic function. For radiation, two effects have to be considered: with decreasing distance x less radiation perpendicular to the tree strip reaches the citrus crop, but also during the day the sunlit time frame decreases.

In this simulation, the reduction of radiation is higher than the reduction of wind speed at 9 and 17 m distance from the tree strip. For the other distances the wind reduction is stronger.

Figure 8.3 shows the water input [mm d^{-1}], which is the same for all the simulations. The nearly year-round irrigation can be seen as horizontal line sections at 6 mm d^{-1}. To this value, the natural precipitation is added. The natural precipitation reaches up to 63 mm d^{-1}. In 2015, 2016, and 2018, most natural precipitation occurs in the winter months June and July. In 2017, the autumn months March and April as well as the late winter/early spring months August and September show more precipitation than May, June, or July. From 2019 to 2021, the daily precipitation is very low, not exceeding 6 mm d^{-1}. This is partly due to a malfunctioning rain gauge. However, in this orchard irrigation is the main source of water input; i.e. the impact on our simulations is neglectable.

Table 8.3 Wind speed [m s^{-1}] and radiation [W m^{-2}] averaged over the simulation period from 12 February 2015 to 1 April 2021 for the different distances from the tree strip. The average wind and radiation attenuation is shown as ratio [%] with respect to the wind speed and radiation from the simulation without the agroforestry model (NoAf)

Distance [m]	Wind speed [m s^{-1}]	Wind ratio [%]	Radiation [W m^{-1}]	Radiation ratio [%]
NoAf	1.97	–	18.75	-
89	1.74	88.4	17.62	94.0
81	1.72	87.3	17.34	92.5
73	1.70	86.1	17.10	91.1
65	1.67	84.7	16.92	90.3
57	1.64	83.2	16.58	88.4
49	1.60	81.4	16.32	87.0
41	1.56	79.2	16.06	85.6
33	1.51	76.7	15.25	81.3
25	1.45	73.4	14.34	76.5
17	1.35	68.8	12.80	68.2
9	1.21	61.2	10.23	54.6

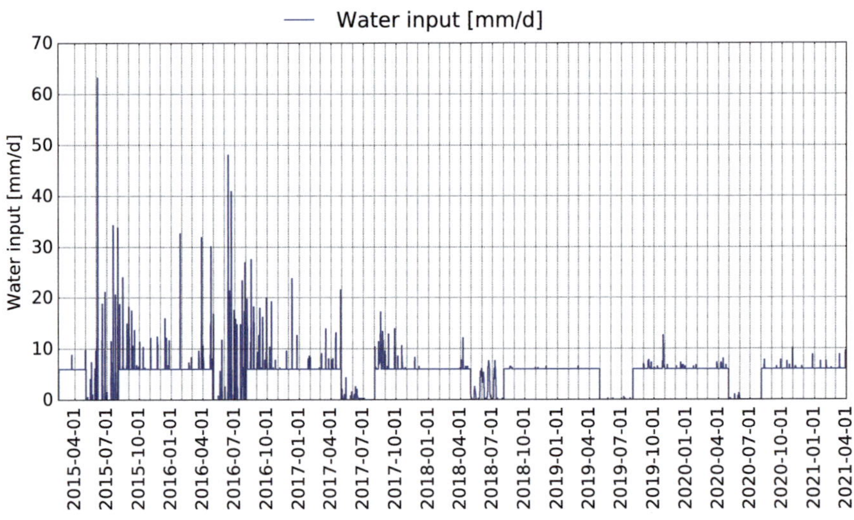

Fig. 8.3 Water input by precipitation and irrigation at Babylonstoren from 12 February 2015 to 1 April 2021

In Fig. 8.4, the simulated fresh biomass is shown without the agroforestry model and for the citrus row nearest to the tree strip. In both simulations, the wood biomass increases from 26,000 kg ha^{-1} at the beginning of the simulation period to about 47,000 kg ha^{-1}. However, the yearly pruning at the beginning of August reduces the wood biomass to 80%. Leaf and root biomass are less than 5000 kg ha^{-1}. The leaf biomass has a yearly maximum in early February and a yearly minimum in mid-July. The root biomass slightly increases throughout the simulation period. Here, there is also a little inconsistency in the model, since root biomass is also reduced by 20% due to pruning which is not realistic in reality. The fruit biomass reaches its maximum at the end of April, when the first fruit harvest occurs and 60% of the fruits are harvested. At the second harvest at the end of May all the remaining fruits are gathered. The simulation without the agroforestry model (Fig. 8.4a) shows yearly maximum fruit biomasses between 50,000 kg ha^{-1} and 55,000 kg ha^{-1}, while no interannual trend can be observed. In 2015 and 2016, the maximum fruit biomass in the citrus row closest to the tree strip (Fig. 8.4b) is similar to the simulation without the agroforestry model. In the following year, the maximum fruit biomass decreases to 40,000 kg ha^{-1} in 2020. These values are rather high, but still in a similar range to Tumbo et al. (2002) who developed and tested a citrus yield monitor and found citrus yields between 9,000 kg ha^{-1} and 54,000 kg ha^{-1} throughout a 5 ha citrus grove in Florida, USA.

For both simulations, the simulated citrus LAI [m^2 m^{-2}] and the height c [m] are presented in Fig. 8.5. During the simulation period, the minimum LAI ranges between 1 m^2 m^{-2} and 1.5 m^2 m^{-2}, and the maximum LAI reaches up to 6.5m^2 m^{-2}. The differences between the LAI of the two simulations only occur in the two months before the LAI peak and do not exceed 0.2 m^2 m^{-2}. These values are similar to those of Delalieux et al. (2008), who measured LAI values between 2.2 m^2 m^{-2} and 11 m^2 m^{-2} at a citrus orchard near Wellington, Western Cape, South Africa. Raj et al. (2020) compared LAI values from drone measurements and from ceptometers as ground truth at a citrus growing area in Central India. Their ground measurements vary from 0.5 m^2 m^{-2} to 3.7 m^2 m^{-2}. The tree heights of the two simulations strongly differ from 2017 onwards. While the simulation without the agroforestry model reaches maximum heights of 4.8 m, the simulation of the citrus row closest to the tree strip shows heights of more than 5 m from 2018 to 2020. This is in accordance with the modelled wood biomasses, which is also lower in the simulation without the agroforestry model. The pruning to a height of 3 m is evident in both simulations.

While the fruit biomass grows stronger in the simulation without the agroforestry model, the height and the wood biomass increase less in comparison with the simulation at 9 m distance from the tree strip. Hence, the Treedyn model's reaction to less radiation and wind speed is a shift from fruit growth to wood growth and height increase. This is indeed a well-known behaviour of trees in forests, which try to obtain more radiation by outreaching the height of other trees. Raveh et al. (2003) also found the same behaviour in an experiment, where they artificially shaded young citrus trees, which reduced radiation between 35 and 55%. As a result, the final tree height was up to 35% larger than non-shaded control trees.

Figure 8.6 shows the simulated volumetric soil water contents [cm^3 cm^{-3}]. Without the agroforestry model (Fig. 8.6a), they range between 0.16 cm3 cm^{-3} and 0.35 cm^3 cm^{-3}. The minima and maximum are more distinct at 30 cm soil depth than at 60 cm or 90 cm soil depth. In all depths, the soil water contents tend to decrease in a similar manner between the beginning of May and the end of July when no irrigation takes place. However, this is compensated in 2015, 2016, and 2018 by

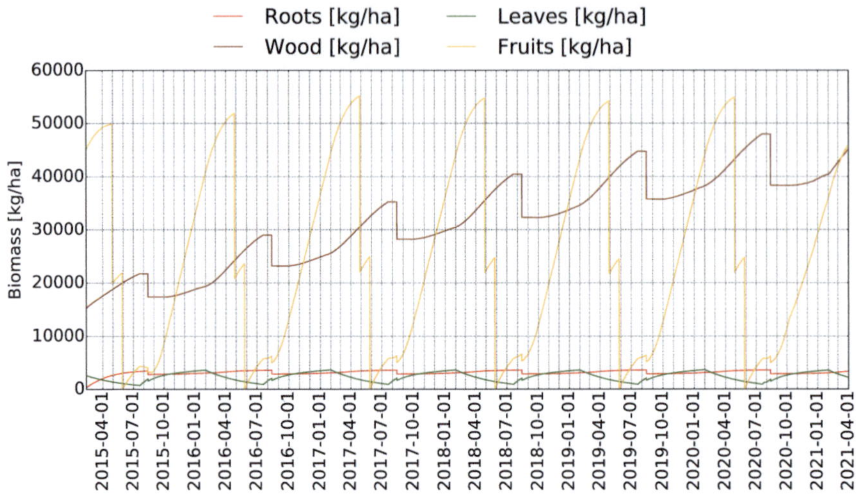

(a) No agroforestry

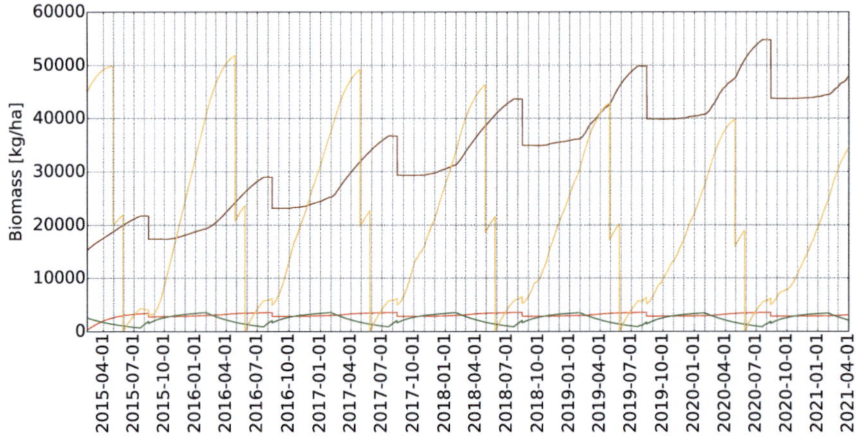

(b) 9 m distance

Fig. 8.4 Simulated fresh citrus biomass at Babylonstoren from 12 February 2015 to 1 April 2021 **a** without the agroforestry model and **b** with the agroforestry model representing the citrus row nearest to the tree strip

natural precipitation. From the beginning of September to the end of March, the soil water contents constantly decrease despite irrigation. This is the period where the citrus trees consume most of the applied water. The irrigation water compensates to a certain extent the decrease of the soil water content at 30 cm soil depth, but not at the deeper soil layers. Hence, the soil water contents at 60 cm or 90 cm soil depth are about 0.05 $cm^3\ cm^{-3}$ lower than at 30 cm depth. From May to July the simulation of the citrus row nearest to the tree strip (Fig. 8.6b) shows a similar trend in comparison with the simulation without the agroforestry model. However, in the

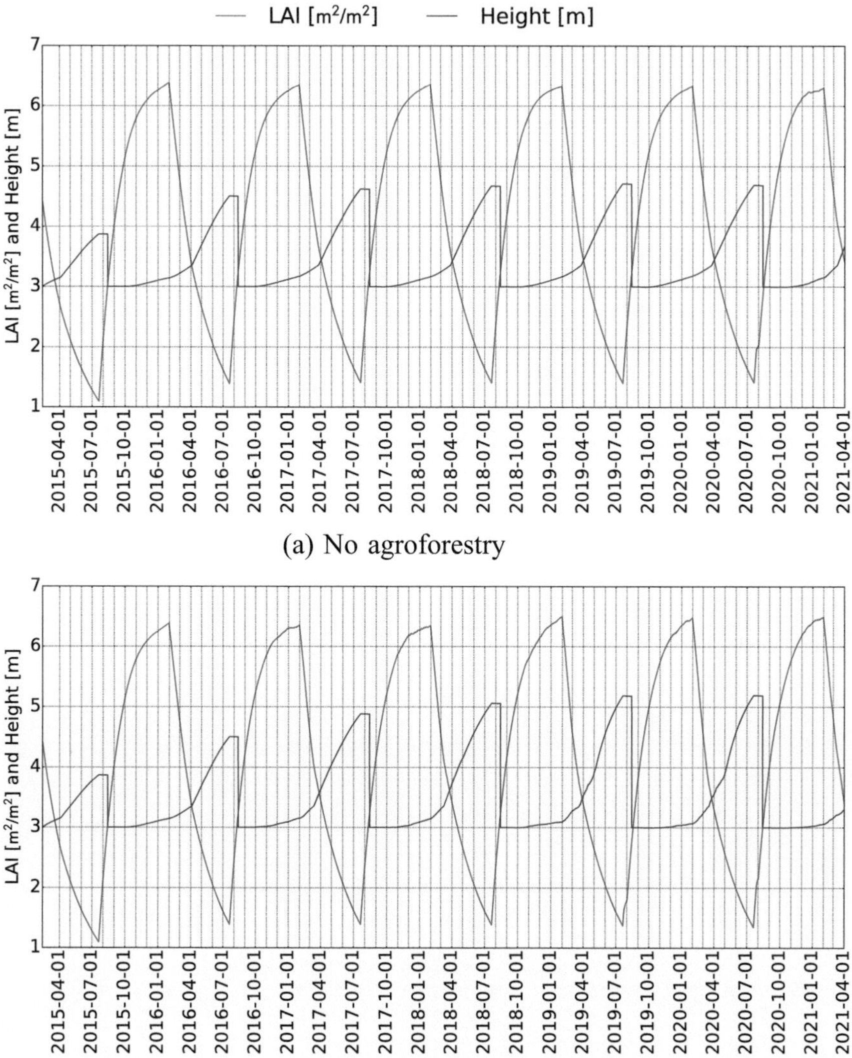

Fig. 8.5 Simulated citrus leaf area index LAI and height at Babylonstoren from 12 February 2015 to 1 April 2021 **a** without the agroforestry model and **b** with the agroforestry model representing the citrus row nearest to the tree strip

agroforestry simulation, the soil water contents are about 0.02 $cm^3\ cm^{-3}$ higher. During the main growing period from September to March, the differences between the two simulations are considerable: the soil water contents in the simulation with the agroforestry model do not differ as much throughout the three soil depths. Due to reduced radiation and wind speed, the citrus trees absorb less water from the soil.

Thus, the water contents lie around 0.30 cm^3 cm^{-3} in all the three soil depths. This is between 0.05 cm^3 cm^{-3} (30 cm depth) and 0.10 cm^3 cm^{-3} (60 cm and 90 cm depth) higher than in the simulation without the agroforestry model.

The cumulative soil water balance [mm] is shown in Fig. 8.7. The soil water input from precipitation and irrigation reaches almost 11,500 mm throughout the simulation period of about 6 years. The water content in the whole soil profile varies between 356 and 620 mm in the simulation without the agroforestry model and between 393 and 630 mm in the simulation with the agroforestry model. The water contents in the soil profile of both simulations approach each other in the winter months. In the other seasons, there is more water in the soil in the simulation with the agroforestry model, which is in line with the soil water contents shown in Fig. 8.6. The simulation of the citrus row closest to the tree strip is characterized by higher percolation and lower actual evapotranspiration in comparison with the simulation without the agroforestry model. The final amounts of percolated water are 4577 and 1488 mm, while the actual evapotranspirations sum up to 6826 mm and 10 039 mm. Especially in the simulation with the agroforestry model, percolation seems to be

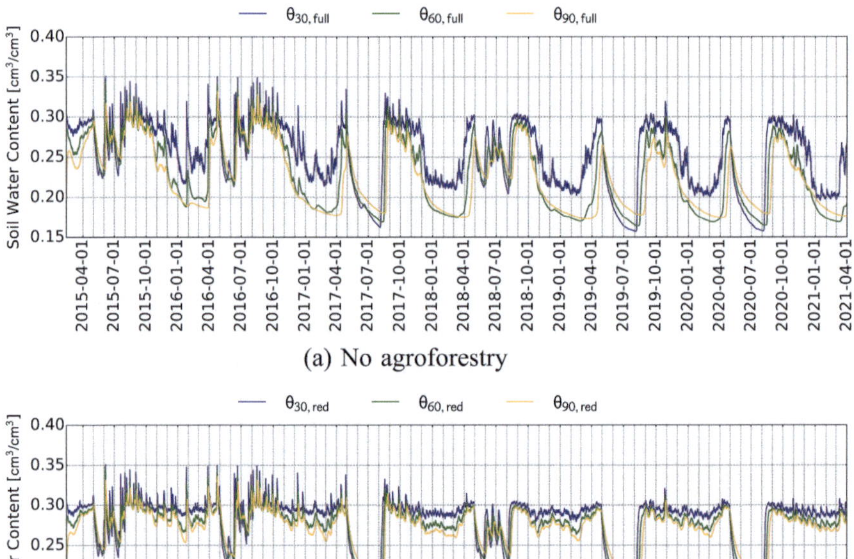

(a) No agroforestry

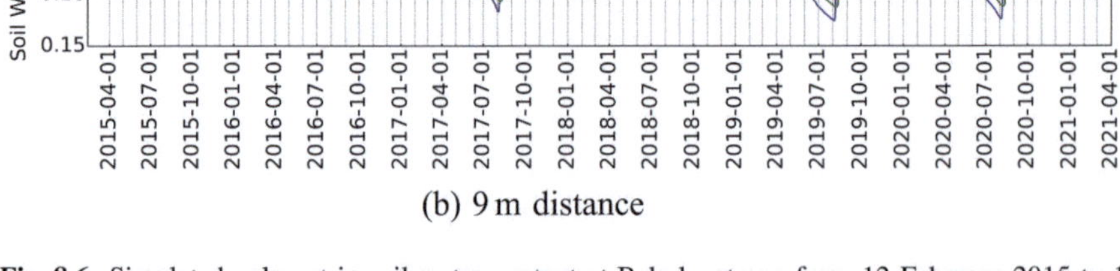

(b) 9 m distance

Fig. 8.6 Simulated volumetric soil water content at Babylonstoren from 12 February 2015 to 1 April 2021 **a** without the agroforestry model and **b** with the agroforestry model representing the citrus row nearest to the tree strip

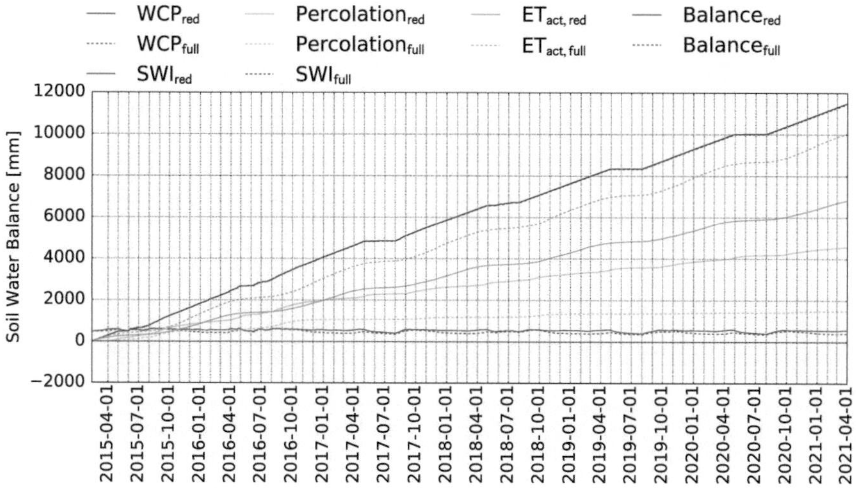

Fig. 8.7 Simulated cumulative soil water balance from 12 February 2015 to 1 April 2021. The solid and dashed lines denote simulations with and without the agroforestry model. The simulations with the agroforestry model represent the citrus row nearest to the tree strip. WCP stands for the water content in the whole soil profile, SWI for the soil water input, and ETact for the actual evapotranspiration

high. However, even if the water percolates out of the described model soil profile, in reality this water may still be reached by some roots. Additionally, the irrigation amount was the same for all model runs; i.e. it was not adapted to the actual citrus water demand, which farmers would typically do.

8.4 Simulation Annual Water Deficit with and Without Irrigation

The yearly potential and actual transpiration of twelve simulations with the agroforestry model and one control simulation without the agroforestry model is presented in Fig. 8.8. In all years, the control simulation exhibits the highest yearly potential transpiration (Fig. 8.8a) between 1668 mm a^{-1} (2020) and 1845 mm a^{-1} (2017). With decreasing distance from the tree strip, the potential transpiration decreases due to the lower solar radiation and wind speed. The actual transpiration (Fig. 8.8b) exhibits a similar pattern, but the differences between the simulations are smaller, since the actual transpiration additionally considers the plant available water in the soil. The actual transpiration in the control simulation lies between 1542 mm a^{-1} (2019) and 1659 mm a^{-1} (2016). The maximum reduction in actual transpiration due to the agroforestry model is 33.3% (9 m distance from the tree strip in 2016).

At 9, 17, and 25 m distance, the yearly potential and actual transpiration do not differ in 2016 and 2018. At 9 m distance, the difference between the potential and

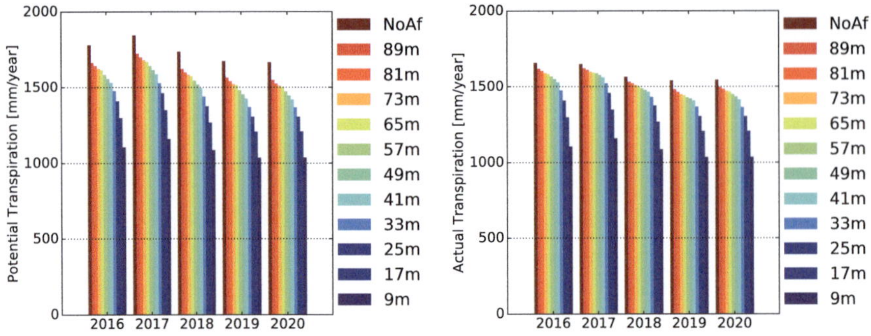

(a) Potential transpiration (b) Actual transpiration

Fig. 8.8 Simulated annual **a** potential and **b** actual transpiration at Babylonstoren from 2016 to 2020. The distances from the tree strip represent different simulations with the agroforestry module, and NoAf is the simulation without the agroforestry model

actual transpiration is lower than 1 mm a^{-1} in the other simulated years. At 17 m distance from the tree strip, the simulated actual transpiration is reduced between 18.3% (2017) and 21.9% (2020) in comparison with the undisturbed simulation. Similar reductions are found for the evapotranspiration (not shown), since soil evaporation only plays a minor role in this simulation. Coussement et al. (2018) also found lower evapotranspiration of crops next to a tree strip in comparison with crops not being under the influence of a neighbouring tree strip. Our simulated reductions of evapotranspiration are higher than those found by Veste et al. (2024) (15.1%) at the same study site. The reason lies in the different study approaches: Veste et al. (2024) used a Penman–Monteith approach with a crop coefficient to calculate the evapotranspiration, which is simpler than a complex soil–plant–atmosphere model like Expert-N, which considers various processes, such as plant growth or soil water flow and uptake. For a wheat field with alley-copping systems in Germany, Kanzler et al. (2019) found that the potential evapotranspiration was up to 25% reduced in comparison with a field without tree rows.

Figure 8.9a shows the yearly transpiration deficit or citrus water demand, which is the difference between the potential and the actual transpiration. At the four rows, which are closest to the tree strip, almost no transpiration deficit occurs. In all the years, the highest difference between potential and actual transpiration is manifested in the control simulation with a maximum of 194 mm a^{-1} in 2017. Even at the furthest distance from the tree strip (89 m), the citrus water demand is at least reduced by 36.4% or 48 mm a^{-1}. However, this effect of the agroforestry model varies over the years. The maximum effect occurs in 2016, and the minimum effect in 2019. The evapotranspiration deficit (Fig. 8.9b) shows similar patterns, but is slightly larger than just the transpiration deficit.

The reason is that evaporation plays a minor role in these simulations not exceeding 55 mm a^{-1} in comparison with the transpiration, which is always higher than

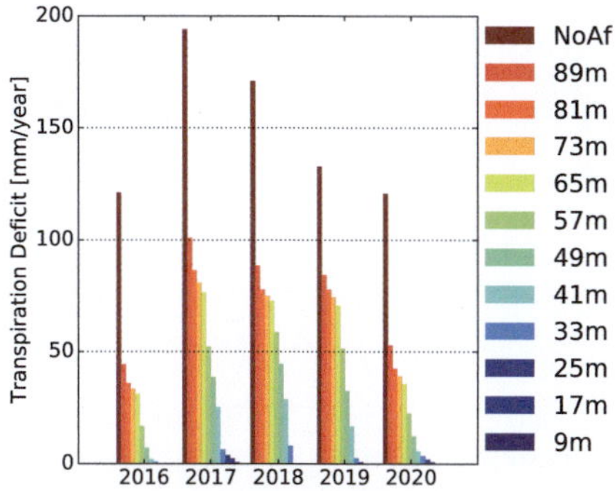

(a) Transpiration deficit

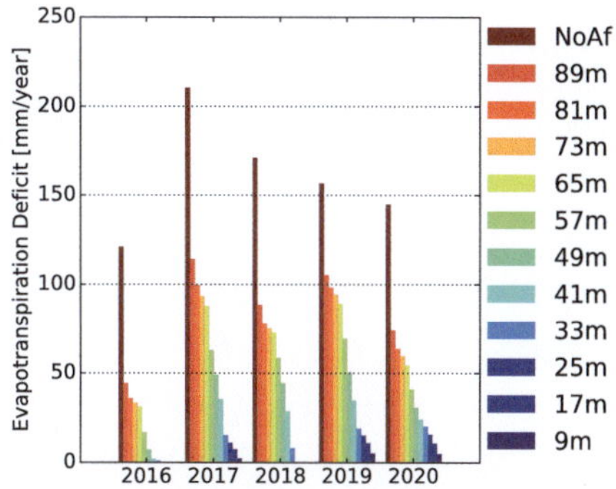

(b) Evapotranspiration deficit

Fig. 8.9 Simulated annual **a** transpiration and **b** evapotranspiration deficit at Babylonstoren from 2016 to 2020. The distances from the tree strip represent different simulations with the agroforestry module, and NoAf is the simulation without the agroforestry model

1000 mm a^{-1}. However, there is some water deficit due to evaporation, since evaporation only occurs in the winter months when the LAI is smaller than 3 m^2 m^{-2} and when there is no irrigation. This can be observed in Fig. 8.10 where in 2017, 2019, and 2020 the evapotranspiration deficit is higher than the transpiration deficit in the winter months. In 2016, the water deficit for all simulations is highest in January, while in 2019 and 2020 the citrus water demand is highest in February. In 2017, 2018, and 2021 the control simulation shows the highest water demand in January, and in the simulations with the agroforestry model, the difference between potential and actual transpiration is highest in February. In the simulations with the agroforestry model, the water deficit occurs later in the growth period. Thus, the tree strip serving as windbreak and for shading may help that the citrus plants are water-stressed later and for a shorter period in comparison with a farming system without tree strips.

Figure 8.11 shows the yearly potential and actual transpiration without any irrigation. In comparison with the potential transpiration with irrigation (Fig. 8.8a), there are only small differences in the years from 2016 to 2018. However, the potential transpiration strongly decreases in 2019 and 2020. In fact, the simulated plants lack a large amount of water, which leads to a loss of biomass and finally these plants die. In 2016, the plants can—despite no irrigation—absorb a substantial amount of water from the soil resulting in yearly actual transpirations between 422 mm a^{-1} and 478 mm a^{-1}). In the following two years, the yearly actual transpirations do not exceed 183 mm a^{-1} and 121 mm a^{-1}. In 2019, between 5.0 mm a^{-1} and 11.5 mm a^{-1} are taken from the soil through transpiration, and in 2020 the actual transpiration is 0 mm a^{-1}. From 2017 to 2019, the actual transpiration slightly increases with decreasing distance from the tree strip, which is contrary to the actual transpiration with irrigation (Fig. 8.8b). When getting closer to the tree strip, there is still more soil water available, and thus more water can be transpired by the plants. In general, in the simulations without irrigation, the amount of actual transpiration is dominated by the soil water availability. In the simulations with irrigation, the amount of actual

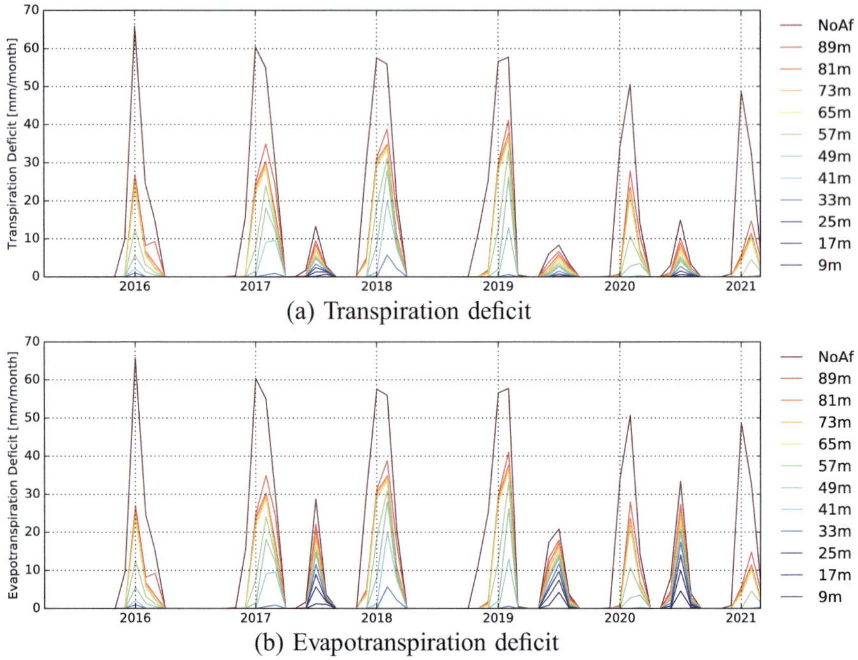

Fig. 8.10 Simulated monthly **a** transpiration and **b** evapotranspiration deficit at Babylonstoren from 2016 to 2020. The distances from the tree strip represent different simulations with the agroforestry module, and NoAf is the simulation without the agroforestry model

transpiration is to a larger extent defined by the potential transpiration, thus by the atmospheric conditions and the plants' leaf area.

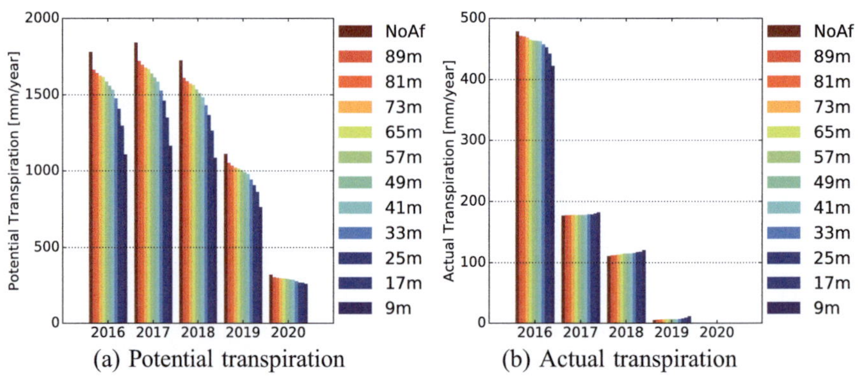

Fig. 8.11 Simulated annual **a** potential and **b** actual transpiration at Babylonstoren from 2016 to 2020 without irrigation. The distances from the tree strip represent different simulations with the agroforestry module, and NoAf is the simulation without the agroforestry model

The transpiration deficit (Fig. 8.12a) without irrigation gives an idea how high the citrus water demand is and what a reasonable irrigation amount would be. For these questions, the transpiration deficit in 2017 and 2018 should be addressed, since 2016 still benefits from the initial soil water availability, while the plants have already died in 2019 and 2020. In the simulation, the citrus trees may die faster than in a reality without irrigation due to the very low rain water input from 2019 onwards. However, the years 2017 and 2018 with correct rain measurements suggest that year after year the soil water availability gets lower, and, in the end, the citrus trees cannot survive without irrigation. In the control simulation without the agroforestry module, the water deficits are 1665 mm a^{-1} (2017) and 1614 mm a^{-1} (2018). At 89 m distance from the tree strip, the corresponding water deficits are 1544 mm a^{-1} and 1499 mm a^{-1}, while at 9 m distance the water deficits are significantly reduced to 981 mm a^{-1} and 965 mm a^{-1}. Thus, the introduction of the tree strip to our model reduces the water demand between 115 mm a^{-1} (89 m distance in 2018) and 684 mm a^{-1} (9 m distance in 2017). In other words, our model suggests that up to 40% of irrigation water may be saved through the introduction of tree strips. In the first three simulated years, the evapotranspiration deficit (Fig. 8.12b) hardly differs from the transpiration deficit. However, from 2019 onwards when the plants shrinked or died, evaporation gets more relevant and thus its deficit also increases.

In 2020, the evapotranspiration deficit ranges from 634 mm a^{-1} to 1039 mm a^{-1} with a share of evaporation between 59% (9 m distance) and 69% (control simulation).

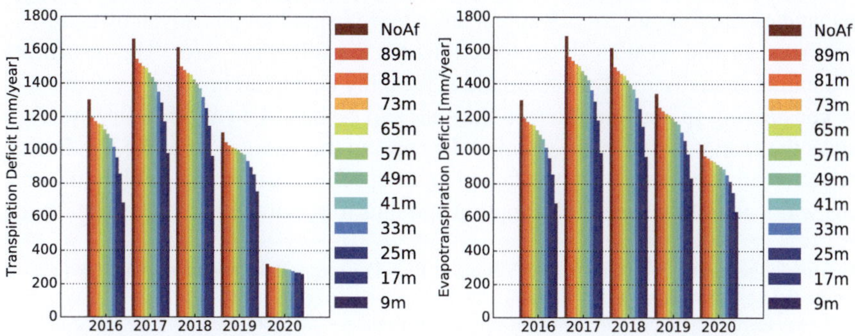

(a) Transpiration deficit (b) Evapotranspiration deficit

Fig. 8.12 Simulated annual **a** transpiration and **b** evapotranspiration deficit at Babylonstoren from 2016 to 2020 without irrigation. The distances from the tree strip represent different simulations with the agroforestry module, and NoAf is the simulation without the agroforestry model

8.5 Conclusions

The Expert-N agroforestry model including the newly developed wind reduction model was successfully tested and applied to a South African windbreak system. While on the one hand, the application of the agroforestry module led to a shift in carbon allocation among the different biomass components leading to lower crop yields and higher wood growth, on the other hand, the simulated water use of the crops was strongly reduced when the agroforestry model was applied. Our model suggests that, depending on the distance from a tree strip, about 20% less water use is achievable in an irrigated agroforestry system. In the non-irrigated system, the crop water demand was up to 40% lower in comparison with the simulation without the agroforestry model. Reducing the water demand is highly desirable especially in regions where the water availability is limited and/or where the price of water is high. Hence, the simulations support the establishment of agroforestry systems.

Applying Expert-N without irrigation, the total crop water demand or deficit can be estimated. In practice, it might also be beneficial to irrigate as close as possible to the crop roots. In this case, it is reasonable to only consider the plant water demand and to neglect the soil water demand. This is only possible with a complex soil–plant–atmosphere model such as Expert-N, while a pure Penman–Monteith approach may overestimate the required irrigation amount. The annual water deficit may help calculate the daily amount of required irrigation water. However, monthly output values may lead to more exact results considering seasonal variations. In principle, Expert-N also delivers daily or even sub-daily output values. Even water demand predictions are conceivable if an Expert-N simulation is driven by historical weather data and weather data predictions for the following days. The possibility to conduct spatially resolved simulations is highly beneficial in all ecosystems (e.g. by considering spatial variations in the soil) and especially in agroforestry systems. Since the crop water demand strongly depends on the distance from the windbreak, irrigation amounts may also be adapted to the specific position in a field. In the end, this will lead to less water consumption and thus to a more sustainable agriculture.

Acknowledgements We thank Ian de Villiers at Babylonstoren, Paarl for support and permission for the field experiments.

The research was funded in the frame of the FarmImpact—project (grant no. FKZ 01LZ1711A-F) as part of "CLIENT II—International Partnerships for Sustainable Innovations" funding initiative and of BonaRes-SIGNAL Sustainable intensification of agriculture through agroforestry, both funded by the Federal Ministry of Education and Research (BMBF).

References

Biernath C, Gayler S, Bittner S, Klein C, Högy P, Fangmeier A, Priesack E (2011) Evaluating the ability of four crop models to predict different environmental impacts on spring wheat grown in open-top chambers. Eur J Agron 35(2):71–82. https://doi.org/10.1016/j.eja.2011.04.001

Bohn FJ, Frank K, Huth A (2014) Of climate and its resulting tree growth: Simulating the productivity of temperate forests. Ecol Model 278:9–17

Bossel H (1996) TREEDYN3 forest simulation model. Ecol Model 90:187227

Brandle JR, Hodges L, Zhou XH (2004) Windbreaks in North American agricultural systems. Agrofor Syst 61–62:65–78

Brisson N, Mary B, Ripoche D, Jeuffroy M H, Ruget F, Nicoullaud B, Gate P, Devienne-Barret F, Antonioletti R, Durr C, Richard G, Beaudoin N, Recous S, Tayot X, Plenet D, Cellier P, Machet J-M, Meynard J M, Delécolle R (1998) STICS: a generic model for the simulation of crops and their water and nitrogen balances. I. Theory and parameterization applied to wheat and corn. Agronomie 18:311–346

Campbell GS (1987) Soil physics with BASIC - transport models for soil-plant systems. Dev Soil Sci 14(3):252254

Coussement T, Maloteau S, Pardon P, Artru S, Ridley S, Javaux M, Garr (2018) A tree-bordered field as a surrogate for agroforestry in temperate regions: Where does the water go? Agric Water Manag 210:198–207

Delalieux S, Somers B, Hereijgers S, Verstraeten W, Keulemans W, Coppin P (2008) A near-infrared narrow-waveband ratio to determine Leaf Area Index in orchards. Remote Sens Environ 112:11

Dupraz C (2002) Tree-crop interaction models: State of the art report (Technical Report No. Deliverable D.1.1 of the SAFE European Research Contract QLK5-CT-2001–00560). INRA, Montpellier

Garca-Barrios L, Mayer-Foulkes D, Franco M, Urquijo-Vsquez G, Franco-Prez J (2001) Development and validation of a spatially explicit individual-based mixed crop growth Model. Bull Math Biol 63(3):507–526

Gayler S, Priesack E (2007) Platho—a simulation model for resource allocation in the plant-soil system. Technical report. https://www.researchgate.net/publication/227945282_PLATHO_a_simulation_model_of_resource_allocation_in_the_plant-soil_system

Graves AR, Burgess PJ, Palma JHN, Herzog F, Moreno, Bertomeu M, Dupraz C, Liagre F, Keesman K, van der Werf W, de Nooy A. Koeffeman, van den Briel JP (2007) Development and application of bio-economic modelling to compare silvoarable, arable, and forestry systems in three European countries. Ecol Eng 29(4):434–449

Hansen S, Jensen HE, Nielsen NE, Svendsen H (1990) DAISY: Soil plant atmosphere system model. NPO Report No. A 10. The National Agency for Environmental Protection, Copenhagen, p 272

Heinlein F, Biernath C, Klein C, Thieme C, Priesack E (2017) Evaluation of simulated transpiration from maize plants on lysimeters. Vadose Zone J 16:1–16. https://doi.org/10.2136/vzj2016.05.0042

Hutson JL, Wagenet RJ (1992) LEACHM: Leaching estimation and chemistry model: A process-based model of water and solute movement, transformations, plant uptake and chemical reactions in the unsaturated zone. Version 3.0. 93–3. Ithaca, NY, USA

Jacobs SR, Webber H, Niether W, Grahmann K, Lüttschwager D, Schwartz C, Breuer L, Bellingrath-Kimura SD (2022) Modification of the microclimate and water balance through the integration of trees into temperate cropping systems. Agric for Meteorol 323:109065

Kanzler M, Böhm C, Mirck J, Schmitt D, Veste M (2019) Microclimate effects on evaporation and winter wheat (*Triticum aestivum* L.) yield within a temperate agroforestry system. Agrofor Syst 93:1821–1841. https://doi.org/10.1007/s10457-018-0289-4

Klein C, Biernath C, Heinlein F, Thieme C, Gilgen AK, Zeeman M, Priesack (2017) Vegetation growth models Improve surface layer flux simulations of a temperate grassland. Vadose Zone J 16(13). https://doi.org/10.2136/vzj2017.03.0052

Kraft P, Rezaei E, Breuer L, Ewert F, Große-Stoltenberg A, Kleinebecker T, Seserman D, Nendel C (2021) Modelling agroforestry's contributions to people—a review of available models. Agronomy 11

Luedeling E, Smethurst PJ, Baudron F, Bayala J, Huth NI, van Noordwijk M, Ong CK, Mulia R, Lusiana B, Muthuri C, Sinclair FL (2016) Field-scale modeling of tree-crop interactions: Challenges and development needs. Agric Syst 142:51–69

Mobbs DC, Lawson GJ, Friend Andrew D, Crout Neil MJ, Arah Jonathan RM, Martin H (1999) HyPAR model for agroforestry systems: technical manual: model description for version 3.0
Mualem Y (1976) A new model for predicting the hydraulic conductivity of unsaturated porous media. Water Resour Res 12:513–522. https://doi.org/10.1029/WR012i003p00513
Nimah MN, Hanks RJ (1973) Model for estimating soil water, plant, and atmospheric interrelations: I. Description and sensitivity. Soil Sci Soc Am J 37(4):522–527
Palma JHN, Graves AR, Bunce RGH, Burgess PJ, de Filippi R, Keesman KJ, van Keulen H, Liagre F, Mayus M, Moreno G, Reisner Y, Herzog F (2007) Modeling environmental benefits of silvoarable agroforestry in Europe. Agric, Ecosyst & Environ 3:320–334
Parton WJ, Ojima DS, Cole CV, Schimel DSA (1994) General model for soil organic matter dynamics: Sensitivity to litter chemistry, texture and management. In: Bryant RB, Arnold RW (eds) Quantitative modeling of soil forming processes. pp 147–167. https://doi.org/10.2136/sssaspecpub39.c9
Pretzsch H, Grote R, Reineking B, Rotzer T., Seifert (2008): Models for forest ecosystem management: A European perspective. Ann Bot 101(8):1065–1087. https://doi.org/10.1093/aob/mcm246
Priesack E, Gayler S (2009) Agricultural crop models: Concepts of resource acquisition and assimilate partitioning. In: Lüttge U, Beyschlag W, Büdel B, Francis D (eds) Progress in botany, vol 70. Springer, Heidelberg, pp 192–222. https://doi.org/10.1007/978-3-540-68421-3_9
Priesack E, Gayler S, Hartmann HP (2006) The impact of crop growth sub-model choice on simulated water and nitrogen balances. Nutr Cycl Agroecosyst 75(1–3):1–13. https://doi.org/10.1007/s10705-006-9006-1
Raj R, Suradhaniwar S, Nandan R, Jagarlapudi A, Walker J (2020) Drone-based sensing for leaf area index estimation of citrus canopy. In: Jain K, Khoshelham K, Zhu X, Tiwari A (eds) Proceedings of UASG 2019. UASG 2019. Lecture notes in civil engineering, vol 51. Springer, Cham, pp 79–89. https://doi.org/10.1007/978-3-030-37393-1_9
Raveh E, Cohen S, Raz T, Yakir D, Grava A, Goldschmidt EE (2003) Increased growth of young citrus trees under reduced radiation load in a semiarid climate. J Exp Bot 54(381):365–373. https://doi.org/10.1093/jxb/erg009
Rosenfeld M, Marom G, Bitan (2010) Numerical simulation of the airflow across trees in a windbreak. Bound-Layer Meteorol 135:89–107
Schaaf T (1998) Integration von Modellansätzen zur Bodenbearbeitung und Düngung in den Baukasten für Stickstoffsimulationsmodelle EXPERT-N. Disseration Zentralstelle für Agrardokumentation und–information (ZADI), Bonn
Schaap MG, Leij FJ, van Genuchten MT (2001) ROSETTA: a computer program for estimating soil hydraulic parameters with hierarchical pedotransfer functions. J Hydrol 251(3–4):163–176. https://doi.org/10.1016/S0022-1694(01)00466-8
Sheppard JP, Bohn Reckziegel R, Borrass L, Chirwa PW, Cuaranhua CJ, Hassler SK, Hoffmeister S, Kestel F, Maier R, Mälicke M, Morhart C, Ndlovu NP, Veste M, Funk R, Lang Seifert F, du Toit TB, Kahle HP (2020) Agroforestry: an appropriate and sustainable response to a changing climate in Southern Africa? Sustain 12(17):6796. https://doi.org/10.3390/su12176796
Skamarock WC, Klemp JB, Dudhia J, Gill DO, Barker D, Duda MG, Huang X-Y, Wang W, Powers JG (2008) A description of the advanced research WRF version 3: Technical Reort 06 2008
Šimůnek J, Šejna M, Huang K, Genuchten MT van (1998) The HYDRUS code for simulating the one-dimensional movement of water, heat, and multiple solutes in variably-saturated media. Version 6.0. 144. Riverside, CA., USA
Tumbo SD, Whitney JD, Miller WM, Wheaton TA (2002) Development and testing of a citrus yield monitor. Appl Eng Agric 18(4):399–403
van Genuchten MT (1980) A Closed-form equation for predicting the hydraulic conductivity of unsaturated soils. Soil Sci Soc Am J 44(5):892–898. https://doi.org/10.2136/sssaj1980.03615995004400050002x

Van Noordwijk M, Lusiana B (1999). WaNuLCAS, a model of water, nutrient and light capture in agroforestry systems. In: Auclair D, Dupraz C (eds) Agroforestry for sustainable land-use fundamental research and modelling with emphasis on temperate and mediterranean applications. For Sci 60:217–242 Springer, Dordrecht. https://doi.org/10.1007/978-94-017-0679-7_14

van Ramshorst JGV, Siebicke L, Baumeister M, Moyano FE, Knohl A, Markwitz C (2022) Reducing wind erosion through agroforestry: a case Study Using large eddy simulations. Sustain 14(20):13372. https://doi.org/10.3390/su142013372

van der Werf W, Keesman K, Burgess P, Graves A, Pilbeam D, Incoll LD, Metsellaar K, Mayus M, Stappers R von Keulen H, Palma J, Dupraz C (2007) Yield-SAFE: a parameter-sparse, process-based dynamic model for predicting resource capture, growth, and production in agroforestry systems. Ecol Eng 29(4):419–433 https://doi.org/10.1016/j.ecoleng.2006.09.017

Veste M, Littmann T, Kunneke A, Du Toit B, Seifert T (2020) Windbreaks as part of climate-smart landscapes reduce evapotranspiration in vineyards, Western Cape Province, South Africa. Plant Soil Environ 66:119–127. https://doi.org/10.17221/616/2019-pse

Veste M, Sheppard JP, Abdulai I, Ayisi KK, Borrass L, Chirwa PW, Funk R, Kapinga K, Morhart C, Mwale SE, Ndlovu NP (2024) The need for sustainable agricultural land-use systems: benefits from integrated agroforestry systems. In: von Maltitz GP, Midgley GF, Veitch J, Brümmer C, Rötter RP, Viehberg FA, Veste M (eds) Sustainability of Southern African ecosystems under global change: science for management and policy interventions. Ecological studies, vol 248. Springer International Publishing, Cham, pp 587–623. https://doi.org/10.1007/978-3-031-10948-5_21

Walter IA, Allen RG, Elliott R, Itenfisu D, Brown P, Jensen ME, Mecham B, Howell TA, Snyder R, Eching S, Spofford T, Hattendorf M, Martin D, Cuenca RH, Wright JL (2005) The ASCE standardized reference evapotranspiration equation

Wilson JD (1987) On the choice of a windbreak porosity profile. Bound-Layer Meteorol 38:37–49

Wilson D (1985) Numerical studies of flow through a windbreak. J Wind Eng Ind Aerodyn 21:119–154

Wright B, Stuhr K (2002) Windbreaks: An agroforestry practice. Agroforestry Notes 25

Chapter 9
Shelterbelts for Livestock Protection: Planning and Design in a High Wind Speed Environment at the West Coast, Western Cape, South Africa

Thomas Littmann, Daniel Littmann, and Maik Veste

Abstract The reduction of near-ground wind speed by design windbreaks was modeled based on long-term wind statistics derived from on-site wind measurements and reanalysis data for different planning purposes. In the high wind energy environment of the Western Cape West Coast Peninsula near Yzerfontein, windbreak planning should significantly reduce wind speeds within a farm enclosure for ostriches susceptible to wind chill stress. The model results for three hedgerows (*Populus* or comparable, height 10 m, width 6 m, length 250 m, spacing 100 m) indicate an overall reduction of wind speed of 45% for the lee zone behind a windbreak and in the middle about 50 m behind it. The results imply that the relatively small invest in 750 running meters of hedgerow planting covering 0.5% of the enclosure creates a comfort zone of 2.5% of the enclosure area which should be sufficient for the entire ostrich stock.

Keywords Windbreak · Wind speed · Landscape structure · Ostriches · Coastal area

9.1 Introduction

The reduction of near-ground wind speed by windbreaks may fulfill several application purposes. Saving irrigation water by significant reduction of crop evapotranspiration is a main feature especially in semiarid environments like the Western Cape

T. Littmann (✉) · D. Littmann
DLC–Dr. Littmann Consulting, Ennepetal, Germany
e-mail: littmann@littmann-consulting.com

M. Veste
CEBra–Center for Energy Technology Brandenburg e.V., Cottbus, Germany

Department of Molecular Botany, Institute of Biology, University of Hohenheim, Stuttgart, Germany

M. Veste et al. (eds.), *Trees in South African Agriculture*, Innovations in Landscape Research, https://doi.org/10.1007/978-3-032-00993-7_9

Fig. 9.1 **a** Farm Slangkop near Yzerfontein embedded in the hilly landscapes of the West Coast, Western Cape, South Africa

of South Africa (Chap. 5) but may also be relevant in terms of resilience against climate change in temperate climate (Chaps. 10 and 11). However, the benefits and planning of windbreaks largely depend on the climatic setting and their specific uses of shelterbelts. The climate of the West Coast of South Africa (Fig. 1.1) is semiarid Mediterranean, mild and without extremes, but with strong seasonal winds, which are mostly southerlies in summer and northerlies in winter. Due to the semiarid climate and the sandy soils, the commercial farming relies here on the production potatoes, rooibos tea, table grapes, and wheat. Crop irrigation plays here an important role for agriculture (Archer et al. (2009). In areas characterized by such a strong wind climate like the West Coast of the Western Cape, a reduction of wind speed is not only beneficial for crop production (Veste et al. 2020) and wind erosion, but also for husbandry. Besides traditional sheep farming, commercial farming of ostriches has become more and more important, and South Africa is the premier producer of ostrich products worldwide (Snyder 2020). Even that they are living in open landscapes, ostriches (Fig. 9.1) are highly sensitive to wind chill (Deeming 1997; Adams and Revell 1998). For the planning of the specific shelterbelts at the West Coast, a comprehensive wind analysis was conducted.

9.2 Study Site

Slangkop Farm is located about 10 km northeast of Yzerfontein, Western Cape Province, north of the R315 and east of the R27. The farm area is hilly terrain with several hilltops reaching elevations of 208 m (Slangkop southeast of the farm) and 160 m to 150 m (hilltops on the western side), and the locally highest elevation is 250 m which was chosen as the site of the wind measurement, adjacent to the existing wind turbines. From the measurement site, the terrain slopes gently toward the northerly sectors to 80 m at Swartwater and toward the southerly and southeasterly sectors to 100 m along the R315 and to 120 m at the farmhouses. Thus, the wind mast location is freely exposed to all directions. The terrain structure is additionally modified by two smaller dry valleys going west and southwest toward the coastline at Yzerfontein.

9.3 Wind Measurements

The installed measurement (Fig. 9.2b) included wind speed and wind direction at 10 m and wind speed at 2 m (sensor: RM Young wind sentry); ambient temperature, relative humidity, and precipitation were also measured at 2 m. Data are sampled at 1 Hz and are stored as 10 min averages. The measurement period was March 4, 2015, to April 6, 2016 (13 months) and is sufficient for further data analysis and the generation of a statistical basis for small scale wind field modeling. Data availability over the measurement period was 100% in all cases. We did not apply further long-term scaling of measured wind speeds as the purpose of the measurement is to provide input data for estimating the relative effects of windbreak obstacles to near-ground air flow and not a detailed investigation of local wind power potential.

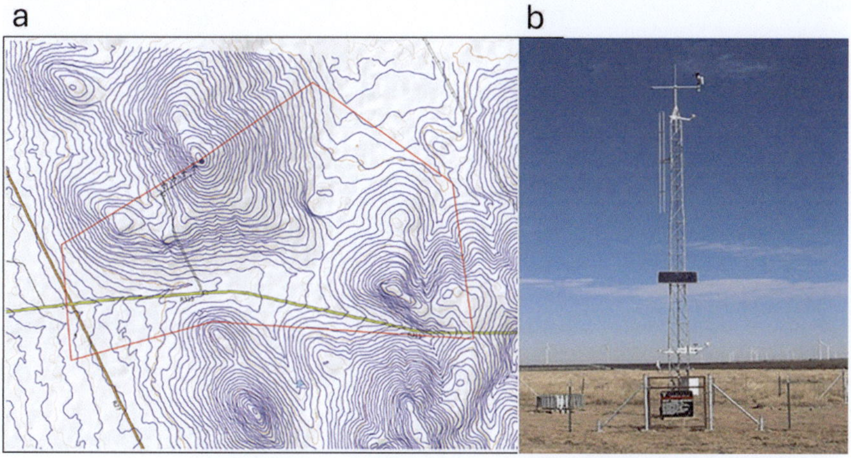

Fig. 9.2 **a** Topography, farm limits of Slangkop farm, and location of the wind measurement (dot), **b** wind measurement mast with wind measurements at 2 and 10 m height

9.4 Results of Wind Measurements

The characteristic wind data at Slangkop Farm are shown in Table 9.1. Mean wind speed at 10 m above ground over the measurement period was 6.5 m s-1 with a median value of 7.35 m s^{-1} (Weibull A parameter). Higher wind speeds occur over the summer months when inland heating leads to low pressure systems and local stronger airflow from the sea (Chap. 4). Mean maximum wind speed at 10 m was 22.7 m s^{-1} while the frequency distribution of wind speed indicates 34% of the total dataset to be within the low wind range up to 5 m s^{-1}, 52.6% are between 5 and 10 m s^{-1}, 12.4% are within the range of 10 and 15 m s^{-1}, and a few cases showed very high 10 min averages (1%). This distribution is well within the range of a typical Weibull frequency distribution but indicates a strong local wind field (Fig. 9.3). The siting of the mast on top of the western ridge with topographical forcing and speed up along the south-facing slope results in considerable turbulence intensity at the relatively low measuring height of 10 m (12.77%).

The wind direction distribution over twelve 30° sectors shows about 48% of all cases to come from south and south-southeast followed by south-southwest (9.6%) while all other directions show frequencies of up to 5%. The main wind directions also show the highest individual wind speeds, indicating long periods of strong sea breeze over the year (an exception is the north-northwest with an average speed of 7.1 m s^{-1}). This implies a clear focus on west–east orientation of windbreaks sheltering the site from the prevailing and heaviest winds (Figs. 9.3 and 9.4).

Table 9.1 Wind characteristics at Slangkop farm

Parameter	Unit	
Mean wind speed	m s^{-1}	6.50
Max. mean wind speed	m s^{-1}	22.69
Standard deviation	m s^{-1}	3.97
Turbulence intensity > 3 m s^{-1}	m s^{-1}	12.77
Weibull A parameter	m s^{-1}	7.35
Weibull k parameter		2.24

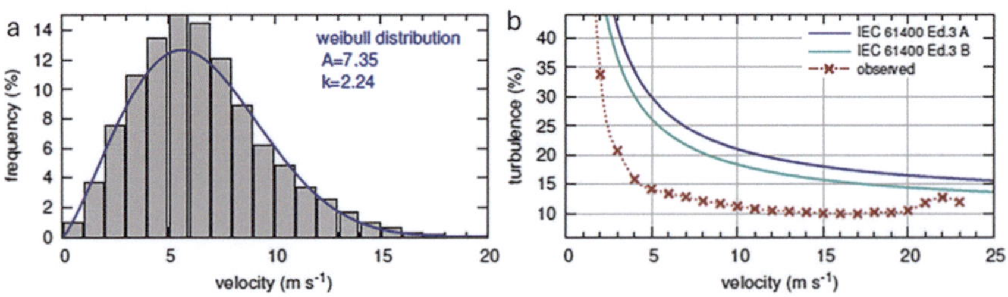

Fig. 9.3 Wind statistics of Slangkop wind measurement (anemometer at 10 m). **a** Frequency distribution of wind speed and Weibull distribution fit; **b** turbulence intensity distribution compared to IEC standard limits

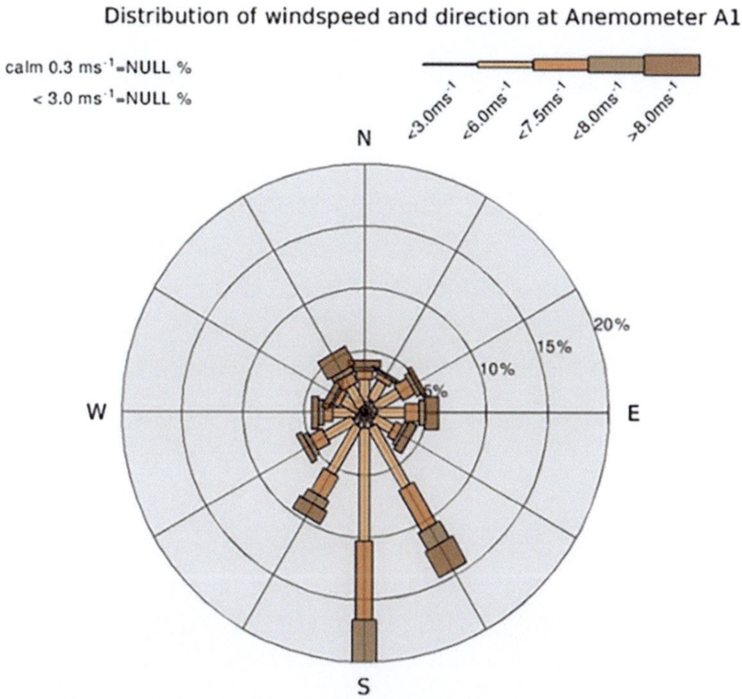

Fig. 9.4 Distribution of wind speed and wind direction, Slangkop Farm

9.5 Windbreak Design

The local wind field at both sites was simulated with EMD WindPRO 3.5 based on hourly ERA 5 reanalysis wind data for 10 m above ground (1990–2020 for the nearest grid nodes) and scaled to the measured wind speed at the respective measurement stations for the overlapping period. The digital terrain model was SRTM 5 m, and the roughness model was CORINE Landcover 2018. The resulting long-term wind speed distributions are representative for the local wind climate on a 30-year timescale. For the initial planning of windbreaks, we focused on the area of farmland with ostrich enclosures.

Three tree windbreaks (*Populus* or comparable) were inserted into one part of the enclosure area as obstacles considering the following parameters:

- Orientation: 260° to 80° from North (i.e., perpendicular to the prevailing wind directions)
- Length: 3 × 250 m
- Height: 10 m
- Width: 6 m
- Porosity: 0.3 on average (aerodynamic porosity indicates the degree of wind still passing through the obstacle)

Fig. 9.5 Siting of the initial windbreak planning. Wind mast: red dot

- Distance between windbreaks: 100 m (Figs. 9.5 and 9.6)

In a first validation approach (Table 9.2), we computed the wind speed at 10 m above ground for 10 reference points around the windbreaks (one reference point in the open space south of the southernmost windbreak with free exposure, one point about 10 m on the leeside, one in the center between two windbreaks, one 10 m on the luv side of the last windbreak, 3 points on the western and eastern sides of the three windbreaks, respectively). Here, the terrain is about 104 m to 115 m lower than the mast position on top of the western hill and slopes slightly from North to South (3°).

The computation of wind speeds at 10 m above ground (i.e., the measurement height at the mast) shows lesser wind speeds along the lower part of the slope and a clear interrelation of elevation and wind speed (Fig. 9.7). The analysis of wind direction distributions shows a shift toward a south-southwesterly component which is an effect of channeling by the small valley head going toward this direction (Table 9.3). The wind speed reductions computed for the reference points located between the three windbreaks are significant already at 10 m above ground with reference to the freely exposed control point T0 in the main wind direction sector in front of the first

Fig. 9.6 Windbreaks and control points

Table 9.2 Computed wind speeds at 10 m above ground

Point	Elevation	Wind speed (m s^{-1}) 10 m without shelterbelt	Wind speed (m s^{-1}) 10 m with shelterbelt	Difference
Mast	248	6.5	6.5	0
T0	127	4.11	4.11	0
T1	135	4.44	2.43	*–45.3*
T2	141	4.55	2.33	*–48.8*
T3	143	4.58	3.47	*–24.4*
T4	135	4,54	4.54	0
T5	141	4.6	4.6	0
T6	147	4.79	4.79	0
T7	137	4.44	4.44	0
T8	134	4.56	4.56	0
T9	140	4.41	4.41	0

windbreak (Fig. 9.8). The overall reduction is about 45% for the lee zone behind a windbreak and in the middle about 50 m behind it. At about 90 m behind the windbreak (and close to the following one), the reduction is significantly less (about 24%). The western and eastern control points do not show any reduction in computed wind speed. The wind speed reductions show a clear concentration on the sectors effectively shielded by the windbreak. With increasing distance from the windbreak, the shielded sectors and thus the effects of the windbreak decrease further until the effect is reduced to the main wind directions.

Based on those findings, we computed the respective wind speeds at 2 m above ground (Table 9.4) which should be the relevant height for the comfort zone for the ostriches. The shielding effect of the windbreaks becomes even more pronounced at this height, especially in the middle part of the field between the windbreaks. At this level, there are also minor effects at the reference point T0 and at control points west and east of the second and third windbreaks (Fig. 9.9).

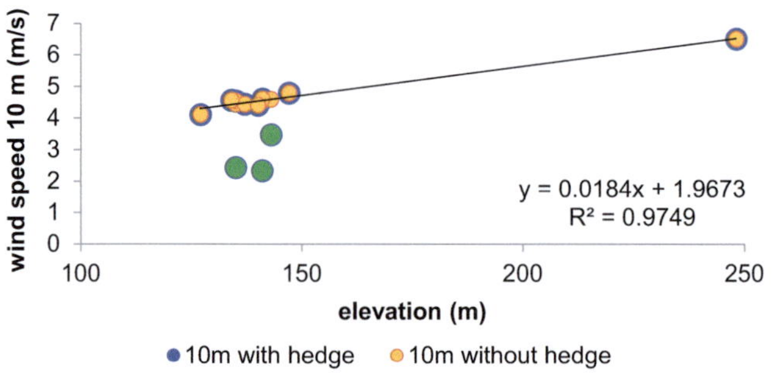

Fig. 9.7 Interrelation of wind speed at 10 m above ground and elevation of control point (without and with windbreaks) Green: T1/T2/T3 between windbreaks

Table 9.3 Sector-wise computed differences in wind speed (m/s) with reference to control point T0

Sector	T1–T0	T2–T0	T3–T0
N	–0.42	–0.37	1.67
NNE	0.86	–0.79	0.75
ENE	–0.90	–0.28	0.39
E	*–2.19*	–0.07	0.07
ESE	*–1.66*	*–1.51*	–0.49
SSE	*–1,60*	*–2.42*	*–1.39*
S	*–2.67*	*–3.00*	*–1.75*
SSW	*–2.35*	*–2.17*	–0.87
WSW	–0.72	–0.64	0.12
W	–0.27	–0.15	–0.23
WNW	–0.65	*–2.27*	–0.27
NNW	–0.80	–0.97	0.79
Total	–1.68	–1.78	–0.64

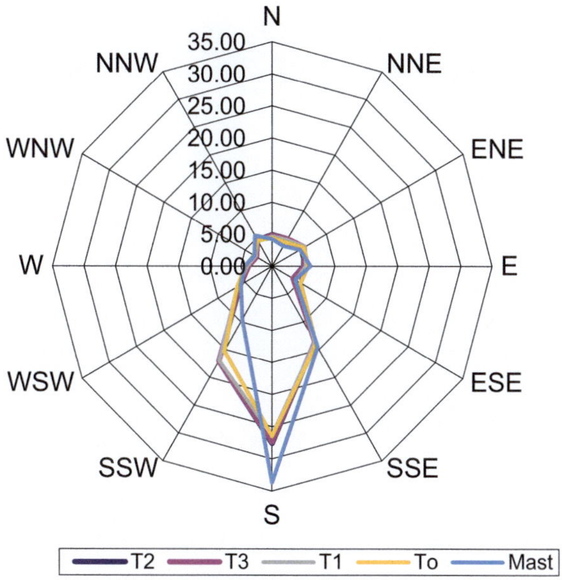

Fig. 9.8 Wind direction distributions compared to the measurement mast. T0 open field, T1/T2/T3 between windbreaks (cf. Fig. 9.6)

Table 9.4 Computed wind speeds at 2 m above ground without and with windbreaks

2m Point	Wind speed (m s^{-1}) (2m) without shelterbelt	Wind speed (m s^{-1}) (2m) with shelterbelt	Difference (%)
T0	2.7	2.6	−3.7 %
T1	2.8	1.5	−46.4%
T2	2.7	1.2	−55.6 %
T3	2.6	1,	−26.9 %
T4	3.0	3.0	0.0 %
T5	3.0	2.9	−3.3 %
T6	3.2	3.1	−3.1 %
T7	2.5	2.5	0.0 %
T8	2.5	2.4	−4.0 %
T9	2.8	2.7	−3.6 %

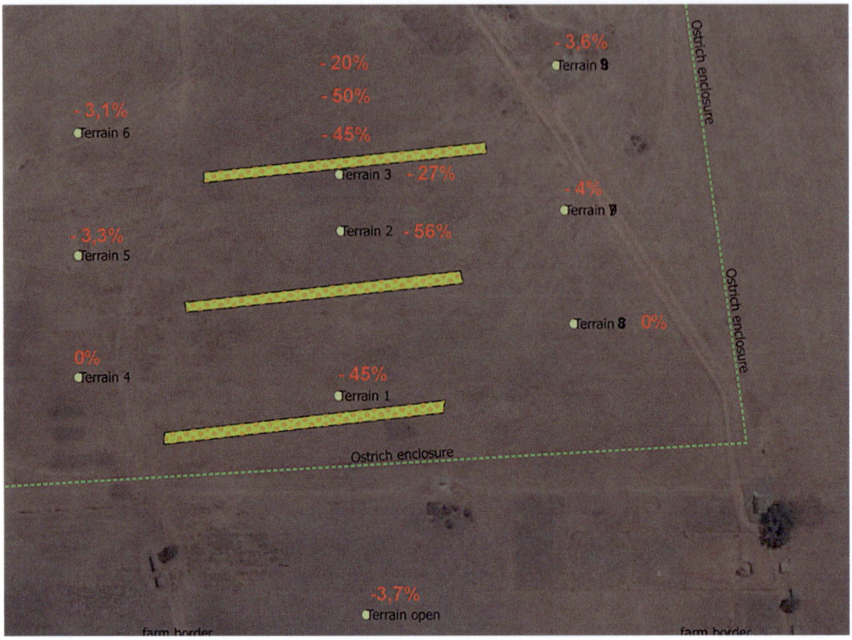

Fig. 9.9 Relative reduction of wind speed at 2 m above ground

9.6 Conclusion

Computations of near-ground wind speed based on a 10 m wind measurement and wind field modeling for Slangkop ostrich farm showed a significant reduction of wind speed at 10 m and at 2 m above ground by 25% to 55% in the field area between windbreaks. The immediate effect of windbreak planting is an effective sheltering over 50 m to 70 m on the leesides behind the windbreaks. A row distance of about 80 m to 100 m is recommended. The results imply that the relatively small investment in 750 running meters of hedgerow planting covering 0.5% of the enclosure creates a comfort zone of 2.5% of the enclosure area which should be sufficient for the entire ostrich stock. It may be concluded that the benefits and planning of windbreaks largely depend on the climatic setting, especially on the necessity of saving irrigation water over a hot summer season with low rainfall as in the Western Cape, and on special applications such as reducing wind chill stress for livestock in high wind energy environments such as the Westcoast Peninsula of the Western Cape. The application of windbreaks preventing mechanical wind damage of crops like grapes and fruit was found feasible in high wind energy environments (Chaps. 3 and 6) of the Western Cape but is ambiguous in low wind energy environments (Chap. 10). The choice of windbreak type (hedgerow or technical) will depend on available space, shading, and the additional water demand of hedgerows (Chap. 5).

Acknowledgements The research was supported by the National Research Foundation (NRF) and the Department of Agriculture, Forestry and Fisheries (DAFF) of South Africa.

References

Adams, J, Revell, BJ (1998) A review and feasibility study of opportunities in the EU; School of Management, Harper Adams University College, Newport, UK. https://macaulay.webarchive.hutton.ac.uk/livestocksystems/feasibility/ostrich.htm

Archer E, Conrad J, Münch Z, Opperman D, Tadross M, Venter J (2009) Climate change, groundwater and intensive commercial farming in the semi-arid northern Sandveld, South Africa. J Integr Environ Sci 6(2):139–155. https://doi.org/10.1080/19438150902916589

Deeming DC (1997) Effect of climatic conditions on the behaviour of adult ostriches (Struthio camelus) in Britain. Anim Welf 6:349–356

Snyders M (2020) Perceptions about commercial ostrich farming: views of consumers, farmers and secondary stakeholders. Master thesis of Agricultural Sciences, Stellenbosch University. https://scholar.sun.ac.za/handle/10019.1/109073

Veste M, Littmann T, Kunneke A, Du Toit B, Seifert T (2020) Windbreaks as part of climate-smart landscapes reduce evapotranspiration in vineyards, Western Cape Province, South Africa. Plant Soil Environ 66:119–127. https://doi.org/10.17221/616/2019-pse

Veste M, Sheppard JP, Abdulai I, Ayisi KK, Borrass L, Chirwa PW, Funk R, Kapinga K, Morhart K, Mwale SE, Ndlovu NP, Nyamadzaw G, Nyoka BI, Sebola P, Seifert T, Senyolo MP, Sileshi GW, Syampungani S, Kahle H-P (2023) The need for sustainable agricultural land-use systems: benefits from integrated agroforestry systems. In: von Maltitz GP, Midgley GF, Veitch J, Brümmer C, Rötter RP, Viehberg, FA, Veste M (eds) Sustainability of southern African ecosystems under global change: Science for management and policy interventions. Ecological Studies 248, Springer Nature, Heidelberg. https://doi.org/10.1007/978-3-031-10948-5_21

Chapter 10
Weather Types Controlling Wind and Evapotranspiration in a Temperate Vineyard—A Transfer Application to a Low Wind Speed System

Thomas Littmann, Tobias Recke, Daniel Littmann, and Maik Veste

Abstract The vineyard Wolkenberg is located in a temperate climate with low wind speeds, but gusts up to 14 m s^{-1} were occasionally measured. Analysis of measurement data aimed at the identification of typical combinations of microclimatic parameters controlling evaporative processes within the vineyard. The statistical approach combines results from cluster analysis and from principal component analysis (PCA). For upscaling and prognostic purposes, daily objective weather types as defined by the German Met Service (DWD) were analyzed. Three PCA clusters were derived that show the main two constellations in which high evapotranspiration can occur in the stand (spring wind-dependent, summer dependent on saturation deficit). In both cases, there is a significant correlation with high-pressure systems in a marginal position, which usually lead to NW or SW flow with more or less wind. The local wind field was modeled to design a windbreak system to reduce wind impacts and evapotranspiration. In this case, a vertical artificial shading net was implemented in model runs. The water balance over the vegetation periods was generally negative. In years without longer summerly dry spells, the irrigation water demand over the vegetation period should be rather low (20–80 mm) and probably even lower in case of deep-rooting systems like grapevine.

Keywords Vineyard · Water balance · Design of windbreaks · Post-mining site · Microclimate

T. Littmann (✉) · T. Recke · D. Littmann
DLC–Dr. Littmann Consulting, Ennepetal, Germany
e-mail: littmann@littmann-consulting.com

M. Veste
CEBra–Center for Energy Technology Brandenburg e.V., Cottbus, Germany

Department Molecular Botany, Institute of Biology, University of Hohenheim, Stuttgart, Germany

M. Veste et al. (eds.), *Trees in South African Agriculture*, Innovations in Landscape Research, https://doi.org/10.1007/978-3-032-00993-7_10

10.1 Introduction

Like South Africa, climate change and the associated droughts will have a significant impact on agriculture in Brandenburg in the coming decades. The region is characterized by relative dry climate conditions in combination by morainic sandy soils with a low soil water storage capacity. Therefore, climate change (increasing temperature, spring and summer droughts) will increase vulnerability of the agricultural and natural systems (e.g., decreasing ground water tables, drought stress, fire risk, productivity losses) (Reyer et al. 2012). The ongoing climate change will affect the crop productivity (Brill et al. 2024; Pohanková et al. 2025). The climate drivers and changes in cropping rotation and cultivated area will increase the irrigation demands in near future (Drastig et al. 2016; Mirschel et al. 2020). In Brandenburg, the main reason is that, at approximately the beginning of the 2000s, the frequency of occurrence of droughts in spring and early summer began to increase (DWD 2014). Furthermore, landscape structure with their treeless agricultural steppes is particularly influenced by the wind, which have negative implication on soil erosion (Deumlich et al. 2006; Funk et al. 2023; Chap. 11).

This motivates the development of adaptation and mitigation measures (Reyer et al. 2012). The integration of fast-growing trees and agroforestry systems (Veste and Böhm 2018) are meanwhile wide discussed as a measure for adaptation to climate change with positive effects on the local microclimate and to increase the economic resilience of agricultural production (Kanzler et al. 2019; Seserman et al. 2019; Riekötter and Hassler 2022; Thiesmeier 2024). A study by Lang et al. (2018) in Rhineland-Palatinate, Germany, showed that nitrogen availability of grapevines was favorable and water relations were improved in a vineyard-agroforestry system with oaks and poplar, whereas wine quality was similar when grown with trees or without. Trees were able to reduce water and nitrogen losses without negative effects on wine quality.

In addition to the meso-scale weather conditions, local conditions such as topography control the wind field, evapotranspiration, and the local water balance of the cropped area. Interrelation of microclimate parameters and weather types is important information for crop management. Such upscaling may facilitate the definition of conditions leading to higher or lower evapotranspiration because weather types are reported on a regular basis and may be also used as a basis for water demand forecasting. Based on precise climate analyzes and local wind conditions, crop management measures and the design of wind protection measures can also be developed. Using the example of a newly created vineyard in the post-mining landscape in Lower Lusatia (Brandenburg, Germany), we show what the wind conditions are like in this open landscape and how wind protection affects the local wind field.

10.2 The Wolkenberg Vineyard in a Newly Created Landscape

The Lower Lusatia region is characterized by several lignite mining areas that are still in operation. New landscapes are being developed here by reclamation of the land surface and transferred to new land uses and changing the landscape structure (Dilly et al. 2010; Gerwin et al. 2023). The topography of the landscape is also being redesigned during the restoration of the post-mining sites. The former ridge of the Geisendorf-Steinitz terminal moraine has been recreated on the 600 ha restoration area at the Welzow opencast mine. In addition to the Geisendorfer Berg, it also includes the Steinitzer Berg (165 m high) and the Wolkenberg (135 m high). At Wolkenberg it was also possible to create an optimized landscape for wine cultivation (Dilly et al. 2007, Fig. 10.1). The vineyard is situated along a south-southwest-oriented slope with an average slope angle of 11°. Seven vine varieties are spread over 26,000 vines oriented north–south on an area of 6 ha.

Due to a more continental character of the climate, the Lower Lusatian region is among the driest regions in Germany with a negative climatic water balance. The region is characterized by a humid continental climate (Köppen-Geiger classification Dfb) according to Peel et al. (2007). The nearest climate station Cottbus (Deutscher Wetterdienst—DWD, ID 880) shows a mean annual temperature of 10.0 °C (1991–2020) with the mean temperature during the coldest month (January) of 0.8 °C and the mean temperature during the hottest month (July) of 19.8 °C. The mean annual precipitation is 566 mm (1991–2020), whereby the mean precipitation in January is 41.5 mm and the mean precipitation in July is 74.8 mm. The availability of water for the plants is limited to rainwater and possible irrigation from a small pond.

10.3 Microclimate Monitoring

Microclimate was monitored with two units (I) WB 1: in the upper western part at 151 m a.s.l. and (II) WB 2 as a reference station at 130 m a.s.l near the irrigation water pond at the southeastern side (Fig. 10.2). The following relevant microclimate parameters were recorded: air temperature and relative humidity, wind speed and wind direction, and global radiation (Fig. 10.3). Additionally, soil moisture was measured with a soil moisture profile probe up to 1 m depth. All microclimate data were recorded on a data logger and transferred daily to an internet-based cloud system.

Fig. 10.1 Wolkenberg vineyard embedded in the restored post-mining landscape of Welzow. Photographs: Maik Veste

10.4 Wind Field Analysis

10.4.1 Local Wind Conditions

At the Wolkenberg, main wind directions are SSW to W (Fig. 10.4). In the WB 2 reference measurement, the highest average wind speeds and maximum wind speeds coincided with these sectors. Average wind speed of 1.58 m s^{-1} is very low (also over the winter months), but gusts of up to 13 m s^{-1} occurred. In the vineyard (WB 1), despite the considerably greater roughness, but at 30 m higher terrain position, the

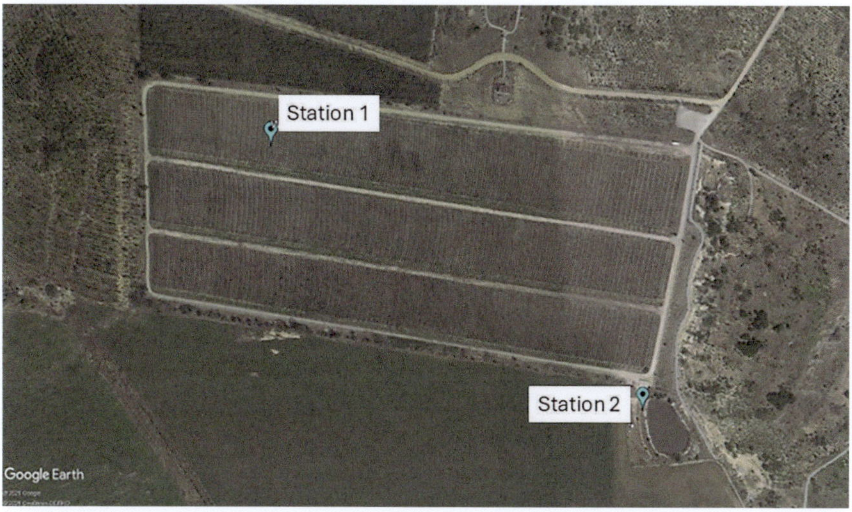

Fig. 10.2 Location map of the microclimate station WB 1—Station 1 and WB 2—Station 2 (Fig. 10.3) at Wolkenberg in the restored post-mining landscape of Welzow (*Source* Google Earth)

Fig. 10.3 Microclimate measurement on the vineyard Wollenberg **a** in the vineyard (WB1), **b** the reference measurements (WB2). Photographs: Maik Veste

mean wind speed was about 16% higher due to the acceleration of airflow uphill from the sectors from SSE to SSW. However, the average maximum wind speeds differ only slightly except for the predominant wind directions. WB 1 shows up to 50% higher mean wind speed than WB 2 with 1.84 m s^{-1} and 1.58 m s^{-1}, respectively, but no significant difference in maximum wind speeds. The following Table 10.1 shows wind statistics for both stations from December 2020 to June 2021.

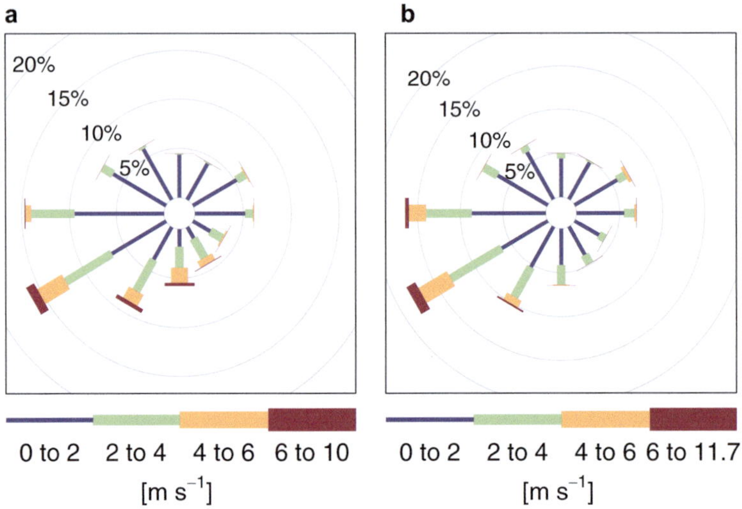

Fig. 10.4 Mean wind speed frequency by wind direction at Wolkenberg in the vineyard (Wolkenberg 1) and the reference station (Wolkenberg 2)

During the day, the highest wind speeds at the WB 2 measurement occur in the afternoon and evening from a southwesterly direction (Fig. 10.5). This indicates frequently occurring thermal effects at the foot of the slope and in the vicinity of the nearby open lignite mine area. In the composition of the upper slope, WB 1 shows the typical leftward rotation of the uphill flow resulting in a broader diurnal distribution of the high wind speed fraction. In this respect, the wind conditions found above the canopy height can significantly influence the evapotranspiration process in the crop stand.

The wind field over the Wolkenberg vineyard is clearly controlled by the topographical setting with an increase of wind speed upslope (Fig. 10.6). Wind field modeling was done with the non-hydrostatic flow model WAsP included in the software package EMD WindPRO version 3.6.377. Long-term wind data for the site (wind speed and wind direction at 100 m above ground) were derived from ERA5-reanalysis data for the nearest grid node (period 1998–2022 in hourly resolution). The model showed mean long-term wind speeds of 2.1 m s^{-1} for the WB 2 reference point and a clearly higher mean wind speed of 3.0 m s^{-1} for WB 1 reference point due to upslope speedup effects. The division of the slope area into the less windy lower part and the windier upper part toward the northern plateau is visualized in a 3D model (Fig. 10.7).

Table 10.1 Measured wind characteristics at the Wolkenberg in the vineyard (Wolkenberg 1) and the reference station (Wolkenberg 2)

Wolkenberg 2 wind direction													
Sector (°)	345–15	15–45	45–75	75–105	105–135	135–165	165–195	195–225	225–255	255–285	285–315	315–345	mean
Frequency (%)	5.0	4.9	7.2	6.9	4.1	4.9	6.5	*10.3*	*18.2*	*16.5*	8.7	7.0	
Wolkenberg 1 (Vineyard)													
Mean wind speed 10 min [m/s]	0.87	0.91	1.44	1.37	1.96	*2.82*	*3.21*	*2.64*	*2.85*	1.84	1.24	0.94	1.84
Max wind speed 10 min [m/s]	3.15	3.17	5.98	5.35	5.51	*7.53*	*7.91*	*10.02*	*8.98*	*7.51*	5.47	3.25	6.15
Wolkenberg 2 (Reference)													
Mean wind speed 10 min [m/s]	1.06	0.85	1.41	1.33	1.32	1.45	*1.73*	*2.21*	*3.0*	*2.25*	1.28	1.08	1.58
Max wind speed 10 min [m/s]	3.98	3.13	6.2	*6.86*	4.31	4.04	4.78	*9.23*	*9.49*	*11.75*	5.95	4.62	6.20
Gust wind speed 10 min [m/s]	5.16	4.41	6.74	*7.57*	5.62	5.27	5.56	*9.77*	*10.97*	*12.94*	7.69	5.42	7.26
Gust wind speed 1 min [m/s]	4.55	6.61	6.6	*8.8*	5.16	5.44	6.21	*10.43*	*13.36*	*11.88*	*8.09*	5.8	7.74

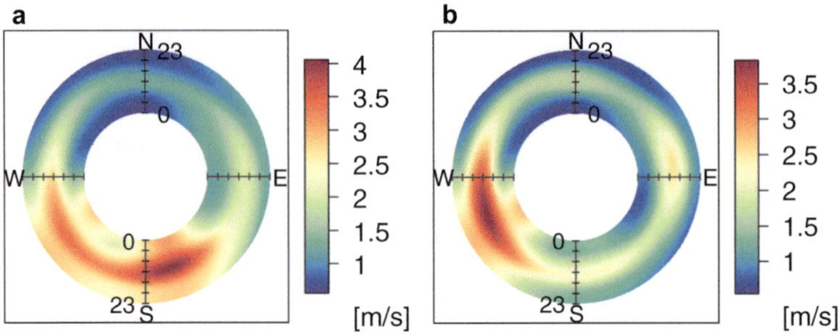

Fig. 10.5 Diurnal course of wind speed within wind direction sectors, Wolkenberg vineyard. **a** At the slope station 1, **b** Reference station 2

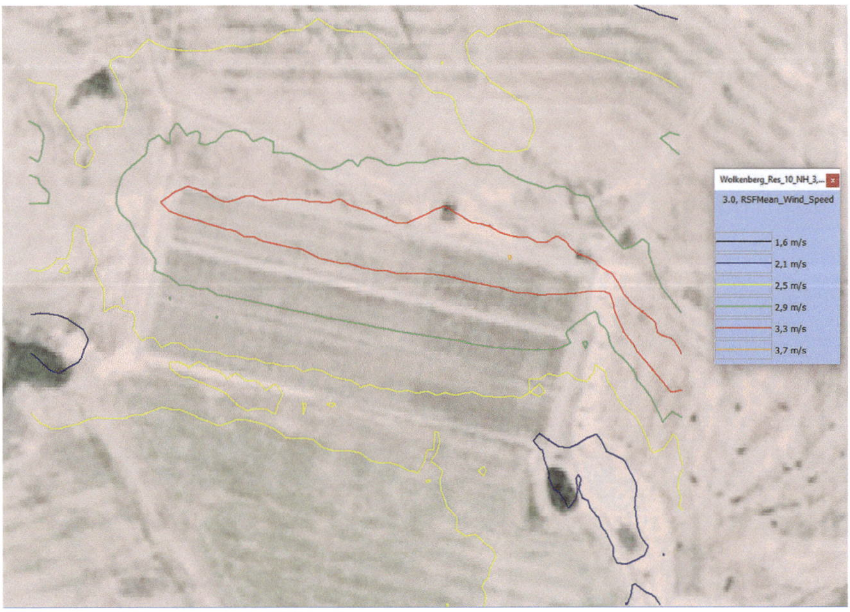

Fig. 10.6 Wind speed isolines in the model area at Wolkenberg

10.4.2 Simulation of a Windbreak Influence

In a further model run, the effect of a windbreak on near-ground wind speed was investigated. As planting of a tree hedgerow and agroforestry with sufficient width would not be feasible because of limited space along the pathway and the adjacent grapevine plantings, a vertical shading net (Chap. 5) was implemented in the simulation as a windbreak along the upper farm path in the western upslope part of the vineyard. The vertical net was defined as 3 m high and with a porosity of 0.2. Compared to the spatial distribution of the wind speed at a height of 3 m without obstacle, a

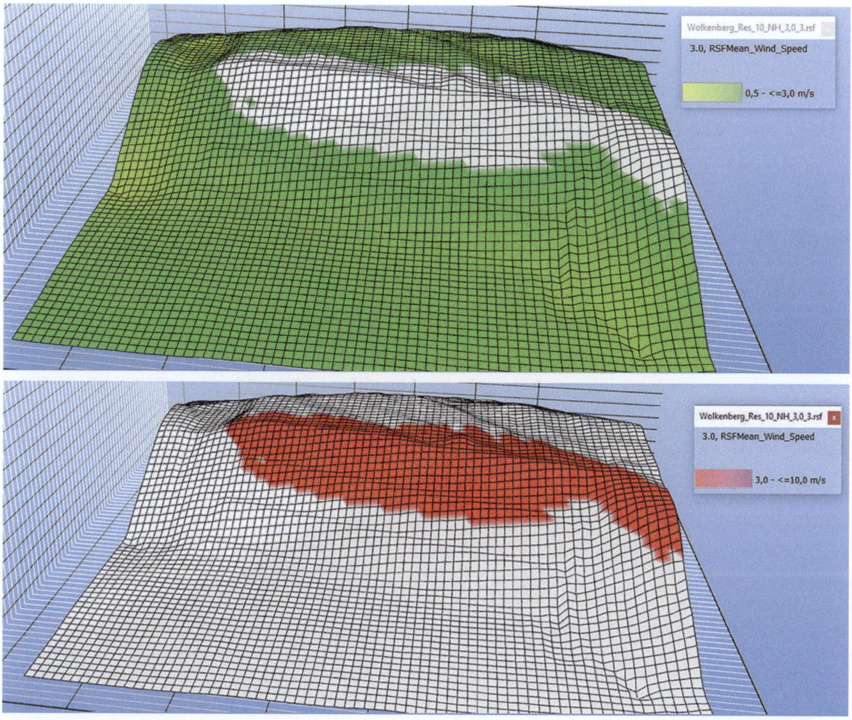

Fig. 10.7 Modeled long-term wind speed at 3 m above ground: **a** wind speed 0.5 m s^{-1} – < 3.0 m s^{-1}, **b** wind speed 3.0 m s^{-1} – < 10.0 m s^{-1}. The upslope part of the vineyard Wolkenberg shows significantly higher wind speed (**b**)

significant reduction in wind speed can be observed at the point of measurement Wolkenberg 1 (33%; from 3.03 m s^{-1} to 2.01 m s^{-1}) and up to the upper edge of the vineyard; there is no change in the eastern half of the upper slope (Fig. 10.8). This result implies that a simple technical measure such as the implementation of a vertical net should significantly reduce wind speed in case of maximum or gust wind speeds. Furthermore, evapotranspiration rates are also clearly reduced by this technical measure for the entire area of wind speed reduction.

On the other hand, shading of the grapevine and the reduced ventilation in a low wind energy environment could trigger negative effects for plant growth and grape quality. Ensuring a well-ventilated bunch zone is a key target of canopy management in viticulture, reducing the spread of fungal diseases. Prolonged leaf wetness can lead to plant diseases caused by fungi, including mildew. Another underestimated factor is the reduced cooling of leaves and increase in temperature. Therefore, the benefit of reducing evapotranspiration in low wind speed system may have counterproductive effects like plant diseases and temperature effects on photosynthesis and wine quality. In order to increase canopy ventilation, mitigate high foliage temperature, and reduce

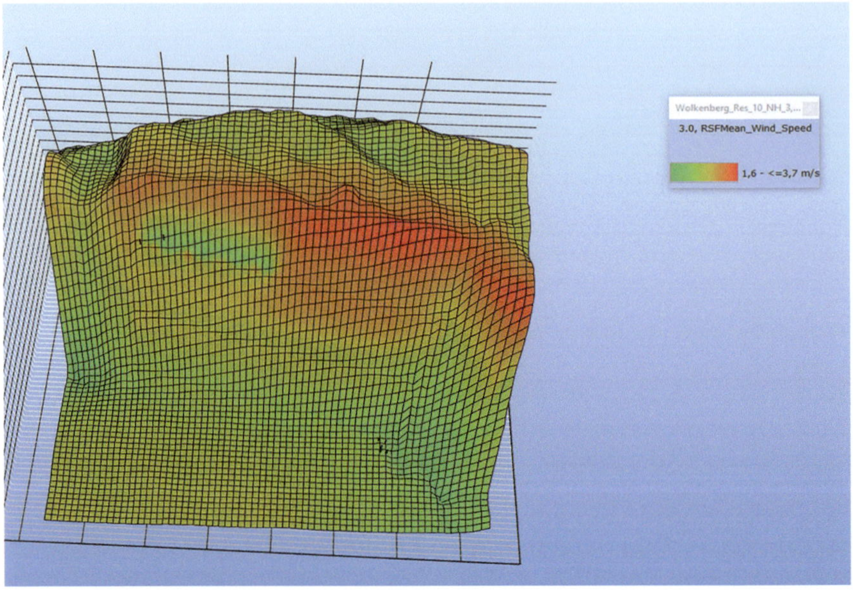

Fig. 10.8 Modification of the local wind field by a windbreak (shading net, black line) in the upper western part of the vineyard

fungal diseases, a row direction parallel to the prevailing wind sector should be preferred (Vendrame et al. 2023).

10.5 Water Balance

For the calculation of a simple climatic water balance (WB) of the vineyard, the measurements of station 1 were used for the period from April 2020 to September 2021:

$$WB = P - ET + \Delta S$$

WB: Climatic water balance [mm].

P: Precipitation [mm].

ET: Evapotranspiration [mm].

ΔS: Change of soil water storage [mm].

Soil water data were derived from TDR measurements (soil profile probe) at the depth of –0.1 m, –0.2 m, –0.3 m, and –1.0 m. Data obtained in vol-% were transformed to $m^3\ m^{-3}$, and the absolute water content Θ was computed for each layer between the measurement depths (Table 10.2). The changes of soil water constant ΔS were expressed as the difference of the entire water column between the surface and the depth of 1.0 m. All WB computations were based on hourly values.

The Fig. 10.9 shows seasonal changes at station 1. The 2020 growing season showed higher precipitation due to two heavy rainfall events in August, but also a high ET due to higher temperatures. Therefore, the soil water storage change was only very small amount, and the water balance was clearly negative. In the 2021 period, there was a significantly lower ET with lower precipitation in cooler weather, so that the positive storage change led to a lower water deficit. The average water balance of the two growing seasons showed a deficit of 50 mm (Table 10.2). Due to the winter half-year with a clearly positive water balance, the annual balance from April 2020 to March 2021 is 64 mm in the excess range. There was no indication or documentation of additional irrigation over the two vegetation periods.

10.6 Microclimatic Patterns

Further analysis of the measured microclimatic data at both measurement stations (parameters wind speed, air temperature, relative humidity, and computed evapotranspiration following the modified FAO approach (Chap. 5)) focused on the identification of similar groups representative for different weather conditions leading to high or low evapotranspiration loss within the vineyard. To compare different methods of grouping algorithms, cluster and principal component analyses were applied.

In a first approach, all measured parameters were subjected to a k_{means}-cluster analysis applying Monte Carlo (bootstrap) gap statistics (500 simulations) and computations of the silhouette coefficient (0.31, which is quite weak after Ester and Sander 2013). The procedure resulted in four clusters. A Kruskal and Wallis H-Test (Table 10.3) showed highly significant (Hedderich and Sachs 2016) differences of the meteorological parameters (Table 10.4) between the clusters, especially for relative humidity, temperature, and evapotranspiration. Based on the cluster analysis, four different clusters could be identified (Fig. 10.10), and their occurrence during the season is shown in Fig. 10.11.

- **Cluster 1** is characterized by cool and moist conditions with low wind speed and evapotranspiration. Over the vegetation period in 2020, it occurred on several days without a clear seasonal pattern.
- **Cluster 2** represents mainly summerly warm and dry conditions with low wind speed but the highest evapotranspiration.
- **Cluster 3** is a frequent summer and fall case with warm and very moist conditions, low wind speed, and below average evapotranspiration.
- **Cluster 4** is mainly a spring case with cool, dry, and windy conditions with high evaporation rates.

Table 10.2 Monthly means and sums of the water balance parameters

Year	Month	TDR	TDR	TDR	TDR	Θ	Θ	Θ	Θ	Sum	Sum	Sum	Sum
		10 cm	20 cm	30 cm	40 cm	0–10 cm	10–20 cm	20 -30 cm	30 -100 cm	P	ET	ΔS	WB
		%	%	%	%	mm	mm	mm	mm	mm	mm	mm	mm
2020	4	1.4	4.5	8.7	12.6	1.4	4.5	8.7	88.2	4.6	70.5	–1.7	–68.0
2020	5	1.6	5.8	9.0	12.6	1.6	5.8	9.0	87.8	35.4	67.1	2.9	–29.9
2020	6	1.9	5.9	8.5	12.5	1.9	5.9	8,5	87.4	44.4	58.8	–5.4	–19.5
2020	7	1.7	5.4	7.7	11.5	1.7	5.4	7.7	80.3	40.6	56.2	–2.3	–25.0
2020	8	2.6	6.2	8.1	11.3	2.6	6.2	8.1	79.2	117.4	60.3	5.5	62.5
2020	9	4.1	8.2	9.5	11.5	4.1	8.2	9.5	80.5	39.0	44.8	7.9	2.1
2020	10	6.2	10.3	10.8	12.8	6.2	10.3	10.8	89.7	98.8	19.3	14.3	93.6
2020	11	6.1	10.4	11.0	13.2	6.1	10.4	11.0	92.2	6.2	13.1	–12.5	–19.4
2020	12	5.1	9.6	10.5	12.7	5.1	9.6	10.5	89.1	9.2	11.9	–2.4	–5.1
2021	1	8.7	11.6	12.1	14.2	8.7	13.3	12.1	99.4	42.6	8.9	44.8	78.4
2021	2	6.4	39.9	13.0	16.9	6.4	35.9	13.0	118.1	15.4	18.5	29.5	26.4
2021	3	5.5	23.6	13.7	17.8	5.5	24.1	13.7	124.6	34.2	38.3	–27.8	–31.9
2021	4	1.8	10.7	13.9	16.8	1.8	10.6	14.0	117.6	27.6	45.8	–3.6	–20.8
2021	5	1.8	11.4	15.6	19.7	1.8	11,4	15.6	138.1	46.2	58.0	10.8	–0.9
2021	6	1.5	7.7	11.9	18.3	1.5	7.7	11.9	134.7	36.2	68.7	–27.2	–59.7
2021	7	1.7	10.0	12.2	17.6	1.7	10.1	12.2	124.3	47.2	57.3	–18.5	–30.1
2021	8	1.7	9.2	11.4	19.3	1.7	9.2	11.4	144.6	20.6	34.5	90.4	88.9
Total period										665.6	732.1	104.9	41.8
Vegetation period 2020										281.4	357.7	7.0	–77.8
Vegetation period 2021										177.8	264.3	51.9	–22.6

(continued)

Table 10.2 (continued)

Year	Month	TDR	TDR	TDR	TDR	Θ	Θ	Θ	Θ	Sum	Sum	Sum	Sum
		10 cm	20 cm	30 cm	40 cm	0–10 cm	10–20 cm	20 -30 cm	30 -100 cm	P	ET	ΔS	WB
		%	%	%	%	mm	mm	mm	mm	mm	mm	mm	mm
Average vegetation period										229.6	311.0	29.4	–50.2
Winter										206.4	110.1	46.0	142.1
Year 04/20–03/21										487.8	467.8	53.0	64.3

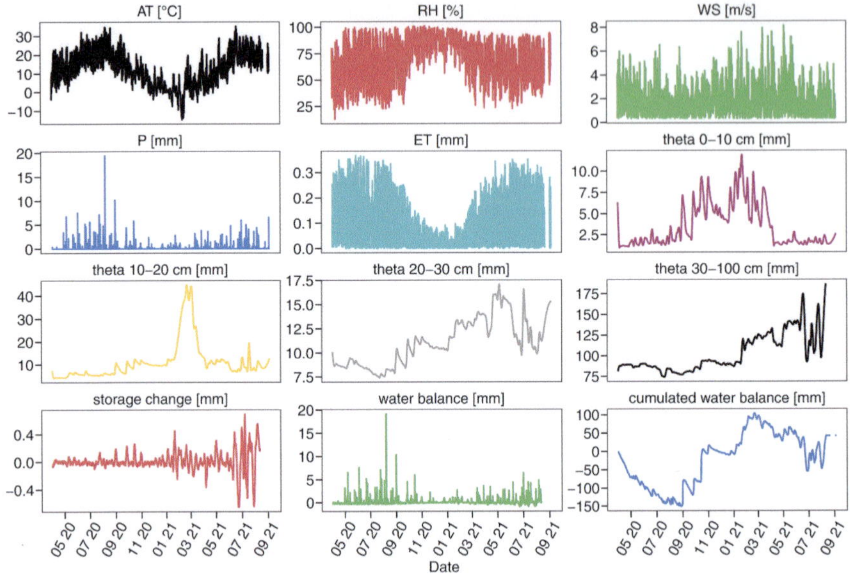

Fig. 10.9 Wolkenberg: Air temperature at 2 m (AT), precipitation (P), relative humidity (RH), wind speed (WS), evapotranspiration, (ET), soil water contents (theta) at 0–10 cm, 10–20 cm, 20–30 cm, soil water storage change, water balance, and cumulated water balance at station 1 from April 2020 to September 2021

Table 10.3 Kruskal–Wallis test for the cluster solution ET: evapotranspiration, Temp: air temperature, RH: relative humidity, WS: wind speed

Parameter	Chi_squared	P-value
ET	136.36	0.0000
Temp	107.20	0.0000
WS	21.97	0.0000
RH	139.97	0.0000

In addition, the measured parameters were subjected to principal component analysis using the R-package "FactoMineR" (Fig. 10.12). The procedure resulted in four principal components (PCs) (Table 10.5). The first two PCs showed Eigenvalues > 1.0 while PC 3 added to a total of 95.9% of explained variance of the dataset. Therefore, finally, three-PC solution shows a clear separation of 2 PCs within the dimension diagram.

The principal component (PC) 1 is characterized by cool and moist conditions with low wind speed and evapotranspiration without a clear seasonal pattern over the vegetation period in 2021. Therefore, it is largely comparable to cluster 1 of the cluster analysis solution. PC 2 is like cluster 4 mainly a spring case with cool, dry, and windy conditions but average evaporation rates. PC 3 represents like cluster 2 mainly summerly warm and dry conditions with low wind speed but the highest evapotranspiration.

Table 10.4 Characteristics of clusters 1–4; Temp in °C, WS in m s^{-1}, ET in mm day^{-1}, RH in %

Parameter	Cluster 1	Cluster 2	Cluster 3	Cluster 4	Overall
min_Temp	10.9	19.2	19.5	8.8	8.8
mean_Temp	18.5	26.5	25.0	16.4	22.2
max_Temp	26.2	34.7	32.1	23.3	34.7
min_WS	0.6	0.9	0.8	0.9	0.6
mean_WS	1.5	1.5	1.5	2.1	1.6
max_WS	3.0	2.2	3.7	4.5	4.5
min_ET	0.0	2.3	0.0	1.4	0.0
mean_ET	0.8	2.9	1.8	2.3	2.0
max_ET	1.8	3.4	2.6	3.1	3.4
min_RH	72.7	37.4	51.7	37.4	37.4
mean_RH	83.2	49.2	63.6	53.1	62.4
max_RH	95.7	61.3	75.5	68.4	95.7

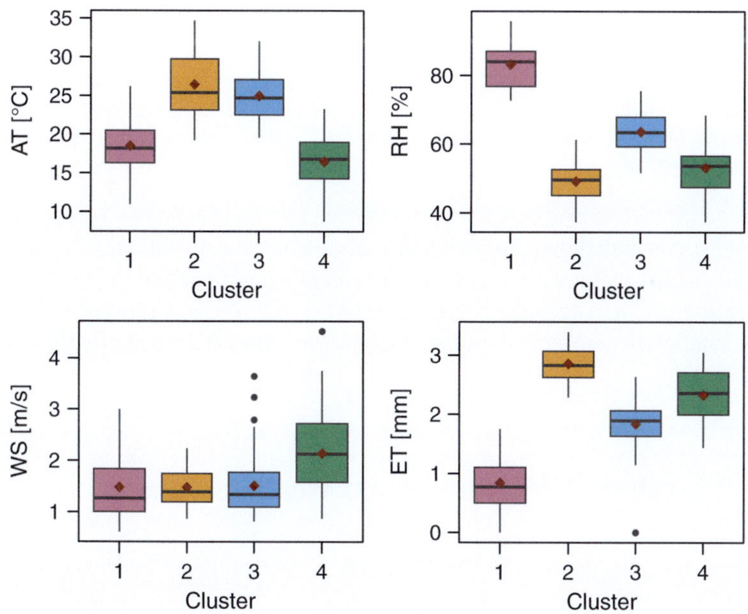

Fig. 10.10 Boxplot of air temperature (**a**), relative humidity (**b**), wind speed (**c**), and evapotranspiration (**d**) of the cluster solution

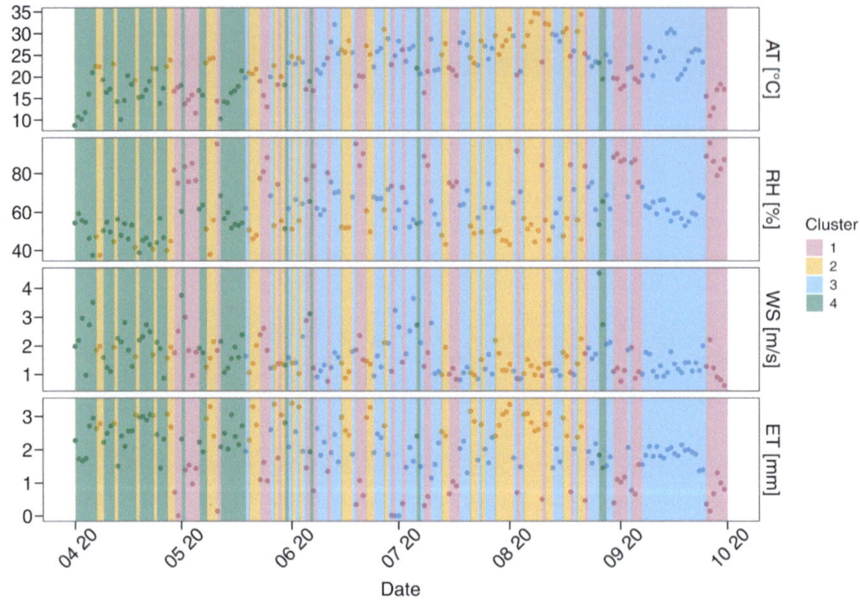

Fig. 10.11 Days with occurrence of four clusters over the vegetation period April–October 2020 and characteristics of parameters. AT: air temperature, RH: relative humidity, WS: wind speed, and ET_0: evapotranspiration

It can be concluded that the principal component solution with only three groupings (Fig. 10.13) shows a more coherent and significant pattern of the microclimatic parameters over the vegetation period 2021 (Fig. 10.14) as compared to the four-cluster solution. Therefore, the three PC groups (clusters) are applied for further analysis.

10.7 Upscaling: Microclimate and Weather Types

For further analysis, the objective weather type classification of the *Deutsche Wetterdienst* (DWD) was applied. Weather types are defined based on four meteorological parameters from 31 × 30 grid nodes over Central Europe: the geopotential height of five pressure layers between 950 and 300 hPa, temperature and relative humidity (precipitable water) on the same layers, and the zonal and meridional wind component on the 700 hPa layer. These data are transformed into indices for cyclonality (cyclonic, anticyclonic on the 950 hPa and 500 hPa layers, respectively), wind direction (five groups), and dry or wet conditions, resulting in 40 combinations (weather types; Bissolli and Dittmann 2001). Figure 10.15 shows the relative frequency of daily weather types within the PC clusters.

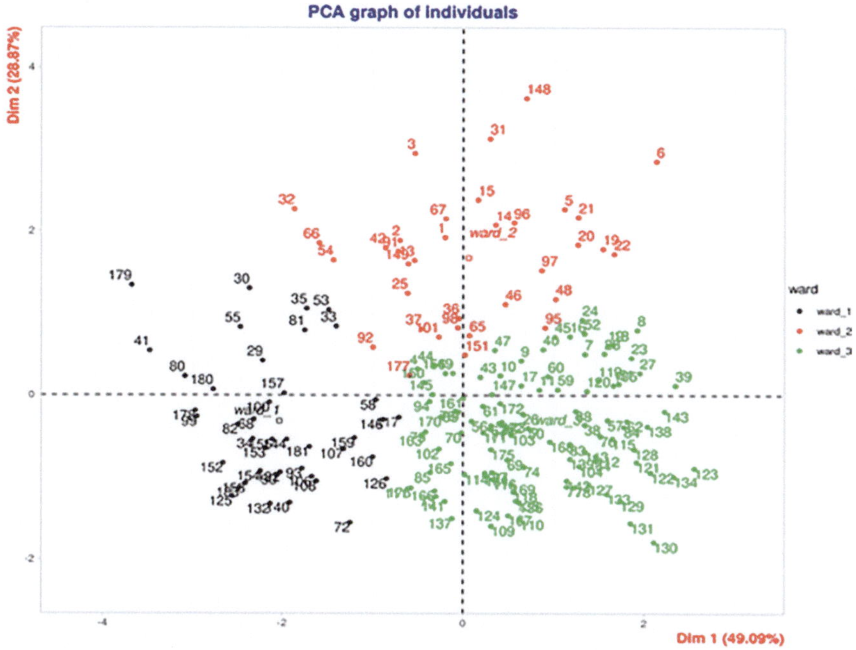

Fig. 10.12 Dimensional distribution of the 3-principal component solution

Table 10.5 Characteristics of principal components

Parameter	PC 1	PC 2	PC 3	Overall
mean_Temperature (°C)	19.0	18.3	24.7	22.2
Mean wind speed (m s^{-1})	1.4	2.6	1.4	1.6
mean_evapotranspiration (mm day $^{-1}$)	0.9	1.95	2.4	2.0
mean_relative humidity (%)	82.4	57.6	56.1	62.4

The three PC clusters (Fig. 10.12) show significant interrelation with several weather types. A chi-square test for the three PC clusters and all weather types with a frequency > 3 days over the vegetation period 2021 showed significance ($p = 0$). Further details were derived from Pearson r-residue ($r = o–e/\sqrt{e}$; o: observed, e: expected) and contribution analysis. High residues contribute largely to the resulting χ^2 values and may be positive (positive interrelation) or negative (negative interrelation). The actual contribution of a given cell of the contingency table of PC clusters and weather types to the chi-square test is computed r^2/χ^2.

PC cluster 1 shows a strong negative correlation with the anticyclonic dry weather type with northwesterly flow (NWAAT), i.e., high pressure over western Europe but positive correlation with autochthonous high- (XXAZT) and low- (XXZZF) pressure cells over central Europe without frontal systems. The contribution of those three

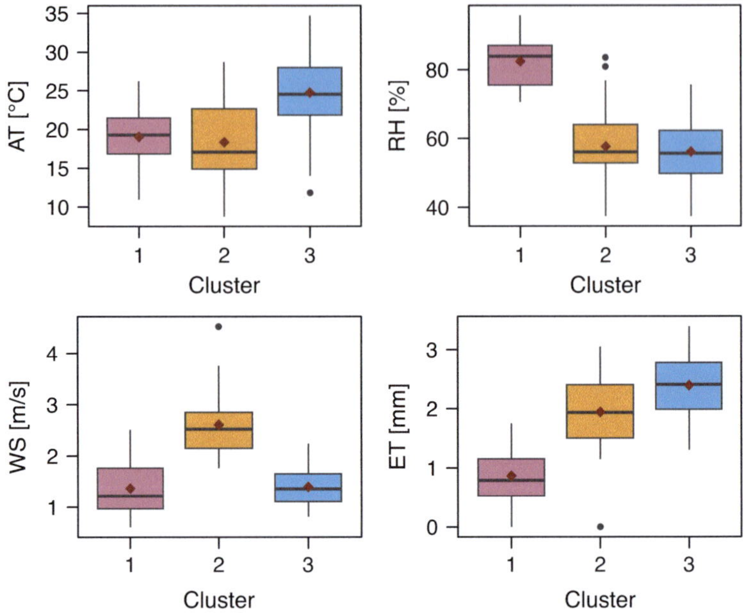

Fig. 10.13 Boxplot of the principal component solution: air temperature (**a**), relative humidity (**b**), wind speed (**c**), and evapotranspiration (**d**)

types within cluster 1 to the chi-square test is significant (16.2%). Windy and dry PC cluster 2 is characterized by northwesterly anticyclonic flow (NWAAT), southeasterly anticyclonic flow from eastern Europe high pressure (SOZAT), or southwesterly frontal systems (SWZZF) which contribute to 15,3% of the chi-square test. PC cluster 3 representing warm and dry summer weather is mainly linked to weak flow types like high pressure over the Baltic area (NOAAT), southwesterly inflow of warm air mass (SWZAF), or autochthonous high pressure (XXZAF). Those three types contribute to 13.1% of the chi-square test.

While the interrelation of near-ground wind speed and cyclonality on the 950 hPa level is not significant, anticyclonic weather types in general correspond to higher temperature, lower humidity, and high evapotranspiration rates over the observation period (Figs. 10.16 and 10.17).

10.8 Conclusions

The wind field over the Wolkenberg vineyard is clearly controlled by the topographical setting with an increase in wind speed upslope. The region can be regarded as low wind speed system compared to coastal regions. The constructions of windbreaks would reduce the wind speed and the evapotranspiration. However, that might lead

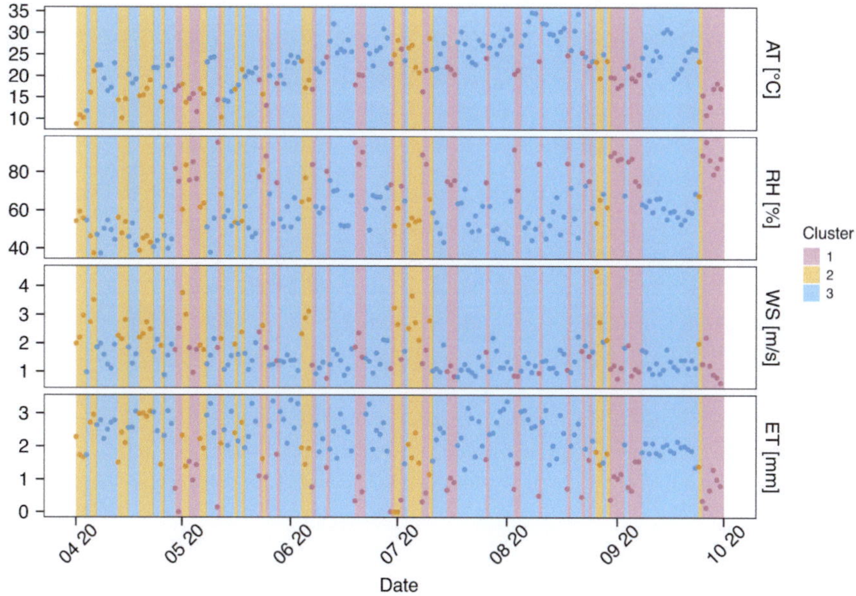

Fig. 10.14 Days with occurrence of three-principal component clusters over the vegetation period April–October 2020 and characteristics of parameters. Air temperature (**a**), relative humidity (**b**), wind speed (**c**), and evapotranspiration (**d**)

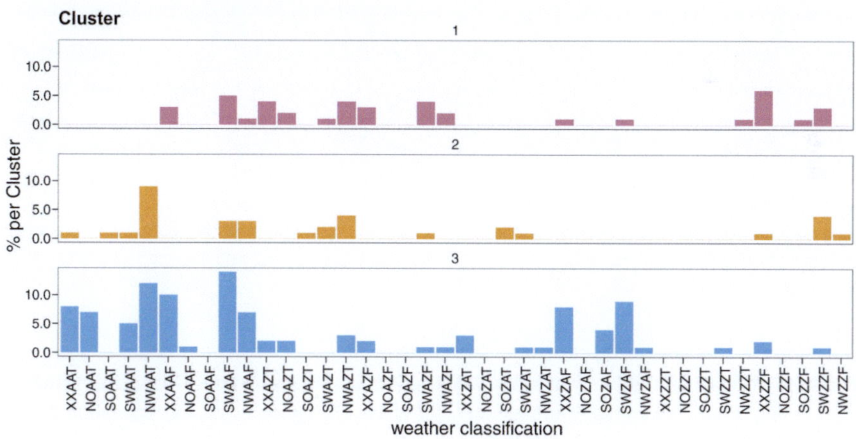

Fig. 10.15 Relative frequency of daily weather types within the microclimatic PC clusters

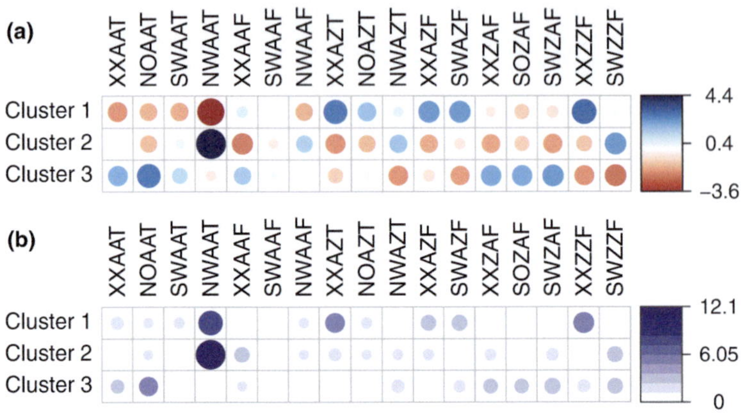

Fig. 10.16 Pearson residues (**a**) and contribution (**b**) of weather types to PC clusters. Blue dots show a positive, while red dots show a negative correlation. The size of the dots represents percentage values

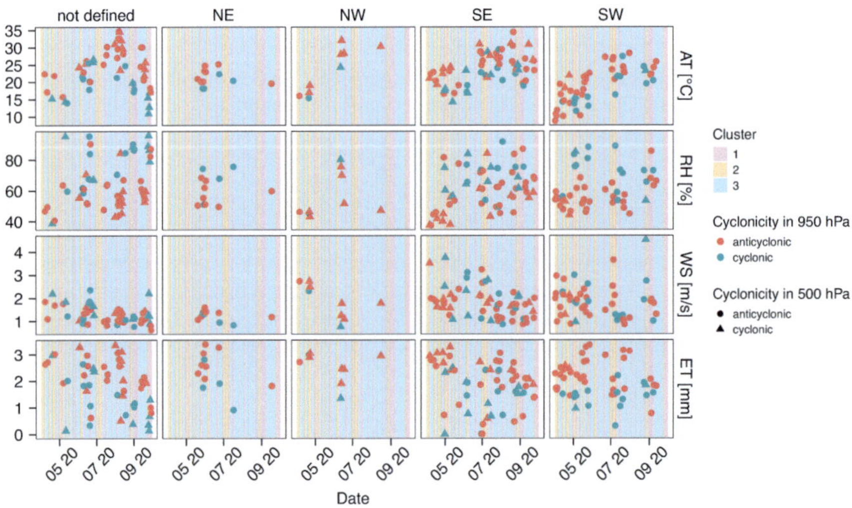

Fig. 10.17 Cyclonality and microclimatic characteristics (air temperature (AT), relative humidity (RH), wind speed (WS), evapotranspiration (ET) of the PC clusters over the vegetation period April–October 2021

to an unknown negative feedback on the canopy ventilation, causing an increase in temperature and fungal diseases. Thus, complex interactions need more investigation for the planning of windbreaks in temperate agroforestry systems.

Computations of a simple climatic water balance over two vegetation periods and the winter in between showed plant available water to be mainly controlled by positive soil water storage by an excess balance of precipitation and evapotranspiration during the prior winter months. The water balance over the vegetation periods was generally negative, with the deficit depending on the prevalence of PC clusters 3 or 1 in the summer half-year. In years without longer summer dry spells, the irrigation water demand over the vegetation period should be rather low (20–80 mm) and probably even lower in case of deep-rooting crops like grapevine.

Analysis of measurement data aimed at the identification of typical combinations of microclimatic parameters controlling evaporative processes within the vineyard. Based on difference analysis, cluster analysis resulted in four clusters, which differ significantly in terms of their characteristics in terms of humidity, temperature, wind speed, and calculated ET. Cluster 1 includes days with low ET, temperature, and wind speed, but high relative humidity, and occurs seasonally non-specific. Cluster 2 refers to dry summer days with high ET and high temperature with low wind, i.e., with a temperature-related high saturation deficit. Cluster 3 shows a similar picture, but the ET is slightly lower, and the humidity is higher. Cluster 4 with high ET was typical for the spring with lower temperature and medium humidity, but higher wind speeds. Thus, the cluster analysis already reveals typical microclimatic situations that lead to higher or lower ET. In spring, when the leaf surface is still small, evapotranspiration dynamics are more likely to be controlled by the wind, in summer by the saturation deficit.

The principal component analysis (PCA) revealed only three groupings based on correlative relationships. PC cluster 1 describes cool, damp, low-wind days with low ET, which included marginal high-pressure systems with cold air influx and precipitation tendency. Cluster 2, on the other hand, is very pronounced with high ET, lower temperature and humidity, but higher wind speeds, especially in spring with a clear anticyclonic northwesterly airflow. The third cluster, on the other hand, refers to warm, dry days with high ET and little wind during the summer months and, like cluster 2, is associated with anticyclonic flow, but here from NW and SW directions.

Thus, the principal component analysis confirms the two constellations in which high ET can occur in the stand (spring wind-dependent, summer dependent on saturation deficit). In both cases, there is a significant correlation with high-pressure systems in a marginal position, which usually lead to NW or SW flow with more or less wind. Further investigation of the correlations with large-scale weather types by a chi-square test confirms the result: The primary anticyclonic NW and SE types clearly occur in PC cluster 2 (high ET) in spring, anticyclonic NO and SW types in cluster 3 (high ET), and in cluster 1 with low ET primarily cyclonic types.

Acknowledgements We thank Bettina Muthmann, Wolkenberg Winery, Welzow and LEAG—Lausitz Energie Bergbau AG, Cottbus for their support and permission for the access to the field site. The research was supported by the Federal Ministry of Education and Research (BMBF) and managed by PT-DLR as part of the joined German-South African project FarmImpact (FKZ 01LZ1711A-F). Because of the SARS-Covid-19 restrictions, we were not able to carry out our planned measurements in South Africa. Instead, the Wolkenberg GmbH gave us the alternative of carrying out our measurements there to the same extent.

References

Bissolli P, Dittmann E (2001) The objective weather types of classification of the German Weather Service andits possibilities of application to environmental and meteorological investigations. Meteorol Z 10(4):253–260

Brill F, Alencar PHL, Zhang H, Boeing F, Hüttel S, Lakes T (2024) Exploring drought hazard, vulnerability, and related impacts to agriculture in Brandenburg, EGUsphere [preprint]. https://doi.org/10.5194/egusphere-2024-1149

Deumlich D, Funk R, Frielinghaus M, Schmidt W-A, Nitzsche (2006) Basics of effective erosion control in German agriculture. J Plant Nutr Soil Sci 169:370–381

Dilly O, Nii-Annang S, Schrautzer J, Breuer V, Pfeiffer E-M, Gerwin W, Schaaf W, Freese D, Veste M, Hüttl RF (2010) Ecosystem manipulation and restoration the basis of long-term conceptions. In: Müller F, Baessler C, Schubert H, Klotz S (eds) Long-term ecological research between theory and application. Springer Science+Business Media, Amsterdam, 411–428. https://doi.org/10.1007/978-90-481-8782-9_28

Dilly O, Zeihser U, Hüttl RF, Kendzia G, Wüstenhagen D, Dähnert D (2007) Perspektiven für einen modernen Weinbau in der Niederlausitz. Forum der Forschung 20, BTU Cottbus, Eigenverlag, pp 89–94

Drastig K, Prochnow A, Libra J, Koch H, Rolinski S (2016) Irrigation water demand of selected agricultural crops in Germany between 1902 and 2010. Sci Total Environ 569:1299–1314. https://doi.org/10.1016/j.scitotenv.2016.06.206

DWD (2014) Deutscher Klimaatlas. http://www.dwd.de/DE/klimaumwelt/klimaatlas/klimaatlas_node.html. Accessed 01 Sep 2024

Ester M, Sander J (2013) Knowledge discovery in databases: Techniken und Anwendungen. Springer, Heidelberg

Funk R, Völker L, Deumlich D (2023) Landscape structure model based estimation oft he wind erosion risk in Brandenburg, Germany. Aeolian Research 62:1000878

Gerwin W, Raab T, Hinz C, Letmathe P, Leuchner M, Roß-Nickoll M, Rüde T, Trachte K, Wätzold F, Lehmkuhl F (2023) Perspectives of lignite post-mining landscapes under changing environmental conditions: what can we learn from a comparison between the Rhenish and Lusatian region in Germany? Environ Sci Eur 35:36. https://doi.org/10.1186/s12302-023-00738-z

Hedderich J, Sachs L (2016) Angewandte Statistik- Methodensammlung mit R. Springer, Heidelberg. https://doi.org/10.1007/978-3-662-45691-0

Kanzler M, Böhm C, Mirck J, Schmitt D, Veste M (2019) Microclimate effects on evapotranspiration and winter wheat (*Triticum aestivum* L.) yield within a temperate agroforestry system. Agrofor Syst 93:1821–2184. https://doi.org/10.1007/s10457-018-0289-4

Lang CP, Merk N, Geilfus C-M, Graeff-Hönninger S, Simon J, Rennenberg H, Zörb C (2018) Interaction between grapevines and trees: effects on water relations, nitrogen nutrition, and wine. Arch Agron Soil Sci 65:224–239

Mirschel W, Wieland R, Luzi K, Groth K (2020) Model-based estimation of irrigation water demand for different agricultural crops under climate change, presented for the Federal State of Brandenburg, Germany. In: Mirschel W, Terleev V, Wenkel K (eds) Landscape modelling and decision support. Innovations in landscape research. Springer, Cham. pp 311–327. https://doi.org/10.1007/978-3-030-37421-1_16

Peel MC, Finlayson BL, McMahon TA (2007) Updated world map of the Köppen-Geiger climate classification. Hydrol Earth Syst Sci 11:1633–1644. https://doi.org/10.5194/hess-11-1633-2007

Piotrowicz K, Ciaranek D (2020) A selection of weather type classification systems and examples of their application. Theoret Appl Climatol 140:719–730. https://doi.org/10.1007/s00704-020-03118-2

Pohanková E, Hlavinka P, Kersebaum, KC, Nendel C, Rodríguez A, Balek J, Balkovič J, Dubrovský M, Hoogenboom G, Moriondo M, Olesen JE, Pullens JWM, Rötter RP, Ruiz-Ramos M, Shelia V, Skalský R, Hoffmann MP, Takáč J, Thaler S, Eitzinger J, Dibari C, Ferrise R, Leolini L, Bohuslav J, Bláhová M, Fischer M, Trnka M (2025) Climate change impacts on two European crop rotations via an ensemble of models. Eur J Agron 164:127456. https://doi.org/10.1016/j.eja.2024.127456

Reyer C, Bachinger J, Bloch R, Hattermann F F, Ibisch P L, Kreft S, Lasch P, Lucht W, Nowicki C, Spathelf P Stock, Welp M (2012) Climate change adaptation and sustainable regional development: a case study for the Federal State of Brandenburg, Germany. Regnal Environ Chang 12:523–542. https://doi.org/10.1007/s10113-011-0269-y

Riekötter N, Hassler M (2022) Agroforestry systems in wine production—mitigating climate change in the Mosel Region. Forests 13:1755

Seserman D-M, Freese D, Swieter A, Langhof M, Veste M (2019) Trade-off between energy wood and grain production in temperate alley-cropping systems: An empirical and simulation-based derivation of land equivalent ratio. Agriculture 9(7):147. https://doi.org/10.3390/agriculture9070147

Vendrame N, Tezza L, Pitacco A (2023) Evolution of turbulent flow characteristics in a hedgerow vineyard during the growing season. Agric for Meteorol 328:109251

Veste M, Böhm C (eds) (2018) Agrarholz—Schnellwachsende Bäume in der Landwirtschaft. Springer Spektrum, Heidelberg. https://link.springer.com/book/10.1007/978-3-662-49931-3

Veste M, Littmann T, Kunneke A, Du Toit B, Seifert T (2020) Windbreaks as part of climate-smart landscapes reduce evapotranspiration in vineyards, Western Cape Province, South Africa. Plant Soil Environ 66:119–127. https://doi.org/10.17221/616/2019-pse

Thiesmeier A (2024) Comparing the economic performance of poplar-based alley cropping systems with arable farming in Brandenburg under varying site conditions and policy scenarios. Agrofor Syst 98:1507–1522

Chapter 11
Spatial Analysis of Wind Shelterbelt Effects: A Model for Regional Planning

Roger Funk, Lidia Völker, Florian Kestel, Maik Veste, and Tobias Hahn

Abstract Windbreaks as part of the landscape structures influence the local wind field. To plan and analyze these windbreaks, it is necessary to precisely record the different landscape structures. The height and spatial arrangement of the landscape elements determine their wind protection effect, which was determined from a digital elevation model based on aircraft-based laser scanning. The ArcGIS toolbox 'WERA' allows a detailed analysis of the landscape structure on wind erosion processes. The presented method combines heights of LE and the directional transport capacities of erosive wind speeds to analyze the spatial wind speed reduction by windbreaks or other landscape elements. Combined with the soil-derived erodibility map, a wind erosion risk map can be derived. The analysis can be performed at the field, landscape, or regional level. This is an important planning tool for windbreaks to minimize wind erosion and evapotranspiration.

Keywords Landscape structure · Wind velocity reduction · Wind erosion · Potential evapotranspiration · Water saving

11.1 Introduction

Agriculture and food security are facing major challenges due to global change worldwide. This puts increasing pressure on the efficient use of limited land and resources to ensure stable agricultural production (Reimund et al. 2011) and avoid conflicts with other livelihoods and economies (Snyder et al. 2015). This is being exacerbated

R. Funk (✉) · L. Völker · F. Kestel
Leibniz Centre for Agricultural Landscape Research (ZALF E.V.), Müncheberg, Germany
e-mail: rfunk@zalf.de

M. Veste
CEBra—Center for Energy Technology Brandenburg e.V., Cottbus, Germany

Institute of Biology, Department Molecular Botany, University of Hohenheim, Stuttgart, Germany

T. Hahn
Orchard Research Station Müncheberg (OBVS), Müncheberg, Germany

M. Veste et al. (eds.), *Trees in South African Agriculture*, Innovations in Landscape Research, https://doi.org/10.1007/978-3-032-00993-7_11

by climate change, particularly in the form of rising temperatures, changes in precipitation patterns, heat waves, and droughts, which will ultimately lead to water scarcity for crops and increased yield uncertainty (Brune 2016; IPCC 2021; von Maltitz et al. 2014). This particularly affects the semi-arid regions where evaporation exceeds the amount of precipitation. Protecting the soils against degradation by erosion and saving water are therefore an absolute necessity for successful and efficient agriculture. All available potentials should be used for this purpose, preferably those that generate positive effects through their presence alone. This applies to the landscape structure, as it is well known that a structured landscape has wind-reducing, shading, and cooling effects. So, landscape structures combine many positive effects (Chap. 14) by reducing soil erosion, creating safe habitats for plants and animals and thus increase biodiversity, connecting biotopes or simply determining the esthetics of a landscape (Vigiak et al. 2003; Pásztor et al. 2016; Sheppard et al. 2020).

Wind erosion is a gradual soil degradation process characterized by the removal of the finest and lightest particles, such as the clay and silt fractions and the soil organic matter (SOM). Over time, this leads to a deterioration in the physical and chemical properties of the soils (Goossens 2004; Borelli et al. 2017; Nerger et al. 2017). This is a particular problem in semi-arid regions with agricultural use, but temperate latitudes are also increasingly confronted with this problem (Reich et al. 2001; Zolina et al. 2013). The classification of the wind erosion risk of larger areas is usually based on the spatial distribution of soil types and wind velocities (e.g., for South Africa Kestel et al. 2023). The effect of vegetation on the wind erosion risk is included as displacement height and aerodynamic roughness length, but only related to the place of its presence (Funk and Reuter 2006; Borelli et al. 2014; 2017). The fact that the influence of forests, hedges, shrubs and other landscape structures can be up to 40 times of their height is hardly considered in large-scale analyses (Hagen et al. 1981; van Eimern 1964; Vigiak et al. 2003).

The water balance of soils is primarily dependent on intrinsic factors such as soil texture, soil organic matter content, or bulk density, which determine the water holding and water transport capacities directly at the site (Horton et al. 2016; Jacobs et al. 2022; Veste et al. 2020). In heterogeneous landscapes, the spatial pattern of soils, topography, and land use determine spatial differences in water availabilities (Schmidt et al. 2004). However, in the landscape context, also other factors contribute to this, which can be attributed to micrometeorological effects, mainly related to evapotranspiration (Allen et al. 1998, Chap. 5, 10). This is particularly relevant for landscape elements as small forests, shelterbelts, trees, or shrubs that influence directional meteorological parameters such as wind velocity or solar radiation and results in windward and leeward effects or sunny and shady sides in close proximity to each other. This causes further spatial differences of temperature, humidity, evaporation or dew formation, and different growth conditions for crops between both sides which go far beyond the effects at the location. Many micrometeorological measurements have been carried out around windbreaks in recent decades (Caborn 1957; Plate 1971; Schwartz et al. 1995; Cleugh 1998; Cleugh and Hughes 2002; Vigiak et al. 2003; Böhm et al. 2014; Kanzler et al. 2019; Weninger et al. 2022), quantifying these effects which can result in different growing conditions and yields of agricultural crops finally (Baker et al. 2016, 2021; Schmidt et al. 2019; Chap. 13).

In the case study, we illustrate the influence of landscape elements on the spatial variability of the directional meteorological parameter wind velocity and how that affects the reduction of the wind erosion risk and the potential evapotranspiration. Our analysis is intended for larger regions and was therefore integrated into a GIS (Funk and Völker 2024). The ArcGIS toolbox "WERA" should help to identify problem areas of wind erosion within the region and to illustrate differences in water availability between or within fields in a quantitative way. The approach can be transferred also to other regions like Southern Africa, where wind erosion is a major threat for agriculture and land uses (Kestel et al. 2023; Baade et al. 2024) for regional and local planning.

11.2 Toolbox "WERA"

Our methodology is based on the ArcGIS toolbox "WERA," which was developed to assess the influence of landscape elements on the wind erosion risk by shadowing effects (Funk et al. 2023; Funk and Völker 2024). The model calculates the reduction of the wind erosivity based on the frequencies of wind velocities above a certain soil specific threshold and their directional distributions. For this purpose, the location and height of all landscape elements are required. This information can be obtained from maps, on-site measurements, or remote sensing methods as laser scanning.

Since the reduction in wind speed not only affects wind erosion, but also has an impact on other micrometeorological parameter as the evapotranspiration, this approach can also be used for this purpose. The aim here was to show how the wind-reducing effects of LE influences this parameter in its spatial variability and to estimate water-saving effects. In this way, shorter periods of time or individual events can be calculated.

11.2.1 Landscape Structure

The landscape structures (Fig. 11.1) for the purposes of this study are needed as information about height and spatial arrangement of each landscape element (LE). We derived a landscape structure model from an aircraft-based laser scanning campaign for the digital elevation model (DEM), where the surface and all objects on it were scanned in a grid with a horizontal resolution of 1×1 m and a vertical resolution of $\pm$ 0.25 m (LGB 2020). The laser scanning resulted in two layers, the digital surface model (DSM), including everything with a certain height, and the derived digital elevation model (DEM), representing the land surface after eliminating all landscape elements as buildings or vegetation. The height of any landscape element can be received by subtracting the DEM from the DSM. This step eliminates topography and the resulting new layer includes only the measured landscape elements on a flat surface. Additionally, all height values inside of agricultural land were set to zero using the structures of a digital field block cadastre (DFBK, LGB 2023), because temporary silos or hay bales left in the fields were also measured by the laser scanning during the overflight.

The decrease in wind speed behind windbreaks is well studied and shows a distinct decrease on the leeward site at first, which then gradually increases back to the original value with increasing distance. One of the few mathematical descriptions of this curve can be found in the WEPS model (Hagen and Fox 2020). The equation calculates the decrease and increase of the wind speed behind the LE as multiples of its height (Eq. 11.1, Fig. 11.3). Also the porosity (p) of the windbreak can be entered as an additional parameter (Eqs. 11.2–11.5). An average value is sufficient for the evaluation of long-term effects; if the method is used on an event-related basis, it is

Fig. 11.1 Landscape elements at the study site of orchard research station, Müncheberg, Brandenburg

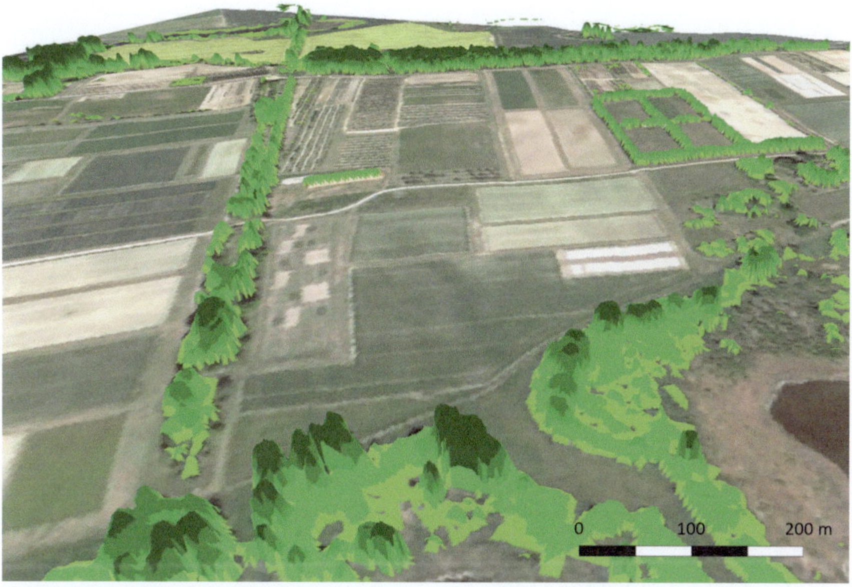

Fig. 11.2 Landscape unit with all scanned landscape elements at the study site of orchard research station, Müncheberg, Brandenburg

possible to react to the different foliage in winter or summer.

$$fu = 1 - \exp(-m * x^2) + n * \exp(-0.003(x + s)^t) \quad (11.1)$$

where

fu wind reduction factor.

x distance from the structure element in multiples of height.

The coefficients m, n, s, and t depend on porosity (p) and are calculated as follows

$$m = 0.008 - 0.17p + 0.17p^{1.05} \quad (11.2)$$

$$n = 1.35(-0.5p^{0.2}) \quad (11.3)$$

$$s = 10(1 - 0.5p) \quad (11.4)$$

$$t = 3 - p \quad (11.5)$$

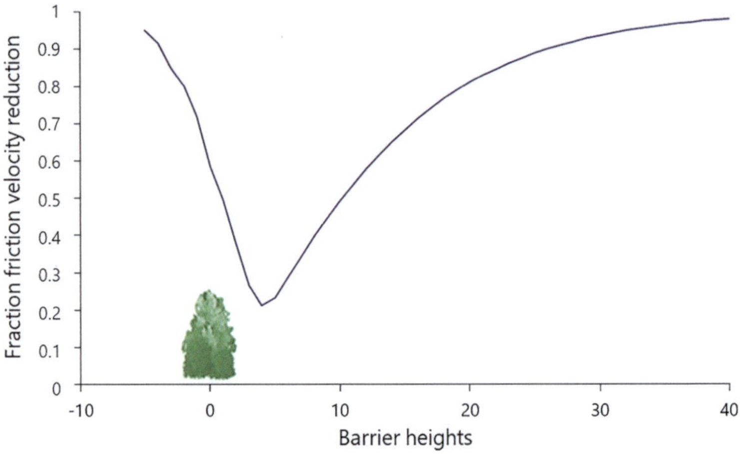

Fig. 11.3 Wind velocity reduction in front and behind a wind barrier in multiples of the barrier heights calculated with Eqs. 11.1–11.5 for a porosity of 40 percent

11.2.2 Meteorological Data

Our study uses hourly data of temperature, precipitation, global radiation, air humidity, wind velocity, and wind directions available from 1991 to 2023. The data were obtained from DWD climate station in Müncheberg (DWD Climate Data Centre, DWD station 3376, 14.12 E, 52.52 N). The data were used to calculate the wind erosivity, the sheltering distance behind the LE for all wind directions, and the potential evapotranspiration (PET) for each hour of that period.

11.2.2.1 Derivation of Parameters Influencing Wind Erosion

The wind erosivity and the wind directions are the basis to derive an effective sheltering distance behind any LE for all measured directions. First, the frequency distribution of measured winds (u) above a threshold wind velocity (u_t) and the transport capacity q (Eq. 11.1) were used to calculate a weighted transport capacity for each direction (Fig. 11.4).

$$q = (u - u_t)^2 \text{ for } u > u_t \tag{11.6}$$

Generally, the wind-reducing influence of a windbreak is indicated with the 40-fold of its height, which refers to the initial level. If a threshold velocity is set, this distance only applies to this velocity. Higher velocities result in shorter distances, so that exceeding this threshold value reduces the protective distance (Fig. 11.5). From the combination of the relative transport capacities and the effective protection

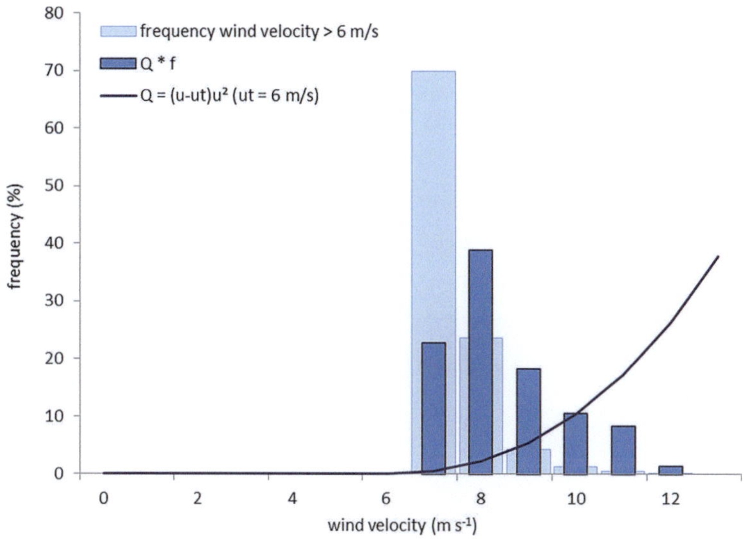

Fig. 11.4 Frequency of hourly wind speeds above the threshold of 6 m s^{-1} (light blue bars); transport capacity of the wind based on the relationship $q = (u - u_t)u^2$ (black line); relative transport capacity q * f (blue bars) (Funk et al. 2023)

distance derived for each wind velocity level, new weighted protection distances can be calculated for each wind direction (Table 11.1).

This is made for each considered wind direction separate. The result is then the effective protective length of a wind barrier in the corresponding direction, expressed in multiples of the height of the LE. In a final step, these effective protection lengths of each direction are then multiplied by the normalized transport capacity of each direction. The direction with the maximum transport capacity is given a factor of 1, all others are reduced accordingly (Fig. 11.6, right).

The ArcGIS Toolbox WERA uses all this data to set virtual shadows that correspond to the effective protection lengths of each LE in all considered directions. To simplify the calculation procedure, the shadows are divided into five protection zones of equal length and their effect is classified from very low to very high (Table 11.2). In the final step, these protection zones are linked to the soil erodibility (Fig. 11.8). This is also available in five levels, resulting in the following the matrix given in Table 11.2.

Derivation of Parameters Influencing the Potential Evapotranspiration

The same meteorological data were used to calculate the hourly potential evapotranspiration (PET) with a modified Penman equation after Wendling et al. (1991):

$$PET = g\left[\frac{G}{410} + (0.5 + 0.54 \cdot u_2)(100 - RH)\,TL/905\right] \quad (11.7)$$

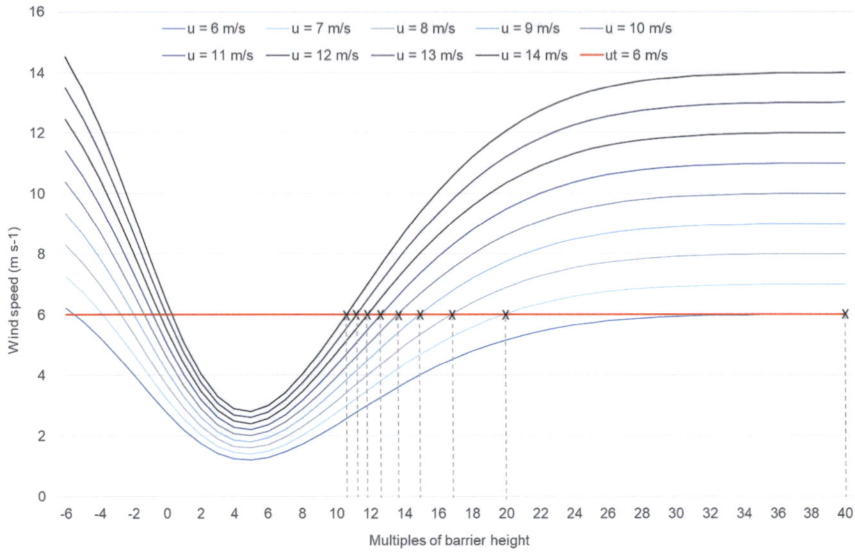

Fig. 11.5 Wind velocity reduction in front and behind a landscape element for wind velocities above a threshold; red line marks the threshold wind speed for wind erosion on sandy soils, the crosses mark the distance from which the wind speed becomes erosive again

Table 11.1 Example for the derivation of the effective protection zone behind a landscape element, taking into account the transport capacities of all occurring wind velocities above the threshold

Wind velocity class	(A) Relative transport capacity (%)	(B) Class middle in multiples of height	A * B Σ/100
6–7 m s^{-1}	22.9	28	641.2
7–8 m s^{-1}	38.8	17.5	679.0
8–9 m s^{-1}	18.2	15	273.0
9–10 m s^{-1}	10.5	12.8	134.4
> 10 m s^{-1}	9.6	10	96.0
			18.2

With $g = 2.4 \cdot (t + 22)/(t + 123)$….

G global radiation (J cm^{-2}).

u_2 wind velocity in 2 m height (m s^{-1})

RH relative humidity (%).

TL length of day (h), setting TL = 1 enables the calculation of hourly values.

As the standard height for wind velocity measurements at the DWD stations is 10 m, the wind velocity for the height of 2 m is calculated with

$$u_2 = u_{10} \cdot 4.2/(3.5 + ln10) \tag{11.8}$$

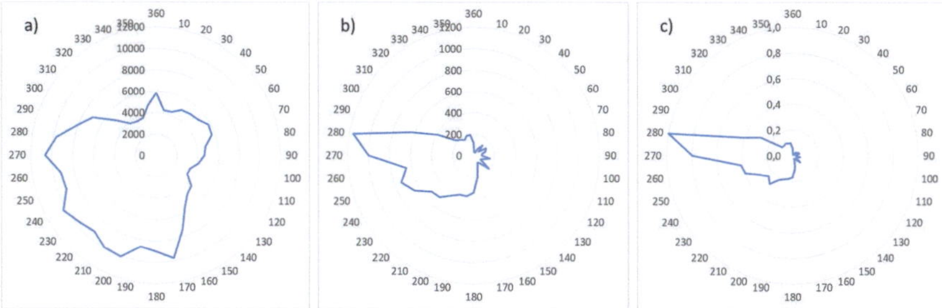

Fig. 11.6 Wind speed distributions: **a** all wind speeds (hours), **b** wind speeds above the threshold of 6 m s^{-1}(hours), and **c** relative transport capacities of each direction (normalized, maximum = 1)

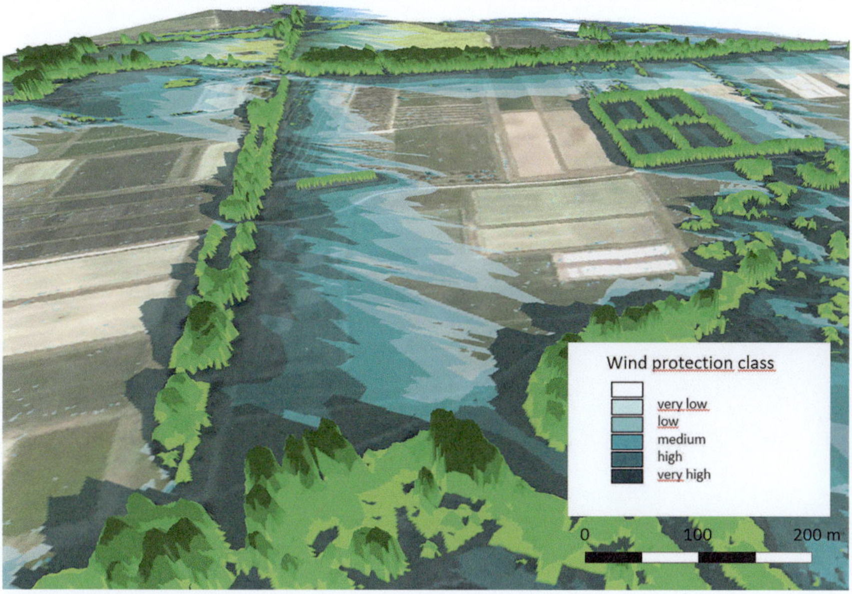

Fig. 11.7 Calculated wind protection zones against wind erosion around landscape elements at the study site of orchard research station, Müncheberg, Brandenburg

but also any other equation for the wind profile is applicable (WMO 2012; Stull 2000).

This hourly PET values were calculated for each of the shadowed zones by reducing the wind speed by the average factors shown in Fig. 11.9. The lengths of the shadows of each direction are related to the frequencies of all measured wind velocities.

Table 11.2 Connection between soil erodibility and wind protection to derive the reduced erodibility by wind shadowing (same colors for the classes are used in Fig. 11.6)

		Wind protection class				
		~~very~~ Very low (1)	~~low~~ Low (2)	~~medium~~ Medium (3)	~~good~~ Good (4)	~~very~~ Very good (5)
Soil erodibility	~~very~~ Very low (1)	0	0	0	0	0
	~~low~~ Low (2)	1	0	0	0	0
	~~medium~~ Medium (3)	2	1	0	0	0
	~~high~~ High (4)	3	2	1	0	0
	Very high (5)	4	3	2	1	0

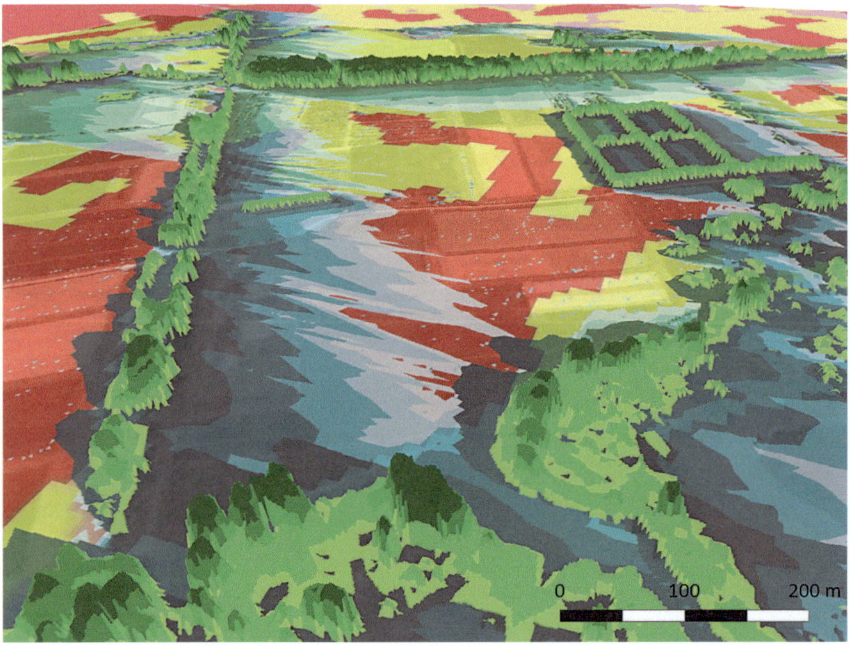

Fig. 11.8 Final combination of the soil erodibility with the wind protection classes for to estimate the reduced soil susceptibility to wind erosion at the study site of orchard research station, Müncheberg, Brandenburg

11.3 Regional Analysis—The Case Study in Brandenburg

11.3.1 The Study Area

The study for the regional analysis is the district Märkisch-Oderland (MOL) located in the Federal State of Brandenburg, Germany. The area of MOL extends east of Berlin to the Polish border and covers 2,127 km^2, of which approximately 60% is

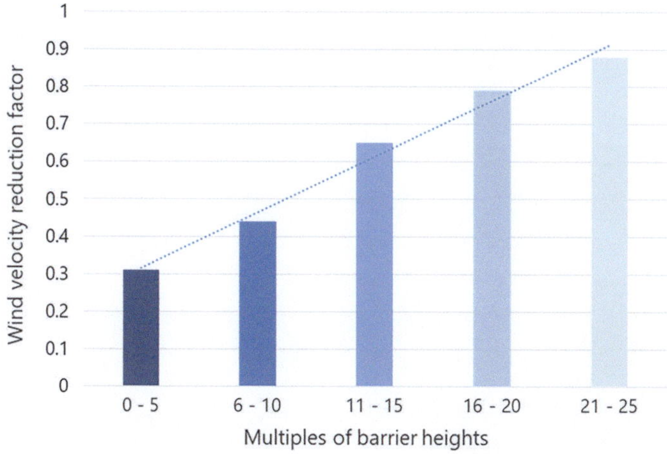

Fig. 11.9 Factors of wind velocity reduction in the zones behind a landscape element, averages of the curves shown in Fig. 11.5

used for agriculture and 23% is covered by forest (Fig. 11.10). The soils originate from the Quaternary period, when large amounts of loose material were deposited during the last Ice Age. In the post-glacial period, water and wind erosion covered large areas with material sorted by the erosion processes. Ground and end moraines, alluvial fans, and aeolian sands are thus the initial substrates of today's soils. As a result, there is a great variability of soils which are immediate adjacent. Soil data are available in a map from the Brandenburg State Office for Mining, Geology and Raw Materials in the scale of 1:10,000 (LGBR 2024).

Climate of the Study Area

The study area climate is characterized by the transition between maritime and continental influences. Especially, the eastern part of Brandenburg with the Oderbruch belongs to the regions with the lowest annual precipitation (P) in Germany, with 533 mm in average with a span from 360 to 760 mm. In the 30-year average (1991–2020), precipitation remains almost constant, with a slight increase of 0.5 mm per year, whereas the last 10 years documents a strong decrease in precipitation of 16 mm per year. In terms of temperatures and daily hours of sunshine, MOL is at the upper limit in Germany (Walter et al. 2006). Thus, the potential evapotranspiration (PET) is also correspondingly high. The 30-year average of PET is 866 mm, ranging with annual values from 740 to 1020 mm. The 30-years trend shows an increase in PET of 77 mm. There is a climatic water deficit (PET > P) from March to October. Based on these climatic conditions, both the protection of the soil from wind erosion and the reduction of evaporation by the LE are of great importance for a sustainable and successful agriculture.

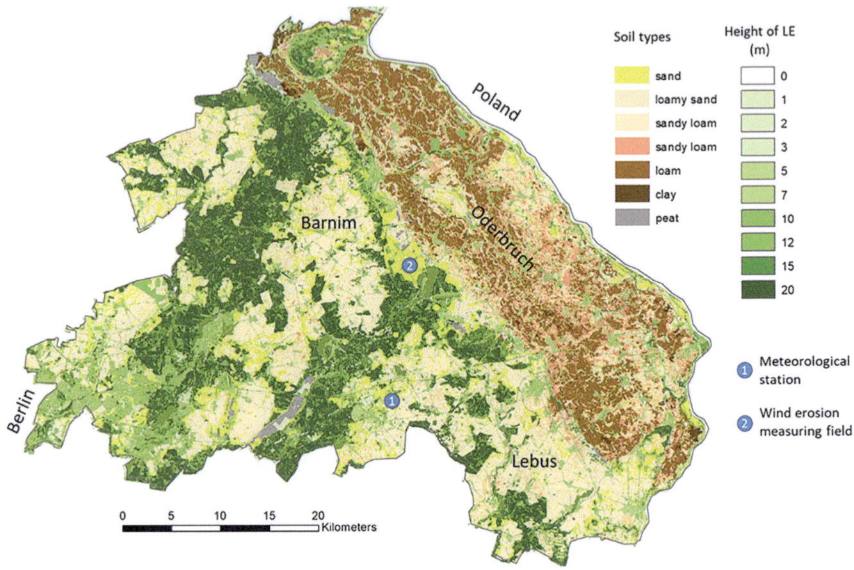

Fig. 11.10 District Märkisch-Oderland (MOL), Brandenburg with the main soil types on arable land and all landscape elements (LE) colored according to their classified heights

11.3.2 Application Wind Erosion

The wind erosion affected area in the district MOL decrease considerably by including the wind-reducing effects of landscape elements (Fig. 11.11). In the comparison to the baseline situation where the erosion risk was only derived from soil data, the wind shadowing effect of the LE results in a reduction of the wind erosion risk area by one third, from 18,740 ha to 12,114 ha. This proves that the inclusion of landscape structures in the determination of the wind erosion risk is an essential part of the process. The ArcGIS Toolbox WERA is now an easy-to-use tool that is available on request.

In addition to the regional analysis, the results can also be used as a planning tool for each individual field to optimize protective measures. The installation of new windbreak hedges can thus be adapted to the real risk potentials and scenarios for their effectiveness can be calculated over certain periods of time. The landscape section used as a case study in this chapter shows that despite the good protection provided by the hedge on the left, a significant proportion of the arable land remains in the highest risk category (Fig. 11.12). It is clear that a hedge running from north to south in the center of the picture would be the best solution here.

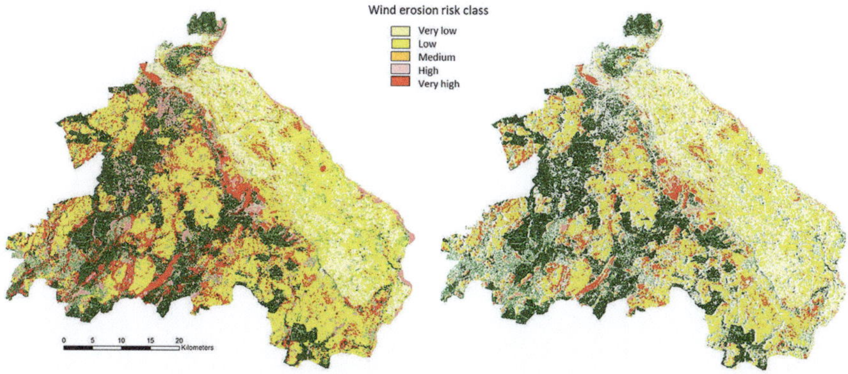

Fig. 11.11 Wind erosion risk in the county Märkisch-Oderland (MOL), Brandenburg Left: based on soil data, right: including wind-reducing effects by landscape structures

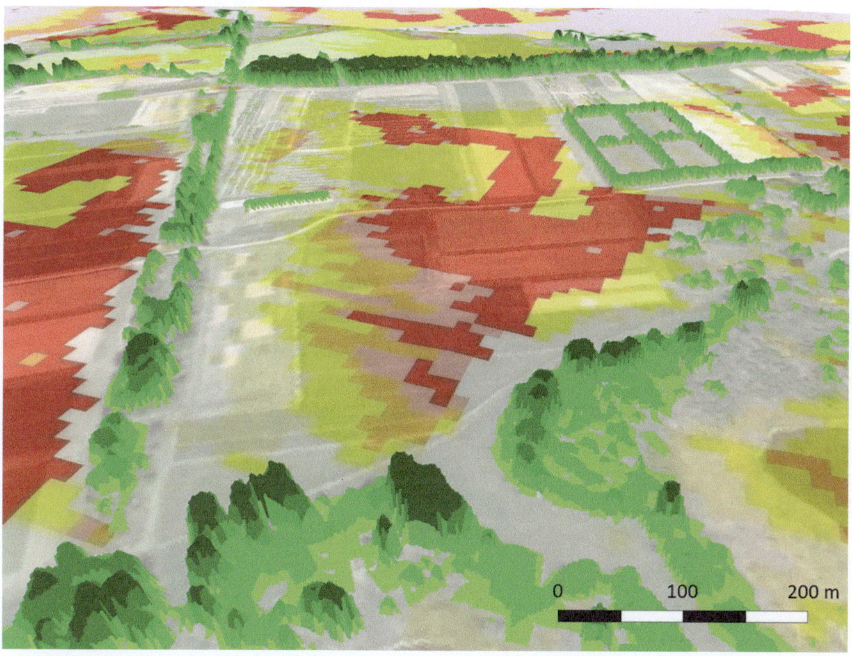

Fig. 11.12 Wind erodibility resulting from the consideration of the wind-reducing effects of shelterbelts at the study site of orchard research station, Müncheberg, Brandenburg (for legend see Table 11.2, Fig. 11.8)

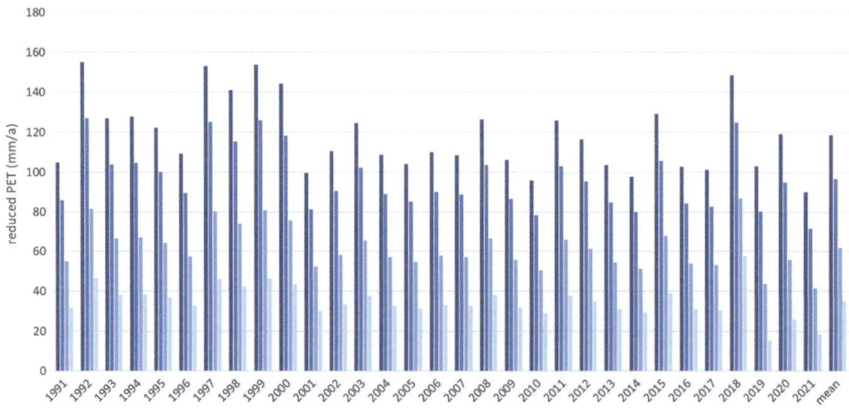

Fig. 11.13 Calculated reduced potential evapotranspirtaion (PET) in the zones of wind reduction behind landscape elements (wind reduction as indicated in Fig. 11.4), dry years: 1992 (418 mm precipitation), 1999 (417 mm), 2018 (390 mm)

11.4 Wind Effects on the Spatial Distribution of the Potential Evapotranspiration (PET)

The reduction in wind speed also has an effect on the calculated potential evapotranspiration PET. This effect is even more pronounced in dry years (Fig. 11.13). During periods of drought, 2–3 days on which the plants still have sufficient water available often decide whether or not there will be serious yield losses. In the 30-years average, the annual PET is reduced to 86% or 118 mm less PET in the zone closest to the LE, corresponding to the 0–fivefold of the height of each LE. With increasing distance and increasing wind velocity, the calculated evaporation increases again until it reaches its initial value again at a distance of the 20-fold of the height. These values are in good agreement with measured values from other authors (Kanzler et al. 2019; Jacobs et al. 2022).

The GIS also allows these effects to be displayed in their spatial distribution (Fig. 11.14).

11.5 Conclusions

Our results support the positive effects of landscape elements on reducing wind erosion susceptibility of soils and as a potential water-saving measure influencing PET. Based on state-of-the-art technologies to measure landscape structures and quantifying the protective effects of wind barriers, a very detailed spatial resolution of the wind erosion risk can be derived. As the wind speed also influences the PET, the same procedure can also be used to calculate the change in PET and to quantify

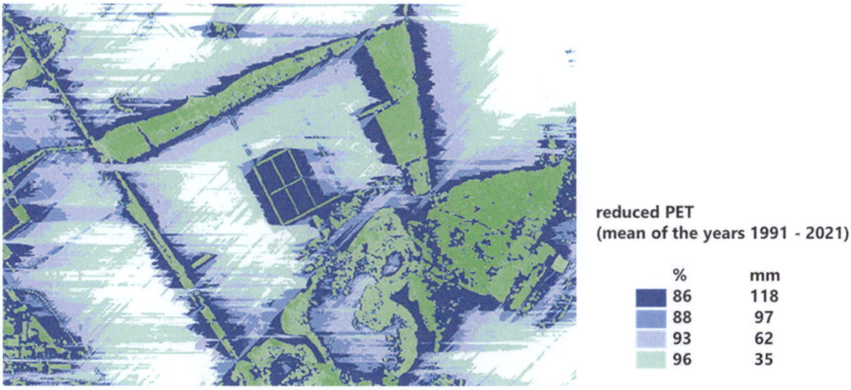

Fig. 11.14 Calculated effects of the reduced wind velocities on the spatial variability of potential evapotranspirtaion (PET)

and visualize the effects at high spatial resolution. This is particularly important in agricultural areas with extended dry seasons, as all means should be used to prevent harmful soil degradation due to wind erosion and to minimize unproductive evaporation. Based on the results, with this powerful planning tool, protection strategies can be better implemented in practice and their positive effects quantified.

Acknowledgements The research was financed by the Federal Ministry of Education and Research (BMBF) and managed by PT-DLR as part of "*Agroforestry in Southern Africa - New pathways of innovative land-use systems under a changing climate*" (ASAP) (FKZ 01LZ1711A-F) within the framework of "*SPACES II—Science Partnerships for the Adaptation/Adjustment to Complex Earth System Processes.*" Because of the SARS-Covid-19 restrictions, we were not able to carry out our planned measurements in South Africa. Instead, the Orchard Research Station Müncheberg (OBVS) gave us the alternative of carrying out our measurements there to the same extent.

References

Allen RG, Pereira L, Raes D, Smith M (1998). Crop Evapotranspiration: Guidelines for Computing Crop Water Requirements. FAO Irrigation and Drainage Paper 56. Rome, Italy: Food and Agriculture Organisation.

Baade J, Aucamp I, Collett A, Eckardt F, Funk R, Glotzbach C, von Holdt J, Kestel F, Knot J, Lombard A, Morgenthal T, Msipa A, Le Roux JJ (2024). Soil erosion research and soil conservation policy in South Africa. In: von Maltitz, G.P., et al. Sustainability of Southern African Ecosystems under Global Change. Ecological Studies. Springer, Cham., pp. 335–368. https://doi.org/10.1007/978-3-031-10948-5_13

Baker TP, Jordan GJ, Baker SC (2016) Microclimate edge effects in a recently harvested forest: Do remnant forest patches create the same impact as large forest areas? For Ecol Manage 365:128–136. https://doi.org/10.1016/j.foreco.2016.01.022

Baker TP, Moroni MT, Hunt MA, Worledge D, Mendham D (2021) Temporal, environmental and spatial changes in the effect of windbreaks on pasture microclimate. Agric for Meteorol 297:1082265. https://doi.org/10.1016/j.agrformet.2020.108265

Böhm C, Kanzler M, Freese D (2014) Wind speed reductions as influenced by woody hedgerows grown for biomass in short rotation alley cropping systems in Germany. Agroforest Systems 88:579–591. https://doi.org/10.1007/s10457-014-9700-y

Borelli P, Balladio C, Panagos P, Montanarella L (2014) Wind erosion susceptibility of European soils. Geoderma 232–234:471–478. https://doi.org/10.1016/j.geoderma.2014.06.008

Borelli P, Lugato E, Montanarella L, Panagos P (2017) A new assessment of soil loss due to wind erosion in European agricultural soils using a quantitative spatially distributed modelling approach. Land Degrad Dev 28:335–344. https://doi.org/10.1002/ldr.2588

Brune M (2016) Urban trees under climate change. Potential impacts of dry spells and heat waves in three German regions in the 2050s. Report 24. Climate Service Center Germany, Hamburg.

Caborn JM (1957) Shelterbelts and microclimate. Forestry Commission Bulletin No. 29, Dep. Of Forestry, Edinburgh University.

Cleugh HA (1998) Effects of windbreaks on airflow, microclimates and crop yields. Agrofor Syst 41:55–84

Cleugh HA, Hughes DE (2002) Impact of shelter on crop microclimates: a synthesis of results from wind tunnel and field experiments. Aust J Exp Agric 42(6):679. https://doi.org/10.1071/EA02005

Funk R, Reuter HI (2006). Wind Erosion in Europe. In Boardman J, Poesen J (eds). Soil Erosion in Europe. Wiley, London.

Funk R, Völker L, Deumlich D (2023) Landscape structure model based estimation of the wind erosion risk in Brandenburg. Germany. Aeolian Research 62:100878

Funk R, Völker L (2024). A GIS-toolbox for a landscape structure based Wind Erosion Risk Assessment (WERA). MethodsX 13: Article 103006.

Goossens D (2004) Net loss and transport of organic matter during wind erosion on loamy sandy soil. In: Goossens D, Riksen M (eds) Wind Erosion and Dust Dynamics: Observations, Simulations, Modelling. ESW Publications, Wageningen, pp 81–102

Hagen LJ, Skidmore EL, Miller PL, Kipp JE (1981) Simulation of the effect of wind barriers on airflow. Trans Am Soc Agric Eng 24(4):1002–1008

Hagen LE, Fox FA (2020) Erosion Submodel of WEPS. In: Tatarko J (Ed.). The Wind Erosion Prediction System (WEPS): Technical Documentation, USDA Agriculture Handbook 727: 85–157. United States Department of Agriculture, Agricultural Research Service, Beltsville, MD.,

Horton R, Horn R, Bachmann J, Peth S (eds) (2016) Essential Soil Physics. Schweizerbart Science Publisher, Stuttgart

Hohenbrink TL, Lischeid G (2014) Texture-depending performance of an in situ method assessing deep seepage. J Hydrol 511:61–71. https://doi.org/10.1016/j.jhydrol.2015.02

IPCC 2021. https://www.ipcc.ch/report/ar6/wg1/downloads/report/IPCC_AR6_WGI_SPM.pdf

Jacobs SR, Webber H, Niether W, Grahmann K, Lüttschwager D, Schwartz C, Breuer L, Bellingrath-Kimura SD (2022) Modification of the microclimate and water balance through the integration of trees into temperate cropping systems. Agricultural and Forest Meteorology 323: Article 109065. https://doi.org/10.1016/j.agrformet.2022.109065

Kanzler M, Böhm C, Mirck J, Schmitt D, Veste M (2019) Microclimate effects on evaporation and winter wheat (*Triticum aestivum* L.) yield within a temperate agroforestry. Agrofor Syst 93:1821–1841. https://doi.org/10.1007/s10457-018-0289-4

Kestel F, Wulf M, Funk R (2023) Spatiotemporal variability of the potential wind erosion risk in Southern Africa between 2005 and 2019. Land Degradation and Development 34(10) 34:2945. https://doi.org/10.1002/ldr.4659

LGB (2020) Landesvermessung und Geobasisinformation Brandenburg - Digital Elevation Model of Brandenburg. https://geobasis-bb.de/lgb/de/geodaten/3d-produkte/, last access: 07.11.2024.

LGB (2023) Landesvermessung und Geobasisinformation Brandenburg - Digitales Feldblock Kataster.

LGBR (2024) Federal Service for Mining, Geology and Resources. Maps of the LBGR: http://www.geo.brandenburg.de/lbgr/

Nerger R, Funk R, Cordsen E, Fohrer N (2017) Application of a modelling approach to designate soil and soil organic carbon loss to wind erosion on long-term monitoring sites (BDF) in Northern Germany. Aeol Res 25:135–147

Pásztor L, Négyesi G, Laborczi A, Kovács T, László E, Bihari Z (2016) Integrated spatial assessment of wind erosion risk in Hungary. Nat Hazards Earth Syst Sci 16(11):2421–2432

Plate EJ (1971) The Aerodynamics of Shelterbelts. Agr. Meteorol. 8:203–222

Řeháček D, Khel T, Kučera J, Vopravil J, Petera M (2017) Effect of windbreaks on wind speed reduction and soil protection against wind erosion. Soil and Water Research 12:128–135. https://doi.org/10.17221/45/2016-SWR

Reich P, Eswaran H, Beinroth F (2001) Global dimensions of vulnerability to wind and water erosion. In: Scott DE, Mohtar RH, Steinhardt GC (eds.) Sustaining the Global Farm. Selected papers from the 10th International Soil Conservation Organization Meeting held May 24–29, 1999 at Purdue University and the USDA-ARS National Soil Erosion Research Laboratory, 838–846.

Rötter RP, Carter TR, Olesen JE, Porter JR (2011) Crop–climate models need an overhaul. Nat Clim Change 1(4):175–177. https://doi.org/10.1038/nclimate1152

Rötter RP, Nkomo, M, Meyer zu Drewer J, Veste M (2024) Agricultural Land-Use Systems and Management Challenges. In: von Maltitz GP, Midgley GF, Veitch J, Brümmer C, Rötter RP, Viehberg, FA, Veste M (eds) Sustainability of southern African ecosystems under global change: Science for management and policy interventions. Ecological Studies 248, Springer Nature, Cham, 551–586. https://doi.org/10.1007/978-3-031-10948-5_20

Sheppard JP, Bohn Reckziegel R, Borrass L, Chirwa PW, Cuaranhua CJ, Hassler SK, Hoffmeister S, Kestel F, Maier R, Mälicke M, Morhart C, Ndlovu NP, Veste M, Funk R, Lang Seifert F, du Toit TB, Kahle HP (2020) Agroforestry: an appropriate and sustainable response to a changing climate in Southern Africa? Sustainability 12(17):6796. https://doi.org/10.3390/su12176796

Schmidt M, Jochheim H, Kersebaum K-C, Lischeid G, Nendel C (2017) Gradients of microclimate, carbon and nitrogen in transition zones of fragmented landscapes - A review. Agric for Meteorol 232:659–671. https://doi.org/10.1016/j.agrformet.2016.10.022

Schmidt M, Nendel C, Funk R, Mitchell MGE, Lischeid G (2019) Modeling yields response to shading in the field-to-forest transition zones in heterogeneous landscapes. Agriculture 9:6. https://doi.org/10.3390/agriculture9010006

Schmidt R, Deumlich D, Reuter HI (2004) From point over regions to the landscape - spatial generalization in the soil landscape research. Archives of Agronomy and Soil Science 50(3):241–257. https://doi.org/10.1080/03650340410001663819

Schwartz RC, Fryrear DW, Harris BL, Bilbro JD, Juo ASR (1995) Mean flow and shear stress distributions as influenced by vegetative windbreak structure. Agric for Meteorol 75:1–22

Snyder RL, Pedras C, Montazar A, Henry JM, Ackley D (2015) Advances in ET-based landscape irrigation management. Agric Water Manag 147:187–197. https://doi.org/10.1016/j.agwat.2014.07.024

Stull RB (2000) Meteorology for Scientists and Engineers. Second Edition, Brooks/Cole, Thomson Learning, USA.

van Eimern J (1964) Untersuchungen über das Klima in Pflanzenbeständen als Grundlage einer agrarmeteorologischen Beratung, insbesondere für den Pflanzenschutz. Berichte des Deutschen Wetterdienstes 96, 76 pp.

Veste M, Littmann T, Kunnecke A, du Toit B, Seifert T (2020) Windbreaks as part of climate-smart landscapes reduce evapotranspiration in vineyards, Western Cape Province, South Africa. Plant Soil Environments 66: 119–127. https://doi.org/10.17221/616/2019-PSE

Veste M, Sheppard JP, Abdulai I, Ayisi KK, Borrass L, Chirwa PW, Funk R, Kapinga K, Morhart K, Mwale SE, Ndlovu NP, Nyamadzaw G, Nyoka BI, Sebola P, Seifert T, Senyolo MP, Sileshi GW, Syampungani S, Kahle H-P (2023) The need for sustainable agricultural land-use systems: benefits from integrated agroforestry systems. In: von Maltitz GP, Midgley GF, Veitch J, Brümmer

C, Rötter RP, Viehberg, FA, Veste M (eds) Sustainability of southern African ecosystems under global change: Science for management and policy interventions. Ecological Studies 248, Springer Nature, Heidelberg, pp 587–623. https://doi.org/10.1007/978-3-031-10948-5_21

Vigiak O, Sterk G, Warren A, Hagen LJ (2003) Spatial modeling of wind speed around windbreaks. CATENA 52:273–288. https://doi.org/10.1016/S0341-8162(03)00018-3

Walter A, Keuler K, Jacob D, Knoche R, Block A, Kotlarski S, Müller-Westermeier G, Rechid D, Ahrens W (2006) A high resolution reference data set of German wind velocity 1951–2001 and comparison with regional climate model results. Meteorol Z 15(6):585–596. https://doi.org/10.1127/0941-2948/2006/0162

Wendling U, Schellin H-G, Thomä M (1991) Bereitstellung von täglichen Informationen zum Wasserhaushalt des Bodens für die Zwecke der agrarmeteorologischen Beratung. Zeitschrift Für Meteorolologie 41:468–475

Weninger T, Gartner K, Riedel S, Scheper S, Michel K (2022) Der Windschutzeffekt von Bodenschutzanlagen am Beispiel Marchfeld. Oesterr Wasser Abfallwirtsch 74:251–257. https://doi.org/10.1007/s00506-022-00851-y

WMO (2012) Technical Regulations, Volume I – General Meteorological Standards and Recommended Practices, 2011 edition, updated 2012, WMO-No. 49

Zolina O, Simmer C, Belyaev K, Gulev SK, Koltermann P (2013) Changes in the duration of European wet and dry spells during the Last 60 Years. J Clim 26:2022–2047

Chapter 12
Farm-Level Financial Analysis of Implementing Tree-Based Windbreaks: A Case Study of South African Mandarin Orchards

Lutz- Heiner Otto

Abstract The financial impact of implementing tree-based windbreaks in South African mandarin orchards was investigated based on a case study from the Western Cape. Windbreaks protect fruit quality and prevent windfall during strong winds but come with establishment costs and land sacrifices to avoid competition with fruit crops. The financial value of windbreaks was assessed using stochastic discounted cash flow techniques. The results show that tree-based systems and agricultural nets outperform open fields in terms of net present value (NPV) and annualized gross margin (AGM). The best windbreak design achieves a mean per-hectare NPV of USD 99,243 over 20 years, while nets have a higher NPV of USD 117,519 and AGM of USD 15,795. Windbreaks also offer shorter discounted payback periods (DPBP) due to reduced orchard costs, while open fields are riskier due to wind damage. Farmers' decisions may be influenced by the internal rate of return (IRR) and DPBP, with nets having a higher NPV but lower IRR compared to windbreaks. Windbreaks offer indirect benefits, like biodiversity enhancement and climate change mitigation, but data limitations prevent further analysis. More research is needed to quantify the broader ecosystem services of windbreaks and their valuation.

Keywords Citrus · Shelterbelts · Sustainable land use · Climate-smart agriculture · Agroforestry · Sustainability

L.-. H. Otto (✉)
University of Hohenheim, Land Use Economics in the Tropics and Subtropics, Stuttgart, Germany
e-mail: lh.otto@tutanota.com

Ecovest Landuse Consulting, Windhoek, Namibia

M. Veste et al. (eds.), *Trees in South African Agriculture*, Innovations in Landscape Research, https://doi.org/10.1007/978-3-032-00993-7_12

12.1 Introduction

Climate change will have drastic implication for farming in the Western Cape (von Maltitz et al. 2024), and operational changes are needed to increase the resilience of commercial fruit farming (Malherbe et al. 2024). Orchards are susceptible to environmental factors, such as wind, which can have detrimental effects on fruit productivity and optical quality (Ali et al. 2021). Strong and gusty winds can cause premature windfall in orchards (Zanamwe 2018). Excessive wind can physically damage fruit, leading to underdeveloped or bruised fruit. Winds can further cause physical damage to fruit trees, especially young or weak trees. This can result in reduced yields and economic losses for fruit growers (Gravina et al. 2011; Cataldo et al. 2013).

Temperature gradients between different regions can affect the movement and strength of air masses, leading to changes in wind patterns. These alterations in atmospheric circulation can be influenced by factors such as changes in land-sea temperature differentials, shifts in weather patterns, and modifications in the distribution of solar radiation. While the overall relationship between temperature increase and wind speed is not linear and can vary in different regions, studies and climate models generally suggest a positive correlation between the two (Otto et al. 2018). Furthermore, extreme events will influence the rainfall in the region and will increase droughts in the region (Chap. 4).

There is evidence that climate change is influencing wind speeds in South Africa. Climate models and observational studies suggest that wind speeds are expected to increase in certain regions of the country as a result of climate change. A study conducted by Kruger and Shongwe (2004) analyzed long-term wind speed trends in South Africa using data from various meteorological stations. The study found a statistically significant increase in wind speeds over the period of 1960 to 2000, particularly in the western and southwestern parts of the country. This increase in wind speeds is consistent with the projections of climate models, which suggest that climate change will lead to stronger and more frequent wind events in these regions (Otto et al. 2018). The coastal areas of the Western Cape Province already experience significant wind speeds, ranging from 5 to 8 m per second ($m\ s^{-1}$) at 10 m above the ground in fruit cultivating regions, and even higher speeds of up to 20 $m\ s^{-1}$ along the ridges. This high wind speed is a critical climatic factor that influences both evapotranspiration and soil erosion dynamics in this region (Veste et al. 2020; von Maltitz et al. 2024).

In the context of wind stress and increased evapotranspiration, the establishment of windbreaks and agroforestry systems can be implemented to mitigate their negative effects on the fruit production (Veste et al. 2023). Windbreaks are rows of trees or shrubs strategically planted around or within orchards to act as protective barriers, reducing wind speed and minimizing its detrimental effects (Chap. 2). They can provide shelter, create a more favorable microclimate, and improve orchard productivity and resilience. Implementing tree-based windbreaks in perennial fruit orchards may thus offer both ecological and economic advantages. However, implementing tree-based windbreaks in new orchards comes at costs. While there are initial establishment costs and ongoing maintenance expenses, land opportunity and competition costs are often farmers' biggest concerns. Further, artificial windbreaks, mainly agricultural nets, have gained attention of citrus farmers in recent years (Chap. 5).

This chapter explores the farm-level impacts of implementing tree-based windbreaks in fruit orchards, focusing on financial approaches under uncertainty and potential returns on investment due to yield benefits. The discussion includes a case study of South African mandarin orchards with empirical data collected in the Western Cape Province in January 2020. Additionally, a scenario analysis on potential increased wind damage as well as a profitability comparison against a net-based structure is provided.

The objective is to develop a practical stochastic simulation model that assesses the long-term financial outcomes of planting new mandarin orchards with tree-based windbreaks or agricultural nets for protection. This approach is designed to assist farmers in making investment decisions when faced with uncertainty, allowing them to compare and prioritize various options for wind protection. The findings of this study can provide valuable insights for both farmers and policymakers when making decisions concerning agricultural practices in the face of a shifting climate.

12.2 Materials and Methods

12.2.1 Study Area

The South African Western Cape Province plays a vital role in the country's citrus industry, particularly in mandarin orchards. Citrus farming, especially for export markets, is known for its high profitability due to the specialized nature of the crop and the ability to achieve premium export prices. The country boasts approximately 1,200 commercial citrus growers, with the majority of production geared toward meeting export demands. Citrus cultivation is concentrated in seven provinces across South Africa, with a total area share of 96,230 hectares in 2020. Among these provinces, the Western Cape ranks third in terms of citrus production, following the Eastern Cape and Limpopo, with a total area of 17,938 hectares (Citrus Growers' Association of Southern Africa 2021). In 2020, the South African citrus industry reached record export volumes, surpassing 2 million metric tons, with a notable surge in soft citrus exports.

The Western Cape region experiences significant south-easterly winds during the spring and summer months (Chap. 4), which can substantially impact fruit yield and quality (Geldenhuys et al. 2023). This highlights the critical need for effective wind protection measures in the area, particularly during the period when fruits are small and immature, spanning from the beginning of November until the end of January (Chap. 7). During this vulnerable phase, the fruits are highly susceptible to wind damage, making it essential to implement robust wind protection strategies, such as tree-based windbreaks, to safeguard the orchards and ensure optimal fruit production and quality.

With the increasing demand for high-value, blemish-free fruit crops like citrus, avocados, peaches, and plums, there has been a growing focus on fruit quality research (Chap. 7), driven by consumers' preferences for visually appealing produce. Consequently, research studies have concentrated on assessing the quality attributes of fruits rather than solely examining the direct yield impacts of windbreaks. The occurrence of premature fruit drop due to strong winds, commonly referred to as windfall, poses a notable concern for the South African fruit industry.

While specific data regarding the removal of tree-based windbreaks by South African farmers is not available, interviews with selected farmers and industry experts indicate a growing interest in alternative windbreak methods, such as nets and artificial structures. These alternatives offer certain advantages, including better control over airflow and protection from specific pests. Additionally, artificial windbreaks may prove more convenient to install and manage compared to traditional tree-based windbreaks, which require ongoing maintenance and management. However, despite the appeal of these alternatives, it is crucial not to overlook the continued importance and potential benefits of tree-based windbreaks in citrus orchards.

12.2.2 Data Collection

A questionnaire was developed to gain insights into the perceptions and practices of fruit and wine farmers in South Africa, particularly in the Western Cape region, regarding windbreaks. The data collection process consisted of two phases: face-to-face semi-structured interviews conducted on farms and online interviews. To obtain an economic inventory of typical farms in the Western Cape, a field research trip was undertaken while the author was affiliated with the University of Stellenbosch. The farm visits took place between the end of 2019 and the beginning of 2020.

The face-to-face semi-structured questionnaire comprised both open-ended and closed questions, divided into two main parts. The first part collected technical–economic data related to company assets, variable costs, income structures, orchard and viticulture management practices, and production practices. The second part focused on open-ended questions aimed at gaining insider perspectives on the use of windbreak hedges, farmers' experiences with windbreaks, and wind damage. The questionnaire was translated into Afrikaans, the dominant language, and was administered with the assistance of scientists from the University of Stellenbosch.

The on-farm interviews were conducted with nine commercial farmers, allowing a relatively high degree of control over the data collection process. The semi-structured nature of the questionnaire enabled the researcher to clarify any ambiguities directly with the respondents during the interviews. The primary purpose of conducting these interviews and using partial budgets was to create a financial model for analysis. Additionally, the open-ended questions provided valuable insights into farmers' management practices and perspectives on windbreaks. These insights complemented the existing literature and informed the development of an online questionnaire, which will be discussed further in the following subchapter.

12.2.3 Investment Indicators

Financial indicators are crucial for assessing the profitability of investment projects, especially when cash flows are received at different time periods. The concept of time value of money is considered to calculate comparable indicators. Money received today is more valuable than money received in the future due to the potential for interest gains. The objective of this study was to comprehensively quantify the costs and benefits associated with windbreak implementation in South African citrus farming in monetary terms. The profitability of windbreak trees planted along the margins of fruit orchards depends on the costs and benefits incurred over time. In the context of fruit farming, the implementation of windbreaks involves the trade-off of foregone land, as the presence of windbreak trees reduces the available land for fruit production. However, it is expected that the benefits in terms of higher yields and reduced fruit damage will outweigh the costs associated with foregone land and windbreak establishment.

In this study, four investment indicators were evaluated using the discounted cash flow (DCF) approach. These indicators include the net present value (NPV) in USD per hectare, internal rate of return (IRR) as a percentage, annual gross margin (AGM) in USD per hectare, and discounted payback period (DPBP).

Cash inflows and outflows are discounted and summed up over the investment horizon to calculate the NPV of a citrus crop (Mutenje et al. 2019). The NPV approach used in this chapter can be represented by the following formula: Σ (from $t = 0$ to n) $[(R_t\text{-}C_t)/(1 + r)^t]$, where C represents the costs, R the returns, t the time of the cash flow, n the investment horizon, and r the interest rate. The NPV method is commonly employed in investment decisions with cash flows occurring at different points in time and long planning horizons. However, it heavily relies on the chosen interest rate. For typical investment projects, where initial negative cash flows are followed by mostly positive flows, the NPV and the interest rate have an inverse relationship. Higher interest rates lead to lower NPVs, and vice versa.

The IRR determines the discount rate at which the NPV would reach zero, indicating a break-even point for the investment (Mutenje et al. 2019). It represents the maximum interest rate at which a project developer could borrow money to break even. One advantage of IRR over NPV is that it does not depend on a discount rate. However, large expenditures generating negative cash flow may lead to distortions in the analysis.

In uncertain investment environments, the DPBP is an essential indicator (Mutenje et al. 2019). It is a simple measure calculated as the quotient of the initial investment and the discounted net cash flow period. The DPBP determines the time required for an investment to break even at given costs and benefits. A higher discount rate results in a longer time for the accumulative NPV of project cash flows to reach zero. The formula for DPBP is Σ [(R_t—C_t) / (1 + r)^t] from t = 1 to t = DPP = I_0, where I_0 is the initial investment, so that DPBP is the time period when the cumulative discounted cash flows equal the initial investment.

AGM for perennial crops can be calculated from the NPV by converting positive and negative cash flows into an average annual value (Testa et al. 2016; Grala and Colletti 2003). AGM is a valuable financial indicator for comparing different perennial land uses as well as annual alternatives. Farmers often consider the annual income when evaluating and comparing perennial land uses to annual crops (Schweier and Becker 2013). The formula for AGM is (NPV * r)/[1−(1 + r)^(−n)].

These indicators help quantify the profitability of different investment alternatives and take into account the time value of money. The debate over which financial indicator is superior for investment appraisal is ongoing in corporate finance. South African stock-listed firms tend to prefer IRR over NPV when evaluating capital investments, but this may vary based on the firm's size and sector (Maroyi 2011). The NPV is commonly used to calculate the expected return of a potential investment across various domains (Griess et al. 2016).

The choice of financial indicator depends on the specific context and preferences of the investors. The IRR and DPBP are suitable alternatives when considering the time it takes for an investment to become profitable. In the South African context, a shorter payback period may be preferable due to the economic instability. However, the NPV and AGM remains widely used in various investment evaluations in agriculture.

A nominal discount rate of 12% was used as a baseline for all investment analyses. This rate reflects the prevailing commercial lending interest rates at the time of the study and represents the farmers' time preference for money and their expectations from high-risk investments (Steyn 2020).

12.2.4 Uncertainty Parameters

There are two main approaches to modeling in the context of farm systems and financial analyses: deterministic and stochastic modeling. Deterministic models are based on fixed calculations and do not account for uncertainty. They yield a single predetermined outcome under specific controlled circumstances (Feuerbacher et al. 2021; Testa et al. 2016).

On the other hand, stochastic simulation models involve solving the model multiple times by varying the input values of certain parameters. This generates a series of outcomes for the dependent variable, taking into consideration the inherent uncertainty in the system (Troost and Berger 2015). Recent studies using stochastic assessments for land-use cost–benefit analyses include Carauta et al. (2021), Declercq et al. (2020), and DeVincentis et al. (2020).

Financial analysis of investments in fruit farming, particularly when incorporating windbreaks, requires a long-term perspective. Given the limited availability of comprehensive data on the productivity effects of windbreaks in South African orchards, this study opted for stochastic methods to monetize the impacts of windbreaks and assess their profitability. The analyses presented here focus on investment decision-making under uncertainty, and the aforementioned approaches yield probability distributions of investment outcomes. Generating these probability distributions is a crucial step in investment analysis when uncertainty is present (Hildebrandt and Knoke 2011). Monte Carlo simulations are commonly used to derive probability distributions of potential investment outcomes. However, this study follows the approach of Carauta et al. (2021), who utilized a Sobol sequence as a sampling technique. The Sobol sequence ensures a more evenly distributed sampling of the parameter space compared to other techniques such as standard Monte Carlo or Latin hypercube sampling (Tarantola et al. 2012). This sequence allows for the random selection of input parameter values from their respective distributions, enabling repeated investment estimations across multiple uncertainty runs. The uncertain input parameters for orchards and windbreaks, as well as their respective distributions, will be discussed in subsequent sections of this chapter. Sobol simulations were utilized to analyze uncertainties associated with fruit yields, variable costs, and prices, following the approach described by Carauta et al. (2021). Due to limited access to historical data, distributions were assumed based on insights from experts and farmers and employed in the simulations instead of probabilities based on historical variations of inputs. The probability distributions used for the investment analysis are presented in the following chapters below. It was assumed that farmers operate as price takers in the global market, and therefore, correlations between crop yields, prices, and variable costs were not considered. Parameters for all different windbreak protection systems were derived from the same Sobol run, with only the distributions being modified (Carauta et al. (2021). The Sobol simulation runs were conducted using the R statistical software, specifically the "fOptions" package developed by Wuertz et al. (2017).

12.2.5 Costs of Orchards and Windbreaks

Initial Establishment Costs "'Nadorcott' Mandarin"

The study compares the impact of windbreaks on the profitability of "Nadorcott" mandarin (*Citrus reticulata*) orchards (Table 12.1), specifically examining two types of wind protection: tree-based windbreaks and protective netting. These are compared against open field conditions. To assess the financial aspects, annual costs and revenues were aggregated from orchard establishment until the end of the 20-year rotation lifetime, using a generic 1-hectare quadrant field for ease of comparison. However, it is important to note that the methods employed in this study can also be applied to evaluate specific field shapes and sizes.

The costs associated with mandarin orchard cultivation encompassed investment costs for establishing a new grove, while variable costs for fruit production varied depending on the age of the orchard (Tables 12.2 and 12.3). These variable costs

Table 12.1 Characteristics of 1 ha citrus orchard without windbreaks

In row tree spacing (m)	2.5
Between row tree spacing (m)	5
Perpendicular row length (m)	95
Parallel row length (m)	97.5
Fruit trees in perpendicular row (m)	19
Fruit trees in parallel row (m)	39
Total stocking rate (trees ha^{-1})	741

included items such as mineral fertilizer and a chemical spraying program. The establishment costs, yields, and prices utilized in the analysis were derived from structured farm interviews and will be discussed in greater detail in subsequent subsections. The prices and revenues were originally calculated in South African Rand and are presented here in US Dollars, with an exchange rate of 16.4 based on the 2020 average. The prices were averaged across all fruit grade categories, which will be further explained below.

Beefwood Windbreak

Casuarina *cunninghamiana* (locally know as beefwood, is an evergreen tree native to New South Wales and Queensland in Australia (Chaps. 1, 2). It typically reaches a

Table 12.2 Expected point estimate yields, revenues, costs, and cash flow for a 1 ha Nadorcott mandarin orchard in open field conditions

Items (USD ha^{-1})	Years						
	0	1	2	3	4–6	7–18	19–20
Yield expected (%)	0	0	0.1	0.2	0.5	1	0.8
Expected revenues	0	0	3,625	7,270	21,809	39,982	32,713
Establishment cost	11,629						
Variable cost	155	429	1,241	2,482	7,445	13,649	11,168
Cash flow	-11,783	-429	2,394	4,788	14,363	26,333	21,545

Table 12.3 Uncertainty parameters on yield and variable costs of orchards

Parameter	Year	Distribution	Min	Max
Yield orchards	3–4	Triangle	0.6	1.4
	5–7		0.7	1.3
	8–18		0.9	1.1
	19–20		0.8	1.2
Var. cost orchards	1	Triangle	0.8	1.2
	2–5		0.8	1.5
	6–20		0.8	1.5

height of 30 m and has a pyramidal crown with oblong and linear leaves measuring 8 to 10 cm in length. The tree structure and fast-grow makes the species perfect for wind protection (Chap. 5). In the Western Cape Province of South Africa, casuarina trees have been widely planted as windbreaks in citrus orchards in regions characterized by significant temperature increases during summer and mild winter frosts (Chap. 5, Diagne et al. 2013).

Windbreak Designs and Foregone Yield

In this study, three windbreak systems consisting of single row casuarina trees are analyzed. The specific arrangements of these windbreak systems are described below. The primary focus of this research is to assess the profitability of establishing new orchard-windbreak fields and evaluate their impact at the farm level.

- **Design 1: Single row perpendicular:**
  For a one-hectare orchard, it was assumed that a single row of 100 casuarina trees arranged in a 100-meter line perpendicular to the prevailing winds would serve as a windbreak. The associated costs and revenues related to the windbreak are presented in Table 12.4.

- **Design 2:** Single rows perpendicular and parallel:
  In cases where fields experience problematic winds from multiple directions, relying on a single row of wind protection may not provide adequate shielding. To address this, it was assumed that a windbreak consisting of a single row of 100 casuarina trees arranged in a 100-m line perpendicular to the prevailing winds, along with an additional single row parallel to the prevailing winds, would be implemented for a one-hectare orchard. This configuration enables the windbreak to effectively block winds originating from two different directions. However, it should be noted that implementing this type of windbreak would require allocating more land for wind protection, resulting in a reduction of productive area.

Table 12.4 Windbreak designs

Item	Design 1	Design 2	Design 3
Orientation	One-sided	Two-sided	Two-sided + within
Avg. distance between trees (m)	1.3	1.3	1.3
Trees in windbreak	73	137	213
Spacing between windbreak and fruit trees (m)	7	7	7
Fruit trees in orchard	654	592	495
Cost			
Establishment cost (USD)	193	383	574
Annual variable costs (USD)	61	121	182
Pruning costs (every 5 years) (USD)	24	47	52
Effective annual forgone yield (%)	11.79	20.15	33.15

- **Design 3:** Two single perpendicular rows and one parallel
  The influence of wind protection diminishes as the distance from the windbreak increases. In the third design, an extra row of windbreak trees is introduced in the center of the orchard, effectively dividing it into two sections. The parallel row from the previous design is retained. This particular windbreak configuration offers a higher level of wind protection compared to design 2. However, it should be noted that implementing this design would require allocating even more land for wind protection, resulting in a greater reduction in productive area.

In summary, design 1 consists of a single row of windbreaks positioned perpendicular to the prevailing winds. Design 2 provides protection from two directions, and design 3 incorporates an additional row of windbreaks placed within the already protected orchard. In this analysis, all three designs are implemented on a hectare scale, with the windbreaks established concurrently with the orchard. In the results section of this chapter, these three designs will be referred to as D1, D2, and D3.

Opportunity Cost of Foregone Land

Implementing tree-based windbreaks requires dedicating a portion of land within the orchard, resulting in an opportunity cost. The forgone land could have been used for additional mandarin trees or other agricultural activities. The effective annual forgone yield is calculated based on the distance of 7 m from the center of the fruit tree to the center of the windbreak tree, versus a deduction of 2.5 m for agricultural paths which were deducted in the unprotected orchards. The reason for such a distance between fruit trees and windbreak trees is to mitigate competition effects, which may arise between orchards and windbreaks. The assumption is made that the orchard's variable costs are reduced by the same amount as the forgone yield, considering the number of fruit trees in the orchard.

Establishment and Ongoing Maintenance Expenses

Establishment costs for the windbreaks ranged between USD 193 for design 1 and USD 574 for design 3. Ongoing maintenance expenses encompass pruning, irrigation, weed control, and periodic replacement of trees. These costs are influenced by factors like tree growth rate, irrigation requirements, labor availability, and machinery use. All costs in this case studies were included as an uncertainty parameter, with ranges given in Table 12.5.

Table 12.5 Variable costs and uncertainty of windbreaks

Parameter	Year	Distribution	Min	Max
Establishment Variable costs	1	Uniform	0.8	1.2
	2–3		0.8	1.5
	4–20		0.9	1.1

Table 12.6 Agricultural nets

	USD ha^{-1}
Establishment cost	
Net	3,512
Structure	6,799
Machinery	1,293
Labor	3,445
Annual variable cost	
Maintenance	98

Agricultural Nets

As an alternative wind protection option for farmers, fabric netting is included in the model as a counterfactual scenario. The data from Brown (2018) is used to incorporate protective netting into the analysis (Table 12.6). To compare windbreaks with protective netting, the same financial calculations are performed on a per-hectare basis for an orchard covered with permanent closed-sided 20% white shade netting. The establishment costs and maintenance costs are provided below, while the benefits of using nets are discussed in subsequent sections. No opportunity cost of foregone land is included in the netting scenario, as it was assumed that nets do not need additional distance to orchards trees above the 2.5 m for agricultural paths.

The total investment cost for the netting option amounts to USD 15,049 per hectare and an annual variable costs for maintenance purposes of USD 98. The investment cost is relatively high compared to the costs associated with tree-based windbreaks. Additionally, it is assumed that during simulated wind damage events, these nets can also be impacted and damaged, leading to repair costs that are proportional to the severity of yield damage, as described in the corresponding section.

12.2.6 Modeled Protection Benefits of Windbreaks

Production risks in fruit production include sunburn, scarring, windfall, and blemishes (Brown 2018). The advantages of windbreaks will be evident in the gross production value as they contribute to higher total yields (reducing fruit loss from being blown off the tree) and minimize physical damage, leading to increased percentages of high-priced export fruit. In this study, the potential returns on investment in tree-based windbreaks can thus be realized through increased yields and improved fruit quality. The case for increased yields is simple: windbreaks protect orchards from windfall and thus a windbreak-orchard system has higher yield compared to unprotected orchards. The case for fruit quality is more intricate: Producer prices of fruit farming in South Africa generally fall into three categories depending on physical fruit appearance. The highest price category are export prices, which farmers also refer to as pack-out prices. Commercial fruit farmers typically strive for high pack-out rates at harvest, which means, a high proportion of export quality fruit. The two domestic price classes are local fresh fruit prices and local processing prices for low quality fruits.

Prices

The average producer price a farmer can achieve thus depends on the price class ratios the harvested fruit falls into. The average price that a farmer can obtain is dependent on the price class ratios assigned to the harvested fruit. These categories are determined through visual inspections that assess factors such as size and damages. When the fruits meet the optical standards, a higher percentage of the total yield falls into the export grade or Class 1 category. The distribution of fruit across different classes influences the average prices, as assumed in this case study are provided in Table 12.7.

The 2020 producer prices for soft citrus were taken from the national statistics of the Citrus Grower Association of South Africa (Citrus Growers' Association of Southern Africa 2021). They include all soft citrus and do not disaggregate at variety level. Based on farmer and expert opinions, assumptions on average producer price with and without wind protection were made. Furthermore, a range of the lowest and highest prices anticipated for different wind protection options was assumed and triangle distributions for the uncertainty analysis were created. Table 12.8 displays the assumed price classifications for Nadorcott mandarins under open field, three windbreak designs, and side netting.

Table 12.7 Price class distributions

	2020 Prices (16.40 Rand = 1 $)		Price class ratios—assumed packouts				
Price classes optical quality	R/ton	$/ton	Open field	Design 1	Design 2	Design 3	Nets
Class 1	16,387	999	70%	75%	85%	90%	95%
Class 2	6,866	419	20%	15%	10%	5%	4%
Class 3	280	17	10%	10%	5%	5%	1%
		Avg. price ($)	785	814	892	921	966
		Ranges	+ - 10%	+ - 10%	+ - 10%	+ - 10%	+ - 5%

Table 12.8 Expected price ranges and ranges for triangular distributions for Nadorcott mandarins

	Expected prices (USD/t)	Range
Open field	785	0.9 – 1.1
Design 1 (D1)	814	0.9 – 1.1
Design 2 (D2)	892	0.9 – 1.1
Design 3 (D3)	921	0.9 – 1.1
Side netting	966	0.95 – 1.05

The range is expressed as percentage yearly variation in avg. prices. For unprotected orchards and all tree-based designs, ranges of plus or minus 10% are assumed. For side netting, ranges of 5% are assumed.

Windfall

To simulate wind damage events and the resulting yield losses, a combination of discrete and continuous distributions is employed. The occurrence of wind damage events is modeled using a discrete Bernoulli distribution, which assigns values of either zero or one. The baseline scenario assumes severe wind damage events every 5 years.

Table 12.9 displays the modeled windfall, ranging from 0% up to 100% for an unprotected orchard. The assumption was made that windbreak design 1 can experience losses of up to 50%, design 2 up to 30%, and design 3 as well as protective nets have a maximum yield loss of up to 20%. It is assumed that wind damage events do not impact the windbreak trees or nets; hence, no additional maintenance costs are considered.

A year with severe wind damage will affect both the windbreak-orchard system and the open field orchard, albeit with varying levels of fruit losses. To determine the frequency and the severity of yield losses, a discrete distribution is multiplied with a continuous distribution as indicated in the example in Table 12.10.

A discrete distribution which takes values of 0 or 1 (Bernoulli) simulates the year in which a wind damage incident occurs. This is then multiplied with a continuous distributed which simulates the percentage loss of fruit yield, depending on the windbreak design as shown in Table 13.9. While all designs experience the windfall event in the same year, the wind shelter effect is calculated across a 1-hectare windbreak-protected production area and is determined by the severity of the yield loss.

Table 12.9 Windfall as percentage of total yield

Windbreak	Range (%)	Distribution
Open field	0–1	Uniform
Design 1	0–0.5	Uniform
Design 2	0–0.3	Uniform
Design 3	0–0.2	Uniform
Nets	0–0.2	Uniform

Table 12.10 Modeled windfall example for unprotected orchards

	Years					
	1	2	3	4	5	…
Discrete distribution	0	0	1	0	0	0
Continuous distribution	61%	83%	28%	58%	11%	19%

Scenarios on Strong Wind Events Causing Windfall

The baseline scenario assumes the occurrence of severe wind damage events every 5 years. In the scenario analysis, different likelihoods of wind damages were simulated. This includes scenarios where wind damage occurs every year, every third year, and every ten years. Additionally, we will simulate a scenario where there are no windfall events, and the impact of wind protection from windbreaks and nets is only reflected in average prices. These likelihoods are expected to have a significant impact on the feasibility of windbreaks in orchards.

All scenarios involving wind damages to fruit, with and without windbreaks, will be simulated while considering uncertainties at the farm level. Below is a list of all the scenarios that will be tested:

- Baseline: 5-year wind damage events
- Wind damage event every year
- Wind damage event every third year
- Wind damage event every 10th year
- No wind damage events, only price differences

12.3 Results

12.3.1 Baseline Impacts

For the baseline results, a probability of 0.2 is considered for wind damage occurrence, indicating the likelihood of wind damage events happening every 5 years. Yearly undiscounted cash flows, incorporating simulated uncertainty, were calculated based on costs and benefits over time. These cash flows were then discounted to derive the financial indicators presented in Fig. 12.1. The financial indicators of NPV (USD/ha), IRR (%), AGM (USD/ha), and DPBP (years) under modeled uncertainty, comparing an open field scenario to the windbreak designs and nets. Windbreak designs are labeled as D1 to D3, followed by nets and the open field orchard. The discount rate used was set at 12%.

The median and interquartile range (IQR) provide insights into the central tendency and variability of the metrics for different designs. For the open field (OF) unprotected orchard, the median NPV was $91,161, with an IQR of $83,301 to $98,331. The median IRR was 0.49, with an IQR of 0.46 to 0.52. The median DPBP was 4.13 years, with an IQR of 3.71 to 4.38 years. Lastly, the median AGM for the open field design was $12,252, with an IQR of $11,196 to $13,216.

For the windbreak-orchards design system labeled "D1," the median NPV was $92,208, with IQR of $87,539 to $96,586. The median IRR was 0.51, with an IQR of 0.48 to 0.54. The median DPBP was 3.96 years, with an IQR of 3.67 to 4.17 years. Lastly, the median AGM for the D1 design was $12,383, with an IQR of $11,765 to $12,981.

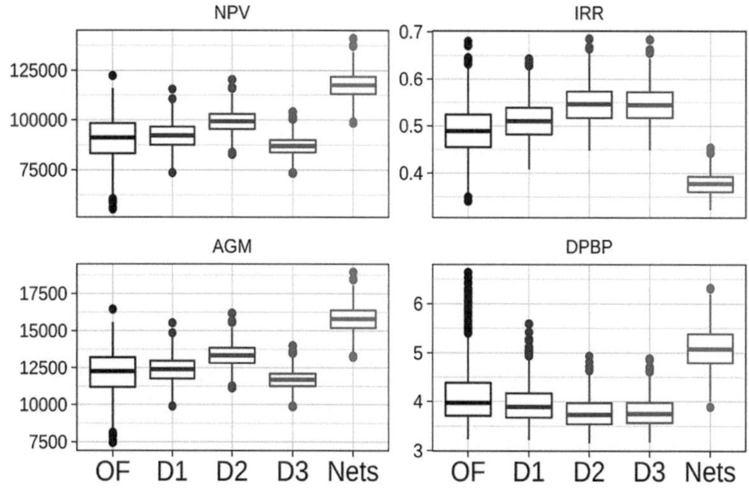

Fig. 12.1 Box plots of financial indicators in the baseline scenario. OF: Open field, D1: Design 1, D2: Design 2, D3: Design 3 (for details see Table 12.4), and nets

For the windbreak-orchards design system labeled "D2," the median NPV was \$99,298, with an IQR of \$95,409 to \$103,069. The median IRR was 0.55, with an IQR of 0.52 to 0.57. The median DPBP was 3.77 years, with an IQR of 3.54 to 3.97 years. Lastly, the median AGM for the D2 design was \$13,332, with an IQR of \$12,823 to \$13,853.

For the windbreak-orchards design system labeled "D3," the median NPV was \$86,975, with an IQR of \$83,778 to \$89,996. The median IRR was 0.54, with an IQR of 0.52 to 0.57. The median DPBP was 3.75 years, with an IQR of 3.57 to 3.98 years. Lastly, the median AGM for the D3 design was \$11,739, with an IQR of \$11,260 to \$12,096.

In the case of the "Nets" for citrus orchard protection, the median NPV was \$117,574, with an IQR of \$4,297, while the median IRR was 0.38, with an IQR of 0.02. The median DPBP was 5.10 years, with an IQR of 0.59, and the median AGM was \$15,804, with an IQR of \$1,206.

These statistics provide an understanding of the central tendencies and ranges of the metrics across different designs, indicating the potential profitability and efficiency of the orchard protection strategies.

12.3.2 Financial Comparisons of Wind Damage Scenarios

To evaluate the financial comparisons of windbreak-orchard systems at different frequencies of windfall events, simulations were conducted following the methods discussed. The financial viability and the practical implementation of windbreaks will rely on the comparative benefits they offer and their influence on investment indicators. The relative change in the AGM serves as a useful metric in assessing the value of a windbreak system under varying assumptions of wind damage frequencies.

Box plots and violin plots are both graphical tools used to depict distributions of numeric data. While violin plots show density curves to represent the data distribution, box plots provide additional information by depicting the quartiles and potential outliers. The width of each density curve in a violin plot indicates the approximate frequency of data points in that region. By combining the two types of plots, a graph can convey both the overall distribution of the data and the key summary statistics. Figure 12.2 presents the relative changes in the agricultural AGM.

In the results provided, the median difference in AGM compared to open field unprotected orchards is reported for different windbreak-orchard designs: D1, D2, D3, and nets. The values represent the relative difference in AGM in dollars per hectare. In the "Baseline" scenario where the likelihood of strong wind damage events with windfall is every five years, the median change in AGM compared to open field

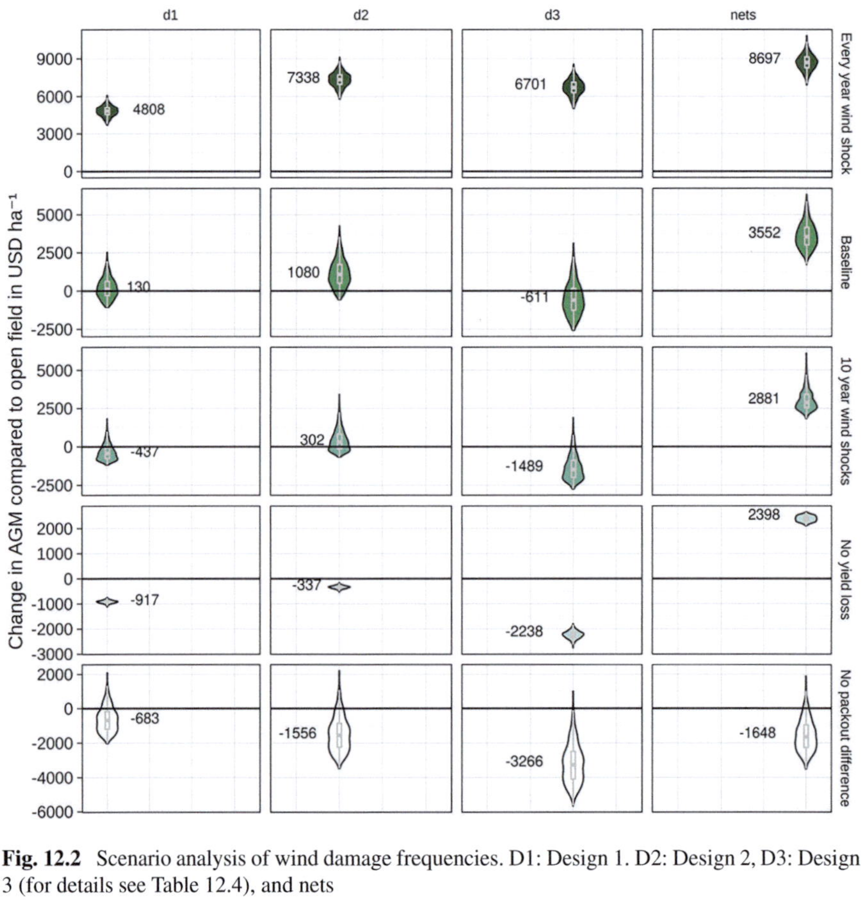

Fig. 12.2 Scenario analysis of wind damage frequencies. D1: Design 1. D2: Design 2, D3: Design 3 (for details see Table 12.4), and nets

unprotected orchards is $130, $1080, -$611, and $3552 for the windbreak-orchard designs D1, D2, D3, and nets, respectively.

When the likelihood of strong wind damage events with windfall is every year, the median changes are as follows: Windbreak design D1 shows a $4,808 higher AGM compared to open fields, D2 is $7,338 higher, D3 is $6,701 higher, and net protected orchards show a median AGM which is $8,697 above open field orchards.

12.4 Discussion

The findings of this study indicate that windbreaks can have a positive impact on the financial performance of citrus orchards in South Africa. When considering the NPV and AGM, windbreak designs 1 and 2, as well as agricultural nets, outperform open field conditions from an investment perspective. The mean per-hectare NPV of the unprotected mandarin orchard is USD 90,392, while the tree-based designs have higher NPVs, with the highest being USD 99,243 over a 20-year period. Nets, however, show even higher NPVs, with a mean per-hectare NPV of USD 117,519 and a mean per-hectare AGM of USD 15,795.

In terms of the payback period, all tree-based systems have a lower DPBP compared to open field orchards, although the differences are small. This is because the investment costs for windbreak trees are offset by the reduction in orchard investment costs due to the utilization of land for windbreaks. It is worth noting that the spread of outcomes for the DPBP is much larger in open field systems due to the potential for complete harvest losses during severe wind damage events.

A notable conclusion from this analysis is that orchards protected by agricultural nets have a higher NPV but a lower IRR or DPBP compared to windbreak trees. The IRR and DPBP are important criteria for farmers' investment decision-making, with some companies in South Africa favoring the IRR over the NPV. The DPBP is particularly suitable in the uncertain economic environment of South Africa as it provides a time period for the investment to amortize. Varying the discount rate while observing the corresponding DPBP allows for quick adjustment and risk analysis of an investment.

The results suggest that agricultural nets have the highest returns but also the highest investment costs, exceeding even the establishment costs of the orchard itself. However, the calculated IRR is well above the current interest rates, indicating that farmers may base their investment decisions on profitability using metrics such as NPV or AGM.

It should be noted that the visual impact of nets on the landscape, especially in the Western Cape Province, where many farms engage in tourism, could be a relevant consideration for farm-level decision-makers (Chap. 14). Farms without tourism activities may prioritize investments in nets due to their higher returns. Agri-environmental premiums and direct payments could be effective policies to incentivize the preservation of trees in productive landscapes.

12.5 Conclusion

This study demonstrates the financial potential of integrating windbreaks into citrus orchards in South Africa. Given the likelihood of increasing wind intensity and frequency in the Western Cape due to climate change, investing in windbreaks can be a viable long-term strategy for protecting orchard productivity.

The choice between tree-based and alternative windbreak systems, such as nets, depends on site-specific conditions, farm management practices, and individual farmer preferences. While net-based systems may offer stronger wind protection, tree-based windbreaks provide additional ecological functions, including habitat provision, soil erosion control, and landscape enhancement—benefits not quantified in this study.

Future research should focus on economically quantifying broader ecosystem services, such as biodiversity enhancement, carbon sequestration, and microclimatic regulation. Potential indirect income sources, including biomass production, carbon credits, and payments for ecosystem services, also warrant further investigation. However, data limitations—particularly regarding Casuarina growth, carbon storage, and irrigation efficiency—restrict a comprehensive economic assessment.

The study's reliance on producer-reported data and assumed wind-damage probabilities presents additional uncertainty. Improved empirical data on wind speeds, damage thresholds, and long-term fruit quality effects would strengthen future analyses. Incorporating temporal variability and uncertainty into financial models will further enhance their relevance for farm-level decision-making.

In summary, while this study primarily assessed windfall reduction as the key benefit of windbreaks, their broader ecological and economic roles highlight their importance for sustainable and resilient agricultural landscapes. Continued interdisciplinary research is needed to fully evaluate and realize these long-term benefits.

Acknowledgements The FarmImpact project was supported by the Federal Ministry of Education and Research (BMBF) and managed by PT-DLR (grant no. FKZ 01LZ1711A-F) as part of the "CLIENT II—International Partnerships for Sustainable Innovations" funding initiative.

References

Ali MM, Yousef AF, Li B, Chen F (2021) Effect of environmental factors on growth and development of fruits. Tropical Plant Biology 14:226–238

Brown R (2018) Effect of permanent shade netting on 'Nadorcott' mandarin tree phenology and productivity (Master's thesis). University of Stellenbosch. http://hdl.handle.net/10019.1/103904

Carauta M, Parussis J, Hampf A, Libera A, Berger T (2021) No more double cropping in Mato Grosso, Brazil? Evaluating the potential impact of climate change on the profitability of farm systems. Agric Syst 190:103104

Cataldo J, Durañona V, Pienika R, Pais PL, Gravina A (2013) Wind damage on citrus fruit study: Wind tunnel tests. J Wind Eng Ind Aerodyn 116:1–6

Citrus Growers' Association of Southern Africa (2021) Industry Statistics 2020 Export Season. Retrieved April 7, 2022, from http://cga.co.za/

Devincentis AJ, Solis SS, Bruno EM, Leavitt AA, Gomes A, Rice S, Zaccaria D (2020) Using cost-benefit analysis to understand adoption of winter cover cropping in California's specialty crop systems. J Environ Manage 261:110205

Declercq R, Loubier S, Condom N, Molle B (2020) Socio-economic interest of treated wastewater reuse in agricultural irrigation and indirect potable water reuse: Clermont-Ferrand and Cannes case studies' cost–benefit analysis. Irrig Drain 69:194–208

Diagne N, Diouf D, Svistoonoff S, Kane A, Noba K, Franche C, Bogusz D, Duponnois R (2013) Casuarina in Africa: Distribution, role and importance of arbuscular mycorrhizal, ectomycorrhizal fungi and Frankia on plant development. J Environ Manage 128:204–209

Hildebrandt P, Knoke T (2011) Investment decisions under uncertainty—A methodological review on forest science studies. Forest Policy Econ 13(1):1–15

Feuerbacher A, Laub M, Högy P, Lippert C, Pataczek L, Schindele SF, Wieck C, Zikeli S (2021) An analytical framework to estimate the economics and adoption potential of dual land-use systems: The case of agrivoltaics. Agric Syst 192:103193

Geldenhuys H, Lötze E, Veste M (2023) Fruit quality and yield of mandarin (Citrus reticulata) in orchards with different windbreaks in the Western Cape, South Africa. Erwerbs-Obstbau 65:959–969. https://doi.org/10.1007/s10341-022-00725-3

Grala RK, Colletti JP (2003) Estimates of additional maize (*Zea mays*) yields required to offset costs of tree-windbreaks in Midwestern USA. Agrofor Syst 59(1):11–20

Gravina A, Cataldo J, Gambetta G, Pardo EA, Fornero C, Galiger S, Pienika R (2011) Relation of peel damage in citrus fruit to wind climate in orchard and its control. Sci Hortic 129:46–51

Griess VC, Uhde B, Ham C, Seifert T (2016) Product diversification in South Africa's commercial timber plantations: a way to mitigate investment risk. Southern Forests 78(2):145–150

Jeong D, Sushama L (2019) Projected changes to mean and extreme surface wind speeds for North America based on regional climate model simulations. Atmosphere 10(9):497

Kruger AC, Shongwe S (2004) Temperature trends in South Africa: 1960–2003. Int J Climatol 24(15):1929–1945. https://doi.org/10.1002/joc.1096

Malherbe W, Biggs R, Sitas N (2024) Comparing apples and pears: linking capitals and capacities to assess the resilience of commercial farming operations. Agricultural Systems, 217, Article 103934, https://doi.org/10.1016/j.agsy.2024.103934

Maroyi V (2011) Capital budgeting practices: A South African perspective (Master's thesis). Wageningen University Department of Social Sciences Management Studies Group

Mutenje MJ, Farnworth CR, Stirling C, Thierfelder C, Mupangwa W, Nyagumbo I (2019) A cost-benefit analysis of climate-smart agriculture options in Southern Africa: Balancing gender and technology. Ecol Econ 163,:126–137

Otto F, Wolski P, Lehner F, Tebaldi C, van Oldenborgh GJ, Hogesteeger S, Singh R, Holden P, Fučkar NS, Odoulami RC, New M (2018) Anthropogenic influence on the drivers of the Western Cape drought 2015–2017. Environ Res Lett 13(12):124010

Schweier J, Becker G (2013) Economics of poplar short rotation coppice plantations on marginal land in Germany. Biomass Bioenerg 59:494–502

Steyn T (2020) Comparing hail risk management strategies through wholefarm multi-period stochastic budgeting for avocado production in South Africa (Master's thesis). University of Stellenbosch

Tarantola S, Becker WE, Zeitz D (2012) A comparison of two sampling methods for global sensitivity analysis. Comput Phys Commun 183:1061–1072

Testa R, Foderà M, Di Trapani AM, Tudisca S, Sgroi F (2016) Giant reed as energy crop for southern Italy: An economic feasibility study. Renew Sustain Energy Rev 58:558–564

Troost C, Berger T (2015) Dealing with uncertainty in agent-based simulation: Farm-level modeling of adaptation to climate change in Southwest Germany. Am J Agr Econ 97(3):833–854

Veste M, Sheppard JP, Abdulai I, Ayisi KK, Borrass L, Chirwa PW, Funk R, Kapinga K, Morhart K, Mwale SE, Ndlovu NP, Nyamadzaw G, Nyoka BI, Sebola P, Seifert T, Senyolo MP, Sileshi GW,

Syampungani S, Kahle H-P (2023) The need for sustainable agricultural land-use systems: benefits from integrated agroforestry systems. In: von Maltitz GP, Midgley GF, Veitch J, Brümmer C, Rötter RP, Viehberg, FA, Veste M (eds) Sustainability of southern African ecosystems under global change: Science for management and policy interventions. Ecological Studies 248, Springer Nature, Heidelberg, pp 587–62

Veste M, Littmann T, Kunneke A, du Toit B, Seifert T (2020) Windbreaks as part of climate-smart landscapes reduce evapotranspiration in vineyards, Western Cape Province, South Africa. Plant Soil Environ 66:119–127

Von Maltitz GP, Midgley GF, Veitch J, Brümmer C, Rötter RP, Viehberg FA, Veste M (eds) (2024) Sustainability of Southern African ecosystems under global change, ecological studies 248, Springer, Cham, https://doi.org/10.1007/978-3-031-10948-5

Wuertz, D, Setz T, Chalabi Y (2017): fOptions: Rmetrics - Pricing and Evaluating Basic Options [R package version 3042.86]

Zanamwe P (2018) The effect of pre-harvest wind damage on post-harvest quality of 'Triumph' persimmon fruit. Afr J Sci Technol Innov Dev 10:753–758

Chapter 13
Yield Impacts and Farm-Level Economics of Tree-Based Windbreaks: Evidence and Research Gaps

Lutz-Heiner Otto and Maik Veste

Abstract This chapter synthesizes contemporary research on the role of windbreaks in enhancing the productivity of annual and perennial crops, emphasizing their influence on adjacent food crop yields and the economic viability of integrating tree-based windbreak systems into agricultural practices. It highlights the complexity of windbreak effects, which are influenced by a multitude of factors, including environmental conditions and windbreak design. Despite the recognized benefits, there is a notable gap in longitudinal studies that document windbreak configurations and their management, underscoring the necessity for comprehensive data collection to better understand their impacts on yield and economic outcomes. The chapter underscores the growing importance of policy support and innovative technological solutions in promoting windbreak restoration and optimization in agricultural landscapes. By providing farmers with decision support tools and leveraging new technologies, the potential for windbreaks to significantly contribute to agricultural productivity and sustainability can be more fully realized. This chapter calls for a systematic approach to windbreak research, advocating for targeted studies that can inform effective windbreak implementation strategies, ultimately enhancing crop yields and providing economic benefits to farmers.

Keywords Sustainable land use · Agricultural landscapes · Climate-smart agriculture · Agroforestry · Agro-ecology · Greening practices · Sustainability · Shelterbelts

L.-H. Otto (✉)
University of Hohenheim Department Land Use Economics in the Tropics and Subtropics, Stuttgart, Germany
e-mail: lh.otto@tutanota.com

Ecovest Landuse Consulting, Windhoek, Namibia

M. Veste
CEBra—Centre for Energy Technology Brandenburg E.V., Cottbus, Germany
e-mail: veste@cebra-cottbus.de

Institute of Biology, Department Molecular Botany, University of Hohenheim, Stuttgart, Germany

M. Veste et al. (eds.), *Trees in South African Agriculture*, Innovations in Landscape Research, https://doi.org/10.1007/978-3-032-00993-7_13

13.1 Introduction

Tree windbreaks and hedges, strategically placed along agricultural field borders, have been a cornerstone of traditional farming practices across various agricultural landscapes for centuries (Schulze et al. 1984, Hansen di Castri, 1992, Leuschner and Ellenberg 2017, Konold 2018). In wind-prone coastal regions, the strategic implementation of windbreaks in from of hedges has proven effective against the progressive wind erosion of sandy soils, thereby enhancing agricultural fields and productivity. Particularly in drylands, the integration of trees and shrubs into windbreak systems plays a pivotal role in combating desertification, curbing soil erosion, and bolstering agricultural production (Breckle et al. 2001; Veste et al. 2006). Beyond their environmental benefits, trees on farmsteads are increasingly recognized for their critical contribution to poverty alleviation and climate change adaptation, underscoring their importance in fostering resilient rural livelihoods (Miller et al., 2021).

The latter part of the twentieth century, however, witnessed a global decline in farmland trees, a trend attributed largely to the push for agricultural intensification (Fischer et al. 2010; Gibbons et al. 2008). This trend has been reversed over the past decade, thanks to a renewed interest in researching and integrating windbreaks within agroforestry systems (Veste and Böhm 2018; Veste et al. 2024), highlighting the critical role of trees in multifunctional agricultural landscapes (Veldkamp et al. 2024, Chap. 14). The recent focus on enhancing biodiversity and sustainability in agricultural production has prompted increased policy efforts aimed at the restoration and propagation of trees as essential components of agroforestry systems (Gibbons et al. 2009; Maron et al. 2010;). In addition to their ecological roles, tree linings serve as a climate-smart, eco-engineering solution designed to adjust the microclimatic boundary conditions conducive to crop production while simultaneously providing valuable wood products. Extensive research has demonstrated the efficacy of tree-based field windbreaks in reducing lee-side wind velocities, thereby offering direct protection to downwind crops from mechanical damage, such as lodging and abrasion (Cleugh et al. 2002; Baker et al., 2018). Despite the proven benefits, the distinction between empirically proven and hypothesized productivity effects of windbreaks on crops necessitates cautious interpretation (Cleugh et al. 1998).

The latter decades of the twentieth century were marked by significant efforts to systematically explore, quantify, and synthesize knowledge regarding the effects of windbreaks on crop production, culminating in seminal reviews and coordinated studies that provided invaluable insights into windbreak, crop, and pasture interactions (Brandle and Hintz 1988a; Baldwin 1988; Kort 1988; Norton, 1988; Cleugh et al. 1998; Nuberg, 1998; Cleugh et al., 2002). Since Nuberg's comprehensive review in 1998, recent literature has predominantly concentrated on ecosystem services, biodiversity, carbon sequestration, farmer perceptions, and the socio-ecological impacts of agroforestry, including windbreaks, signifying a paradigm shift in windbreak research toward sustainability-oriented themes (Miller et al. 2019; Brown et al. 2018; Bentrup et al. 2019; Mayrinck et al. 2019; Brownson et al. 2020; Sheppard et al.

2020; Smith et al. 2021; Weninger et al. 2021; Veldkamp et al. 2023; Veste et al. 2024). Baker et al. (2018) offered a contemporary perspective through a focused review on the quantitative benefits of windbreaks on crops and pasture within Australian conditions, revealing a broad spectrum of tree windbreak effects on crop productivity, with yield effects varying widely from −24% to + 81%. Majaura et al. (2024) analyzed the effect of alley-cropping systems on crop yields in the temperate zone. In this study, the tree rows impact crop productivity in both positive and negative ways. In the competition, crop yield is reduced because of light limitation and reduced soil moisture. Furthermore, crop management contributes also to the yield reduction.

This variability underscores the complexity of windbreak-tree interactions and the challenges inherent in predicting crop growth and modeling economic outcomes based on windbreak implementations.

This chapter aims to provide an updated global overview of windbreak research post-1996, emphasizing studies that offer clear descriptions of windbreak architecture, quantitative yield impacts, and the simulation of economic returns of field windbreaks within agricultural systems. By adopting a comprehensive and flexible approach to examine the current body of knowledge, this review seeks to highlight existing evidence and identify research gaps, thereby contributing to the advancement of windbreak integration within contemporary agricultural practices.

13.2 Background

13.2.1 Primary Functions and Benefits of Windbreaks

Windbreaks play a crucial role in enhancing agricultural productivity, conserving soil, and regulating microclimates, as extensively studied in various research papers. Already Baldwin (1988) highlighted the positive effects of windbreaks on vegetable and specialty crops, noting improvements in crop quality and yield increases of up to 50% due to reduced wind abrasion and earlier ripening. Peri and Bloomberg (2002) reviewed research in southern Patagonia, Argentina, showcasing the differential yield response of crops to wind stress and emphasizing windbreak characteristics such as height, porosity, and orientation in determining efficacy. Additionally, Baker et al. (2018) focused on the quantitative benefits of windbreaks in Australian conditions, identifying key research gaps and the need for accurate predictions regarding impacts on farm productivity. Bird (1998) provided a review on windbreak effects on pastures within temperate grazing systems, noting increased pasture growth and wool production in sheltered conditions. Furthermore, Ucar and Hall (2001) discussed the potential of windbreaks as a pesticide drift mitigation strategy, emphasizing significant reductions in spray drift achieved through their use. Nordstrom and Hotta (2004) highlighted the effectiveness of windbreaks in erosion control, emphasizing their role in reducing soil erosion and dust emissions on croplands in the USA. Lastly, Dix et al. (1995) reviewed the influence of trees on the abundance of natural enemies of insect pests, suggesting that windbreaks can enhance biological pest control efficacy in agroecosystems.

Windbreak function can loosely divide into the following categories:

(I) Wind Reduction: Windbreaks serve as natural barricades, significantly diminishing wind velocity and redirecting airflow. This protective barrier fosters more stable and conducive growing environments for crops, mitigates wind-inflicted structural damage, and enhances the comfort of both humans and animals by buffering against harsh winds.

(II) Erosion Control: By curtailing wind velocity at the surface level, windbreaks play a pivotal role in soil conservation. They prevent the displacement of topsoil, thus combating wind erosion—a critical concern in agricultural zones where soil erosion can significantly impede productivity and lead to ecological harm.

(III) Microclimate Modification: Beyond their role as wind barriers, windbreaks have a profound influence on local microclimates. By modulating wind chill and providing shade, they can moderate temperature extremes throughout the year, leading to improved conditions for crop maturation, extended growing seasons, and overall habitat comfort for a broad range of organisms.

(IV) Biodiversity and Habitat Enhancement: Serving as verdant corridors and refuges within predominantly open landscapes, windbreaks enrich local biodiversity. They support a diverse array of fauna, from avian species to beneficial insects and small mammals, fostering ecological networks essential for pollination, pest control, and the sustenance of various life forms.

(V) Carbon Sequestration: Windbreaks contribute to climate change mitigation through carbon sequestration. The biomass growth within these tree and shrub arrays captures and stores atmospheric CO_2, transforming it into organic carbon—a process that not only combats greenhouse gas accumulation but also enhances the carbon stock within both the vegetation and the soil.

(VI) Soil Health Improvement: The organic matter contributed by the decomposition of leaf litter, fallen branches, and roots from windbreak vegetation acts as a natural soil amendment. This enrichment process improves soil structure, fertility, and water retention, promoting healthier plant growth and contributing to the overall sustainability of agricultural systems.

(VII) Pollination and Pest Control Support: By fostering habitats for a variety of insect species, windbreaks can enhance pollination services for adjacent agricultural fields and orchards. Furthermore, they can contribute to natural pest control by supporting predator and parasitoid species that help manage crop pests.

(VIII) Esthetics and Visual Screening: Windbreaks contribute to the esthetic enhancement of landscapes, offering scenic beauty and a sense of tranquility. They also provide practical benefits such as privacy, noise reduction, and the screening of unsightly views or odors from adjacent roads, industrial sites, or other sources of visual or environmental pollution.

13.2.2 Mechanisms by Which a Tree Windbreak Can Affect Crop Productivity

When designing and implementing tree-based windbreaks for agricultural systems, factors such as tree species selection, spacing, height, and arrangement are considered to optimize their effectiveness. Proper maintenance, including pruning, weed control, and replanting, when necessary, is also crucial for ensuring the long-term success of windbreaks.

Windbreak impacts on crop yields may depend on various exogenous, abiotic, and biotic factors (Fig. 13.1). These factors can be divided into four main abiotic and biotic categories:

(I) Microclimate: The windbreaks are affecting microclimatic conditions directly by, for example, increasing air temperature, relative humidity, radiation, and wind speed. The effects of the wind on evapotranspiration are well investigated and quantified (Baker et al. 2018, Cleugh 1998, Kanzler et al. 2018, Veste et al. 2020; Jacobs et al. 2022). Windbreaks ameliorate the wind effect, i.e., reduce wind speed and the turbulent flux of heat and water between the crop, soil and atmosphere in the sheltered area. Tree strips and hedges are a direct obstacle for the wind, so that the wind speed is directly reduced depending on the height and the porosity. In addition, turbulence of the wind can occur, so that the positive effect of the windbreak is reversed.

(II) Direct shelter: Wind can cause severe damage to fruit trees due to the very high kinetic energy being exerted by wind gusts. Wind damage is the result of either severe irregular wind incidents which lead to uprooting and lodging, or

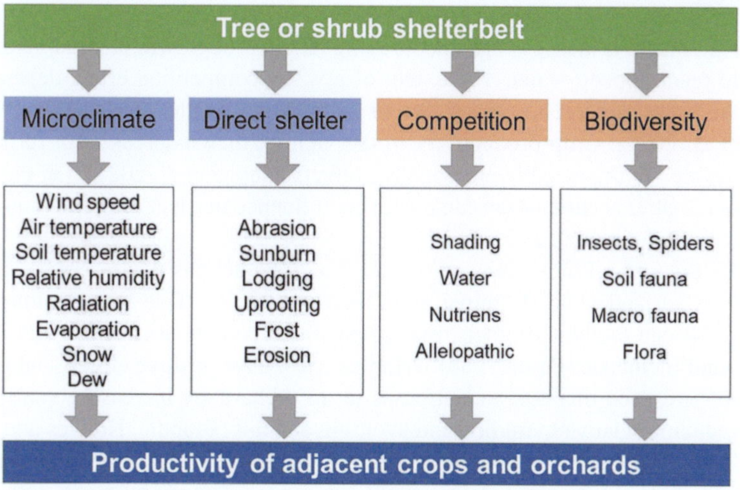

Fig. 13.1 Mechanisms by which a tree windbreak affects abiotic and biotic factors and influences plant productivity

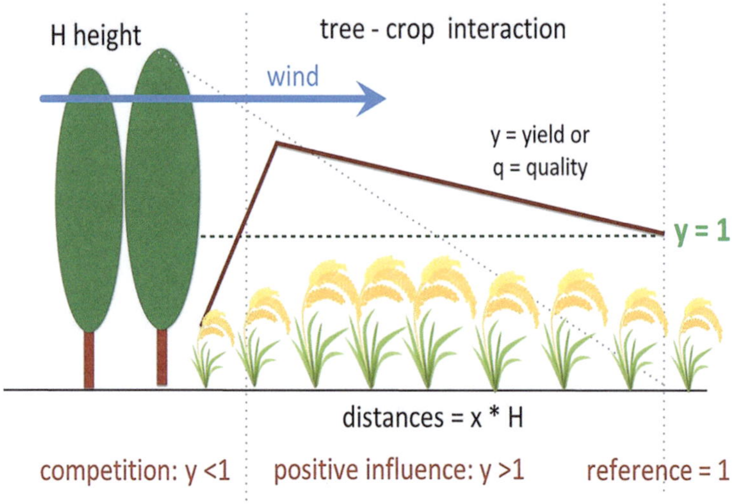

Fig. 13.2 Influence of wind on the tree-crop interactions (yield or quality). Three different zones can be characterized: (I) tree-crop competition with reduced yield, (II) zone of positive influence (increased yield), (III) no influence of wind

persistent winds of intermediate strength. Direct sheltering effects on downwind crops thus include protection from mechanical stress such as lodging and physical abrasion, but also shading and erosion control are direct effects of windbreaks on crop productivity. Furthermore, a reduction in the radiation due to shading close to the trees lowers the temperature and can under certain circumstances increase frost hazard but reduces the risk of sunburn of fruits.

(III) Competition: Nutrient and root water competitiveness can occur when the water table is inaccessible, encouraging lateral root growth of windbreak trees to reach topsoil water. The extent of a water competition effect depends on the ability of trees to reach deeper water levels in the soil profile (Swieter et al. 2021). Crop productivity in the vicinity of windbreaks can further be negatively impacted by allelopathic effects (Zubay et al. 2021). On the other hand, litter accumulation can positively influence nutrient accumulation in the soil (Swieter et al. 2021).

(IV) Biodiversity: Pollination, pests, and fungal pathogens play a critical role in fruit set and yield (Ramirez and Davenport 2016). Tree linings along crop fields can facilitate higher insect populations compared to areas with higher wind incidence (Pasek 1988). Wind can also have positive effects and reduce the presence of pests and diseases in an orchard by increasing ventilation, reducing relative humidity, and avoiding dewfall (Brandle, Hodges and Zhou 2004). On the other hand, the prevailing wind speed affects the flight patterns of insects. Moderately strong winds can reduce damage by insect herbivores by effectively disrupting their flight pattern. In areas with low or no wind, wingless and flying insects tend to assemble more often, and damage is more common (Bentrup et al. 2019).

13.3 Effects of Windbreaks on Food Crops

13.3.1 Tree-Crop Interactions and Feedbacks

Windbreaks have an impact on crop and fruit production. The results show that 32 papers state positive effects, while eight papers show negative effects, and two state no effects on crops. However, this depends on the crop. Table 13.1 gives an overview of papers which had the primary intention of collecting data on windbreaks' effects on crop yields. Windbreak specifics provided by the authors, such as species composition, average tree height, width, orientation, and distance from the windbreak to the hedge are recorded in the column "windbreak design." Windbreak impacts on agricultural productivity stretches to multiples of the tree height; hereafter indicated with a capital H. All analyzed papers mentioned that the influence of trees on crops stretches to multiples of the tree height. This effect is commonly recorded as x H, where x represents a scalar multiplied by the actual tree height (H).

Three different zones can be characterized: (i) tree-crop competition with reduced yield, (ii) zone of positive influence (increased yield), (iii) zone where there is no measurable influence of wind protection to open field conditions anymore. Overall, wind protection and microclimate effects do not always improve crop and fruit productivity. Where losses in the competitive zone outweigh potential benefits, competition management strategies such as root-pruning may mitigate nutrient and water competition. Several studies emphasize the positive effects of reduced wind speed for the measured or modeled evaporation or evapotranspiration and concluded a higher water availability for the crops compared to unprotected fields. However, crop physiology (leaf level), soil water, evapotranspiration, and productivity are complexly intertwined. Field investigations in the Cape Winelands by Veste et al. (2021) showed that the leaf conductance and leaf transpiration was unaffected by windbreaks in citrus trees, however, was lower for the vines located within the windbreak zone than those located away from the windbreak. Quantum efficiency of PSII did not differ between plants located inside or outside of the windbreak zone, both for the citrus and grape vines. Also, the positive effects of wind for heat transfer and cooling needs to be considered, but so far only insufficiently analyzed (Steubing 1968). This is of more relevance under climate change in low wind speed environments to avoid overheating.

Table 13.1 Yield effects of tree windbreaks on annual field crops

Location (Reference)	Crops protected	Tree species in windbreak	Windbreak design	Yield effect
Japan, Hokkaido (Iwasaki et al. 2018)	Maize	*Larix kaempferi Abies sachalinensis*	Four trees in a row, 7m apart, avg. total width of 33 m, avg. height of larchs was 29 and of firs was 16. Porosity given at 78%	Yield decrease in the competitive zone (1 H) and increase in shelter zone (3–5 H)
Germany, Lower Saxony (Swieter et al. 2019)	Winter wheat and winter rapeseed	3 poplar clones *(P. nigra L. × P. maximowiczii, P. maximowiczii × P. trichocarpa, P. koreana × P. trichocarpa)*	North–south oriented alley-cropping system, 10,000 trees/ ha (0.5 m × 2 m) and 1.5m between tree and first crop row, tree height was approximately 5 m in 2015 and 6 m in 2016	Reduced crop yield was observed up to 7 m distance from the tree strip edges, no positive windbreak effect on crop yields could be found
Germany, Brandenburg (Kanzler et al. 2019)	Winter wheat	*Populus nigra L. x P. maximowiczii Henry*	North- south oriented alley-cropping system, four double rows 75cm within rows, and 1.80 m between double rows, tree spacing of 90cm, total of 11 m windbreak. Trees had a height of 3m at start, and grew to 5m during trial	Wheat yields within 3 m distance of the hedgerow were 1% less; yield gains of more than 30% at 3–24 m from windbreak and average crop yield for the entire crop alley was 16% higher compared to the open field
China, Northeast (Zheng et al. 2016)	Maize	Mostly *Populus spp. (*incl. *Populus canadensis, Populus beijingensis, Populus xiaozuanrica, Populus simonii, and Populus pseudo-simonii)*	The average heights of shelterbelts that are < 10, 10–20 and > 20 years of age were approximately 8, 16 and 18 m	In zones with middle/ high climatic potential productivities shelterbelts increased maize yield about 4% and in low potential areas by about 9%

(continued)

Table 13.1 (continued)

Location (Reference)	Crops protected	Tree species in windbreak	Windbreak design	Yield effect
Canada, Quebec (Rivest and Vezina 2015)	Maize	Four windbreak sites: (1) *Picea glauca* (Moench) Voss (2) *Fraxinus pennsylvanica* Marsh. and *Picea glauca* (3) *Pinus sylvestris* L (4) *Larix laricina* (Du Roi) K. Koch and *Picea glauca*	(1) 40 years old, 11 m high, 3.5 m between rows,, north–south orientation while dominant west winds (2) 12 years, 10 m, 2 m, northeast–southwest orientation while dominant west winds (3) 40 years, 13 m, 3.5 m, northeast–southwest while dominant west winds (4) 25 years, 8 m, 1.5 m, northeast–southwest while dominant northwest and west winds	Overall yield for all positions between 0.5 and 20H relative to the 24H position: (1) 6% lower (2) No significant yield effect (3) No significant yield effect (4) 16% higher Across all but one site significant competitive yield decreases were recorded at 0.5 - 1H
Australia, Queensland (Huth et al. 2010)	Wheat and cotton	*Eucalyptus argophloia Blakely*	Four trees in a row, 5 × 5-m spacing, north–south orientation, about 10m height aged 7 years old at start of study	Competition zone to 3.5 H, with high root competition up to 2 H

(continued)

Table 13.1 (continued)

Location (Reference)	Crops protected	Tree species in windbreak	Windbreak design	Yield effect
Italy, Apulia (Campi et al. 2009)	Durum wheat	*Cupressus arizonica* L	Trees were 20 years old, 3 m in height and porosity was estimated equal to 40%	Over a 3 year period, the maximum yield was measured between 2.7H and 4.7H. The area closer than 2.7H to the windbreak was characterized by a production 12% lower than the maximum, while at distances greater than 18H and not influenced by the windbreak any more, production was reduced by 25%
Australia, South (Bennell and Verbyla 2008)	Wheat, barley and faba bean	Predominantly *Pinus halepensis* and some *Eucalyptus spp*	Various sites with reasonably uniform trees, no gaps, height exceeding 5 m, and length of the windbreaks greater than 10 H	Net yield increase of 3.0% for wheat and 19.1% for faba beans and a decrease of 1.1% for barley
Australia, Western and NSW (Oliver et al. 2005)	Wheat, lupins, barley, canola, and peas	32 sites with 8 eucalyptus species (*E. sideroxylon, E. populnea, E. leucoxylon, E. polybractea, Eucalyptus* spp.*, E. microcarpa, E.wandoo, E. camaldulensis*), a pine species (*Pinus radiata*) and tagasaste *(Chamaecytisus proliferus*)	The trees ranged in age from 6 to 16 years and had an outer tree height of 3–23 m. The treebelts had 1–24 rows (median 4 rows) of trees, which were spaced 2–8 m apart (median 3 m) with 2–16 m between the trees within the windbreak	Tree-crop competition was found to be the norm for the 32 sites studied, with an average 7.4% yield loss equivalent to 9.5 m of land lost to production. Yield increases in the sheltered zone were measured at 22 of the sites, but a net yield gain occurred at only 6 sites

(continued)

Table 13.1 (continued)

Location (Reference)	Crops protected	Tree species in windbreak	Windbreak design	Yield effect
Nebraska, USA (Hodges et al. 2004)	Snap beans cultivars 'Strike' and 'Rushmore'	*Fraxinus pennsylvanica L* *Juniperus virginiana L* *Pinus nigra Arnold*	Two rows, avg. height 13–14 m, east–west oriented, with beans planted between 1 and 2H but at least 7.62 m from either end of tree rows	Increase in total and marketable snap bean pod yields by 47% and 50% in a normal rainfall and 64% and 92% in the drier year
Nebraska, USA (Zhang et al. 1999)	Muskmelon	*Fraxinus pennsylvanica L* *Juniperus virginiana L* *Pinus nigra Arnold* *Pinus sylvestris L*	Two rows, avg. height between 11.9 and 12.5 with a width of 8 m and a summer density of 60%	Sheltered plants developed earlier and grew faster, total yields were not significantly affected

13.3.2 Annual Field Crops

Wind protection in agriculture, encompassing buffer strips, shelterbelts, artificial windbreaks, and shade netting, plays a crucial role in enhancing crop productivity through microclimate modification. These measures are integral components of agroforestry systems (Veste et al. 2024) globally, offering a strategic defense against prevailing winds through the strategic placement of tree and shrub rows.

Iwasaki et al. (2018) innovatively employed unmanned aerial vehicles (UAVs) to map the influence of windbreaks, comprising fir and larch trees, on maize yields. Their findings identified a competitive zone where yields decreased within 1 H (height of the windbreak) and a shelter benefit zone, located 3–5 H from the windbreak, where yields increased significantly. The study attributed the yield improvement primarily to the windbreak-induced increase in soil temperature, showcasing the direct impact of microclimatic alterations on crop growth.

In an interdisciplinary cooperation approach, the influence of a temperate alley-cropping systems composed by poplar trees on crop production and environmental factors were investigated under different climate and soil conditions in Brandenburg and Lower Saxony. For the drier site in Brandenburg, Kanzler et al. (2019) demonstrated a positive effect of the shelterbelts on the winter wheat yields with an average of + 16%. Here, the wind reduction leads to a reduced evaporation and increase of the leaf area index in the alleys between the trees. In contrast to the previous investigation Swieter et al. (2019, 2021) showed for Lower Saxony, no winter wheat yield increases compared to the references in open. Even though the trees were north–south oriented, both papers emphasize the reduction of the yields in the competition zone close to the trees, which were influenced by tree height and shading. The calculated land equivalent ratio (LER) at the drier site in Brandenburg is less than 1, whereas in Lower Saxony LER < 1 (Seserman et al. 2019). This variation was likely due to differing regional climate conditions. The studies emphasized the influence of windbreak orientation, tree height, and shading on yield outcomes, particularly noting the reduced yields in competition zones proximal to the trees. A study by Koch et al. (2024) in Germany demonstrates the different influence shelterbelts types (willow short rotation, hedges, and walnuts) on winter crop yields. In contrast to the willow short rotation coppice, walnut and hedgerows did not evoke significant winter crop yield declines in their close proximity. Zheng et al. (2016) utilized remote sensing to assess maize yields across a productivity gradient in Northeast China, discovering that shelterbelts notably enhanced yields in drier, less favorable conditions. This observation aligns with the German studies, highlighting the pronounced benefits of windbreaks in arid areas. Huth et al. (2010) conducted a three-season study in Queensland, Australia, on wheat and cotton adjacent to *Eucalyptus argophloia* windbreaks. They identified a distinct competition zone extending to approximately 3.5 H from the windbreak, within which crop yields significantly declined, illustrating the critical balance between windbreak benefits and competitive stress on adjacent crops. Further research by Bennell and Verbyla (2008) and studies at the University of Nebraska (Hodges et al. 2004; Zhang et al. 1999) have broadened the understanding of windbreak effects, showing varying impacts on different crops ranging from wheat, barley, and faba beans to beans and muskmelons. These studies collectively underscore the nuanced relationship between windbreak presence, microclimatic modification, and agricultural productivity. They highlight the potential for windbreaks to significantly influence crop outcomes, though the degree of impact is contingent upon a multitude of factors including crop type, windbreak design, and environmental conditions.

13.3.3 Fruit Crop Yields and Quality

Underestimated, but of high practical relevance, are the impacts of shelterbelts on fruit quality and fruit damages. However, the methodical review only identified three papers evaluating the impact of tree-based windbreaks on yield in perennial orchards. Fruit trees in zone 0.5–2 H compete with windbreak trees for water, nutrients, and light, and are associated with lower yields and poorer fruit quality (Baldwin 1988; Nuberg 1998). The impact on annual crops has been well documented since 1996 but quantitative data for fruit and nut crops are lacking. The following section gives an overview of the reports on the effect of windbreaks on fruit yield.

Monelos and Peri (1998) reported on the effect of different distances from single-row *Populus nigra* cv. *Italia* windbreaks (25 m height, 40% porosity) on cherry yield in Valdivia, Chile. Significantly higher yields were recorded in the most wind-protected area of the orchard (2H−7H). Peri et al. (2002) conducted studies on the effect of different distances from a single-row poplar windbreak (19 m height, < 15% porosity) on strawberry yield in Santa Cruz, Argentina. Significantly higher yields were recorded in the most wind-protected area of the orchard (1.5H–8H). Furthermore, Peri and Bloomberg (2002) investigated on the effect of wind on cherry yield at different distances from *Populus nigra* cv. *italica* windbreaks in mature, low density (5 × 5 m) orchards in Southern Patagonia, Argentina. In the dense windbreak, significantly higher yields coincided with the point of maximum wind protection (2H). In contrast, in the semi-dense windbreak, peak yield was closer to the windbreak (1.7H) than the point of maximum wind speed reduction (4H). The lowest yields were recorded for both windbreaks at 0.4H and primarily due to root competition and shading by the windbreaks.

An overview of the effects of tree-based windbreaks on the yield of certain fruit crops is given by Norton (1988). However, since 1996, research focused primarily on the effect of wind protection (shelterbelts, artificial windbreaks, and shade netting) on external fruit quality. The demand for high-value, blemish-free fruit crops like citrus, avocadoes, kiwifruit, blueberries, and plums are increasing due to consumers' preference primarily based on external appearance (Stones et al. 2019; Turner and Hardy 2006). This is one of the contributing factors promoting research on fruit quality instead of yield. Important studies by Cataldo (2009) and Gravina et al. (2011) in Uruguay quantified the effects of artificial windbreaks on citrus wind scar damage. There is also an increasing trend for producers to implement shade netting as a tool to optimize fruit production (Stander and Cronjé, 2016) instead of tree windbreaks, with intensive research over the last 10 years being conducted.

In a recent study, conducted in the Cape Winelands (Western Cape, South Africa) single-row *Casuarina cunninghamiana* and *Populus chinensis* windbreaks (9 m and 14 m height respectively) had no significant effect on soft citrus yield and fruit quality during two consecutive seasons (Geldenhuys et al. 2023). However, external fruit quality was significantly affected by the presence of a windbreak, as well as the distance from the windbreak. 'Tango' mandarin (*C. reticulata*) on Carrizo citrange (*C. sinensis* x *Poncirus. trifoliata*) rootstock with an established tree windbreak showed a significantly higher peel wind scar damage percentage (17.7% less export) with progression from the windbreaks. Maximum wind protection and the highest percentage of export quality fruit was observed between 2H–4H from the windbreaks. It was also evident that most of the peel wind scar damage occurred during exposure of fruit during early summer, when fruitlets are the most susceptible to wind damage. Turner and Hardy (2006) reported a significantly higher incidence of severe wind scar damage (37%), over three consecutive seasons, for Washington Navels grown without wind protection compared to an orchard protected by a single-row *Casuarina cunninghamiana* windbreak (10 m height) in South Australia.

Harrison et al. (2013) provided data on 13-year-old satsumas planted behind a windbreak of mainly oak trees and some unspecified underbrush. They concluded that fruit under these windbreaks showed less freeze damage than an unsheltered control orchard. However, no significant yield differences were found between satsumas mandarin growing in a full sun control and under the oak windbreak.

Also, no significant differences in yield of the sheltered and wind-exposed open field grapes cv. Merlot were reported by Pienaar (2005) in a single-row *Casuarina cunninghamiana* windbreaks in the Western Cape production region, South Africa.

13.4 Economic Impacts of Windbreaks at Farm Level

The economic viability of incorporating trees along field margins, known as windbreaks or shelterbelts, hinges on a careful analysis of costs and benefits over time. The costs include not only the initial expenses of establishing these windbreaks—such as the purchase of trees, fuel, herbicides, labor, and irrigation—but also the opportunity costs associated with the land that is no longer available for crop production. On the flip side, the benefits can be substantial, ranging from increased crop yields due to the microclimatic improvements afforded by the windbreaks to reduced irrigation needs and the potential revenue from selling timber or biomass from the trees themselves (Grala 2004). However, the economic outcomes of windbreak systems are not one-size-fits-all. They vary significantly based on location, environmental conditions, and the specific design of the agroforestry system. While models exist for predicting the economic returns of agroforestry systems, their specificity to site and system limits their general applicability, particularly in the context of wind protection (Baker 2018). Table 13.2 presents an overview of published studies in which model windbreaks are independent land management features and not as components of general agroforestry systems. However, most of the studies are related to field crops, while the potential benefits for fruit production is missing, and therefore, a new model for tree-based windbreaks in mandarin orchards was developed to analysis the specific conditions in the Western Cape (Chap. 12).

The economic analysis of windbreak systems is further complicated by their dual role as both agricultural enhancement tools and land management features. For instance, Thevs and Aliev (2021) evaluated the net present values (NPV) of poplar windbreak systems in Kyrgyzstan, revealing that while multi-row systems showed no financial gain, single-row setups could break even or yield minor profits under certain scenarios. This underscores the importance of model selection and assumptions in determining the financial viability of windbreaks. Similarly, Sudmeyer et al. (2012) highlighted the significant opportunity costs associated with tree-crop competition in Eucalyptus-based systems, suggesting that strategic management practices such as wide spacing between trees and root-pruning can enhance profitability. A practical approach to assessing the economic benefits of windbreaks involves calculating the additional yield required to offset the costs associated with the land dedicated to the windbreaks. Grala and Colletti (2003) found that while all scenarios necessitated increased yields for cost recovery, the specifics of windbreak management—such as the lifespan of the trees and the intensity of maintenance—played a crucial role in determining the break-even point. Advanced modeling, such as the Agricultural Production Systems Simulator (APSIM) used in studies by Huth et al. (2010) and Carberry et al. (2002), provides insights into the nuanced effects of windbreaks on microclimate and crop growth. These studies suggest that while windbreaks may not always fully compensate for the reduced cropping area, the yield improvements for certain crops can offset a significant portion of the opportunity costs. Jones and Sudmeyer (2002) employed a partial budget approach to evaluate the economic trade-offs of Pinus pinaster and Eucalyptus globulus windbreaks, illustrating how

Table 13.2 Overview of modeling approaches of economic benefits of tree windbreaks in crop and fruit production systems

Location (Reference)	Windbreak design and assumptions on tree-crop management	Economic model and results
Australia (Sudmeyer et al. 2012)	Three Eucalyptus species (mallee) at 15 trial sites on private farms protecting various crop (wheat, oats, barley, canola, lupins) and pasture rotations Windbreak trees were at least 5 years old, planted in linear belts with at least 48 m between belts	Gross margin analysis was used to calculate break-even distance (where costs of production were equal to returns) and opportunity-cost distance (width of the area alongside the belts over which agricultural income was forgone) Averaged across all site-years, the opportunity cost of competition was equivalent to forgoing agricultural production of about 14 m The mean break-even distance for unharvested mallees over 6 years was 7.5 m Root-pruning of windbreaks further decreased the opportunity cost
Australia (Sudmeyer and Flugge 2005)	Mainly Eucalypt trees as windbreaks in alley systems in which annual cereals, canola and lupins are grown Tree growth and crop response for pruning and coppicing recorded from 3 year experimental trial at 10 sites Effect of root-pruning on annual equivalent return($/ha) from sole-crop and windbreak systems with and without root-pruning affecting tree growth, and with and without damaging wind events thinning and high-pruning root–pruning with ripper and polyethylene root barrie 110-m-wide alleys between mallee hedges or sole-crops, with varying numbers of years with wind damage	The windbreak model by Jones and Sudmeyer was used to calculate the cumulative agricultural production and net income from land with windbreaks as percentages from the same area without windbreaks. The model simulates the effect of root-pruning and/or severe wind damage on crop yield(where severe damage was set as open yield being reduced by 80%).The NPV and AER were calculated for different scenarios With only 3 sever damaging wind events over the 35-year period, windbreak system show a higher AER even if tree growth is reduced by root-pruning Overall the NPV of crops behind windbreaks was greater if the trees were root-pruned

(continued)

Table 13.2 (continued)

Location (Reference)	Windbreak design and assumptions on tree-crop management	Economic model and results
Australia (Huth et al. 2002)	*Eucalyptus argophloia* as windbreak on the edge of a 1 ha hypothetical field where winter chickpea is grown over a 30-year period 5/5 m spacing resulting in 400 trees planted in four rows The trees were sown at the beginning of this simulation in 1955 and they reached a height of approximately 25 m by the end of the simulation in 1984	APSIM simulation framework Cumulative cash flows for the windbreak system do not catch up to the benchmark system until the final year of the investment period when trees are sold No years for which cash flows for the agroforestry system are higher than for the benchmark Increases in production due to crop sheltering are not high enough to account for the reduced cropping area and reduced yield
Western Australia (Jones and Sudmeyer 2002)	*Pinus pinaster* and *Eucalyptus globulus* modeled as if planted in year 0 until year 35 based as windbreaks on continuous cropping lupin–cereal–canola rotation with substantial variability in land management practice Management strategies such as varying the distance between windbreaks, root pruning to modify the competition effect and commercial harvesting of the windbreak are included	A benefit: cost analysis, using a partial budget approach, compared situations with and without windbreaks allowing crop yield to be affected by specified wind damage events in a 36-year period from initial planting of the tree belts If there is no wind damage, windbreaks can cause reductions in NPV of up to 6.5% over 36 years (income loss amounted to a total of $187/ha for continuous cropping in an area protected by P. pinaster windbreaks without root pruning Root pruning along the outside rows of windbreaks reduces economic losses, however still totaling to an annual loss of $9/ha if no wind damages occur In areas subject to frequent strong winds establishing windbreaks would most likely achieve net financial benefits compared to taking losses associated with wind damage Sites with a high risk from damage would benefit from spacing of between 20 and 25 H, while 30–35 H is suited to less susceptible areas Where windbreaks provide an income from timber, the frequency of wind damage necessary for benefits to exceed costs can be reduced

(continued)

Table 13.2 (continued)

Location (Reference)	Windbreak design and assumptions on tree-crop management	Economic model and results
Australia (Carberry et al. 2002)	APSIM simulated yields of maize, wheat and mungbean for transects behind hypothetical windbreaks and different wind scenarios 10 m high windbreak and 100 ha paddock and assuming a 20% yield loss due to tree competition in the 1.0–3.5 H section Ignored the establishment costs of the windbreak, the costs and returns from production of saleable wood, the aggregated benefits of windbreaks over a sequence of years (as opposed to a response on average), the influence of frequent wind-damaging events	Gross margins averaged across all sites and crops, resulting in an average loss of –$0.60/ha.crop Gross Margins of crop-windbreak systems were sensitive to paddock and windbreaks size as well as root pruning Yield benefits from microclimate changes can at least partly offset the opportunity costs of positioning tree windbreaks on farms

(continued)

Table 13.2 (continued)

Location (Reference)	Windbreak design and assumptions on tree-crop management	Economic model and results
Kyrgyzstan (Thevs and Aliev 2021)	Poplar trees combined with wheat, barley, corn, alfalfa, cotton, and rice Hypothetical square shaped plots with single row or multiple rows of windbreaks at all for sides A zone with zero crop yield behind windbreak was assumed Costs of crop cultivation assumed to be equal in windbreak and open field systems	NPV open field versus protected- tree growth and resulting impact on crop yields were modeled after Alemu (2016) Multiple-row systems did not show financial gains, while single row windbreaks were cost-neutral or showed small gains under various yield impact and discount rate scenarios 200 × 200 m windbreak spacing showed the highest financial gains
USA (Grala and Colletti 2003)	Four windbreak scenarios, harvested at 5 lifespans (10 to 50 years) with two management scenarios: intensive and extensive Poplar trees planted in four rows, windbreak is 12.19 m wide, 2691 trees ha^{-1} Mixed windbreak planted in three rows; one row of Poplar, one of silver spruce and one row of mixed shrubs; width of the windbreak is 7.31 m; 1794 trees ha^{-1} Spruce planted in two rows; width of the windbreak is 9.14 m; 717 trees ha^{-1} Spruce planted in four rows; width of this windbreak is 12.19 m; 1077 trees ha^{-1}	Break-even model: annual required additional crop yield that would have to be obtained every year during windbreak lifespan in order to offset windbreak costs Break-even vary significantly across windbreak scenarios, lifespans and lengths of the protected zone Smallest additional maize yields to break even is a mixed windbreak planted with extensive management at low costs for a 50-year lifespan In contrast, a 4-row spruce windbreak planted with in-tensive management at high cost for a 10-year lifespan is the worst scenario because it requires the greatest additional maize yields

windbreaks can become economically beneficial when they mitigate the impacts of wind damage on crop yields. The decision to implement windbreak systems in agricultural practices involves a comprehensive evaluation of economic trade-offs. By considering the potential for increased crop yields, reduced resource inputs, and additional revenue streams, farmers can make informed decisions about integrating windbreaks into their land management strategies. Future research should continue to refine economic models and consider the long-term environmental benefits of windbreaks, such as carbon sequestration and biodiversity enhancement, which may further influence their profitability and sustainability.

13.5 Conclusion

As the agricultural sector increasingly recognizes the multifaceted benefits of integrating trees into farming landscapes —ranging from ecological and climate advantages to public health improvements—the role of windbreaks in enhancing agricultural sustainability and productivity warrants further emphasis (Chap. 14). The exploration of tree-based wind protection in agriculture, despite its long history as a subject of academic inquiry, reveals notable gaps on design for windbreak efficiency and yield and finally on financial outcomes (Table 13.3).

Table 13.3 Research gaps on farm-level impacts of tree windbreaks in crop and fruit production systems

Topic	Gaps	Recommendations
Yield impacts	• Windbreak design—Better understanding of the required physical and compositional characteristics for windbreaks • Fruit yield data missing • Accessability of data • Water use	• Systematically gather primary research data in an international database • Studies on impacts of tree-based wind protection on quality of major fruit crops
Climate change	• Optimization of wind fields under climate change • for optimal heat transfer / cooling for crops • Water use efficienncy of selected tree species	• Experiments and modelling approaches for tree-crop interactions • Crop ecophysilogy
Agroecosystems	• Biodiversity • Multifunctional landscapes	• Further field experiments • Integration in modelling approches
Farm-level economics	• Biophysical modeling • Aggregated benefits—as opposed to a response on average water, soil erosion, in-field shelterbelts also provide other valuable benefits • Availabilty of data to farmers	• Database • Multi-benefit • Climate

The potential of windbreaks to influence agricultural yields and economic returns highlights the need for policies and incentives that encourage their adoption, particularly in scenarios where their presence might not immediately translate to positive yield impacts. Providing financial incentives or compensation could prove pivotal in maintaining and expanding the use of trees in agricultural settings, thereby ensuring resilience and sustainability. A deeper understanding of the specific conditions under which windbreaks thrive—including climate variables, windbreak architecture, and optimal placement—is crucial for maximizing their benefits. Detailed research into the effects of windbreak parameters such as spacing, height, porosity, and species composition, as well as local climatic conditions, is essential for crafting effective wind protection strategies. This knowledge would empower stakeholders to design windbreak systems tailored to their unique environmental and agricultural contexts.

The development of a comprehensive knowledge base on windbreak-crop interactions, accentuated by context- and species-specific insights, could significantly enhance the planning and management of diverse agricultural landscapes. Moreover, the advent of new technologies and methodological innovations offers promising avenues to mitigate the financial and logistical challenges associated with assessing windbreak impacts, thereby enabling more accurate predictions and optimizations of their contributions to agricultural productivity.

Ultimately, advancing our understanding and application of windbreak systems in agriculture necessitates a concerted effort among researchers, policymakers, and practitioners. By prioritizing the integration of ecological principles with agricultural practices, we can move toward more resilient and sustainable farming systems that not only yield economic benefits but also contribute to the health of our planet and its inhabitants.

Acknowledgements FarmImpact project was supported by the Federal Ministry of Education and Research (BMBF) and managed by PT-DLR (grant no. FKZ 01LZ1711A-F) as part of "CLIENT II - International Partnerships for Sustainable Innovations" funding initiative.

References

Alam M, Olivier A, Paquette A, Dupras J, Revéret J-P, Messier C (2014) A general framework for the quantification and valuation of ecosystem services of tree-based intercropping systems. Agrofor Syst 88(5):679–691. https://doi.org/10.1007/s10457-014-9681-x

Baker TP, Moroni MT, Mendham DS, Smith R, Hunt MA (2018) Impacts of windbreak shelter on crop and livestock production. Crop Pasture Sci 69:785. https://doi.org/10.1071/cp17242

Baldwin CS (1988) The influence of field windbreaks on vegetable and specialty crops. Agric Ecosyst Environ 22(23):191–203

Bennell MR, Verbyla AP (2008) Quantifying the response of crops to shelter in the agricultural regions of South Australia. Aust J Agric Res 59:950. https://doi.org/10.1071/ar08188

Bentrup G, Hopwood J, Adamson NL, Vaughan M (2019) Temperate agroforestry systems and Insect pollinators: A Review. Forests 10(11):981. https://doi.org/10.3390/f10110981

Blanco J, Sourdril A, Deconchat M, Barnaud C, San Cristobal M, Andrieu E (2020) How farmers feel about trees: perceptions of ecosystem services and disservices associated with rural forests in southwestern France. Ecosyst Serv 42:101066. https://doi.org/10.1016/j.ecoser.2020.101066
Botina A, B.L., García M, MC, Romero BY (2019) Pre- and post-harvest factors that affect the quality and commercialization of the Tahiti lime. Scientia Horticulturae 257: 108737. https://doi.org/10.1016/j.scienta.2019.108737
Brandle J, Hodges L, Zhou X (2004) Windbreaks in North American agricultural systems. Agrofor Syst 61:65–78
Brandle JR, Hintz DL (1988a) Windbreaks for the future. Agric Ecosyst Environ 22:593–596
Brandle JR, Hintz DL (1988b) In: Windbreak technology: proceedings of an international symposium on windbreak technology, Lincoln, Nebraska, June 23–27, 1986. Elsevier Publishing Company. ISBN 0444430199
Brandle JR, Takle E, Zhou E (2022) Windbreak practices. In: Garrett HE, Jose S, Gold MA (eds) North American Agroforestry. American Society of Agronomy, Wiley, pp 91–125
Breckle S-W, Veste M, Wucherer W (2001) Sustainable Land Use in Deserts. Springer, Berlin Heidelberg
Bird PR (1998) Tree windbreaks and shelter benefits to pasture in temperate grazing systems. Agrofor Syst 41(1):35–54
Brown SE, Miller DC, Ordonez PJ, Baylis K (2018) Evidence for the impacts of agroforestry on agricultural productivity, ecosystem services, and human well-being in high-income countries: a systematic map protocol. Environmental Evidence 7:14. https://doi.org/10.1186/s13750-018-0136-0
Brownson K, Cox C, Padgett-Vasquez S (2020) The impacts of agricultural windbreaks on avian communities and ecosystem services provisioning in the bellbird biological corridor, costa rica. Agroecol Sustain Food Syst 45:592–629. https://doi.org/10.1080/21683565.2020.1838029
Campi P, Palumbo AD, Mastrorilli M (2009) Effects of tree windbreak on microclimate and wheat productivity in a Mediterranean environment. Eur J Agron 30:220–227. https://doi.org/10.1016/j.eja.2008.10.004
Carberry PS, Meinke H, Poulton PL, Hargreaves JNG, Snell AJ, Sudmeyer RA (2002) Modelling crop growth and yield under the environmental changes induced by windbreaks. 2. Simulation of potential benefits at selected sites in Australia. Aust J Exp Agric 42:887–900
Cataldo J, Pienika R (2009) Wind damage study and control on citric fruits. America Conference on Wind Engineering 11:1–11
Cataldo J, Durañona V, Pienika R, Pais P, Gravina A (2013) Wind damage on citrus fruit study: wind tunnel tests. J Wind Eng Ind Aerodyn 116:1–6. https://doi.org/10.1016/j.jweia.2001.008
Cleugh HA (1998) Effects of windbreaks on airflow, microclimates and crop yields. Agrofor Syst 41:55–84. https://doi.org/10.1023/A:1006019805109
Cleugh HA, Prinsley R, Bird PR, Brooks SJ, Carberry PS, Crawford MC, Jackson TT, Meinke H, Mylius SJ, Nuberg IK et al (2002) The Australian national windbreaks program: overview and summary of results. Aust J Exp Agric 42:649. https://doi.org/10.1071/ea02003
Connor DJ, Gómez-del-Campo M, Rousseaux MC, Searles PS (2014) Structure, management and productivity of hedgerow olive orchards: a review. Sci Hortic 169:71–93. https://doi.org/10.1016/j.scienta.2014.02.010
Dix ME, Johnson RJ, Harrell MO, Case RM, Wright RJ, Hodges L, Brandle JR, Schoeneberge MM, Sunderman NJ, Fitzmaurice RL, Young LJ, Hubbard KG (1995) Influences of trees on abundance of natural enemies of insect pests: a review. Agrofor Syst 29(3):303–331
Eugster W (2008) Wind effects. Encyclopedia of Ecology, Elsevier., pp 391–400
Fischer J, Zerger A, Gibbons P Stott J, Law BS (2010). Tree decline and the future of Australian farmland biodiversity. In: Proceedings of the national academy of sciences 107: 19597–19602. https://doi.org/10.1073/pnas.1008476107
Gardiner B, Berry P, Moulia B (2016) Review: Wind impacts on plant growth, mechanics and damage. Plant Sci 245:94–118. https://doi.org/10.1016/j.plantsci.2016.01.006

Geldenhuys H, Lötze E, Veste M (2023) Fruit quality and yield of mandarin (Citrus reticulata) in orchards with different windbreaks in the Western Cape, South Africa. Erwerbs-Obstbau 65:959–969. https://doi.org/10.1007/s10341-022-00725-3

Gibbons P, Briggs SV, Ayers D, Seddon J, Doyle S, Cosier P, McElhinny C, Pelly V, Roberts K (2009) An operational method to assess impacts of land clearing on terrestrial biodiversity. Ecol Ind 9:26–40. https://doi.org/10.1016/j.ecolind.2008.01.006

Gibbons P, Lindenmayer DB, Fischer J, Manning AD, Weinberg A, Seddon J, Ryan P, Barrett G (2008) The future of scattered trees in agricultural landscapes. Conserv Biol 22:1309–1319. https://doi.org/10.1111/j.1523-1739.2008.00997.x

Grala R (2004) An evaluation of the benefits and costs of in-field shelterbelts in Midwestern USA

Grala R, Colletti J (2003) Estimates of additional Maize (*Zea mays*) yields required to offset costs of tree-windbreaks in Midwestern USA. Agrofor Syst 59:11–20

Gravina A, Cataldo J, Gambetta G, Pardo E, Fornero C, Galiger S, Pienika R (2011) Relation of peel damage in citrus fruit to wind climate in orchard and its control. Sci Hortic 129:46–51. https://doi.org/10.1016/j.scienta.2011.03.002

Hansen, AJ, di Castri, F (eds.) (1992) Landscape boundaries, consequences for biotic diversity and ecological flows. Ecological Studies 92, Springer-Verlag, Heidelberg, https://doi.org/10.1007/978-1-4612-2804-2

Harrison MR, Spiers JD, Coneva ED, Dozier W, Woods FM (2013) Orchard design influences fruit quality, canopy temperature, and yield of satsuma mandarin (Citrus unshiu 'Owari'). Int J Fruit Sci 13:334–344. https://doi.org/10.1080/15538362.20748376

Hodges L, Suratman MN, Brandle, JR, Hubbard KG (2004) Growth and yield of snap beans as affected by wind protection and microclimate changes due to shelterbelts and planting dates. HortScience 39: 996–1004. https://doi.org/10.21273/hortsci.39.5.996

Huth NI, Robertson MJ, Poulton PL (2010) Regional differences in tree—crop competition due to soil, climate and management. Crop Pasture Sci 61:763–770. https://doi.org/10.1071/cp09254

Huth NI, Carberry PS, Poulton PL, Brennan LE, Keating BA (2002) A framework for simulating agroforestry options for the low rainfall areas of Australia using APSIM. Eur J Agron 18(1–2):171–185

Iwasaki K, Torita H, Abe T, Uraike T, Touze M, Fukuchi M, Sato H, Iijima T, Imaoka K, Igawa H (2018) Spatial pattern of windbreak effects on maize growth evaluated by an unmanned aerial vehicle in Hokkaido, Northern Japan. Agrofor Syst 93:1133–1145. https://doi.org/10.1007/s10457-018-0217-7

Jacobs SR, Webber H, Niether W, Grahmann K, Lüttschwager D, Schwartz C, Breuer L, Bellingrath-Kimura SD (2022) Modification of the microclimate and water balance through the integration of trees into temperate cropping systems. Agric for Meteorol 323:109065

Jones HR, Sudmeyer RA (2002) Economic assessment of windbreaks on the south-eastern coast of Western Australia. Aust J Exp Agric 42(6):751–761

Kanzler M, Böhm C, Mirck J, Schmitt D, Veste M (2019) Microclimate effects on evaporation and winter wheat (*Triticum aestivum* L.) yield within a temperate agroforestry system. Agrofor Syst 93:1821–1841. https://doi.org/10.1007/s10457-018-0289-4

Konold W (2018) Feldholzanbau im kurzen Umtrieb: ein Rückblick in die Geschichte. In: Veste M, Böhm C (eds) Agrarholz—Schnellwachsende Bäume in der Landwirtschaft. Springer, Heidelberg, pp 17–55

Kort J (1988) Benefits of Windbreaks to field and forage crops. Agr Ecosyst Environ 22–23:165–190. https://doi.org/10.1016/0167-8809(88)90017-5

Koch O, Moore J, Hörl J, Cormann M, Gayler S Lewandowski I, Marhan S, Munz S Pflugfelder M, Piepho H-P, Schneider J, von Cossel M Weinand T, Winkler B, Schweiger AH (2024) Sheltered by trees—long-term yield dynamics in temperate alley cropping agroforestry with changing water availability, 10 September 2024, PREPRINT (Version 1) available at Research Square [https://doi.org/10.21203/rs.3.rs-4861911/v1]

Lado J, Gambetta G, Zacarias L (2018) Key determinants of citrus fruit quality: Metabolites and main changes during maturation. Sci Hortic 233:238–248. https://doi.org/10.1016/j.scienta.2018.01.055

Leuschner C, Ellenberg H. (2017) Forest Edges, Scrub, Hedges and Their Herb Communities. In: Ecology of Central European Forests. Springer, Cham. pp. 747–774 https://doi.org/10.1007/978-3-319-43042-3_11

Majaura M, Böhm C, Freese D (2024) The influence of trees on crop yields in temperate zone alley cropping systems: A Review. Sustainability 16(8):3301. https://doi.org/10.3390/su16083301

Maron M, Dunn PK, McAlpine CA, Apan A (2010) Can offsets really compensate for habitat removal? The case of the endangered Red-tailed Black-Cockatoo. J Appl Ecol 47:348–355. https://doi.org/10.1111/j.1365-2664.2010.01787.x

Mayrinck RC, Laroque CP, Amichev BY, Van Rees K (2019) Above- and below-ground carbon sequestration in shelterbelt trees in Canada: A Review. Forests 10:922. https://doi.org/10.3390/f10100922

Mditshwa A, Magwaza LS, Tesfay SZ (2019) Shade netting on subtropical fruit: effect on environmental conditions, tree physiology and fruit quality. Sci Hortic 256:108556. https://doi.org/10.1016/j.scienta.2019.108556

Miller DC, Mansourian S, Gabay M, Hajjar R, Jagger P, Kamoto JFM, Newton P, Oldekop JA, Razafindratsima OH, Shyamsundar P et al (2021) Forests, trees and poverty alleviation: policy implications of current knowledge. Forest Policy Econ 131:102566. https://doi.org/10.1016/j.forpol.2021.102566

Miller DC, Ordoñez PJ, Brown SE, Forrest S, Nava NJ, Hughes K, Baylis K (2019) The impacts of agroforestry on agricultural productivity, ecosystem services, and human well-being in low-and middle-income countries: An evidence and gap map. Campbell Syst Rev 16:e1066. https://doi.org/10.1002/cl2.1066

Monelos L, Peri P.L. (1998) Incidencia del efecto protector de las cortinas cortaviento en la producción de cerezas (Prunus avium var Bing) en Patagonia Sur. In: Proceeding Primer Congreso Latinoamericano de IUFRO, 22–28 November 1998. Tema 2 (10): Sistemas Integrados de Producción y Desarrollo Rural. International Union of Forest Research Organisations (IUFRO), Valdivia, Chile

Nordstrom KF, Hotta S (2004) Wind erosion from cropland in the USA: a review of problems, solutions and prospects. Geoderma 121(3–4):157–167

Norton R (1988) Windbreaks: Benefits to orchard and vineyard crops. Agr Ecosyst Environ 22–23:205–212

Nuberg IK (1998) Effect of shelter on temperate crops: A review to define research for Australian conditions. Agrofor Syst 41:3–34. https://doi.org/10.1023/a:1006071821948

Oliver YM, Lefroy EC, Stirzaker R, Davies CL (2005) Deep-drainage control and yield: the trade-off between trees and crops in agroforestry systems in the medium to low rainfall areas of Australia. Aust J Agric Res 56:1011–1026

Pasek JE (1988) Influence of wind and windbreaks on local dispersal of insects. Agr Ecosyst Environ 22–23:539–554. https://doi.org/10.1016/0167-8809(88)90044-8

Peri L, Bloomberg M (2002) Windbreaks in southern Patagonia, Argentina: a review of research on growth models, windspeed reduction, and effects on crops. Agrofor Syst 56:129–144

Pienaar JW (2005) The effect of wind on the performance of the grapevine. Master's thesis, stellenbosch university, department of forestry sciences. Retrieved from https://scholar.sun.ac.za/handle/10019.1/50486

Pirkner M, Tanny J, Shapira O, Teitel M, Cohen S, Shahak Y, Israeli Y (2014) The effect of screen type on crop microclimate, reference evapotranspiration and yield of a screenhouse banana plantation. Sci Hortic 180:32–39. https://doi.org/10.1016/j.scienta.2014.09.050

Plante P-M, Rivest D, Vézina A, Vanasse A (2014) Root distribution of different mature tree species growing on contrasting textured soils in temperate windbreaks. Plant Soil 380:429–439. https://doi.org/10.1007/s11104-014-2108-7

Ramírez F, Davenport TL (2016) Mango (*Mangifera indica* L.) pollination: A Review. Sci Hortic 203:158–168. https://doi.org/10.1016/j.scienta.2016.03.011
Rhee JA van (1958) The cropping of fruit trees in relation to windbreak protection. Neth J Agric Sci 6: 11–17. https://doi.org/10.18174/njas.v6i1.17719
Rivest D, Vézina A (2015) Maize yield patterns on the leeward side of tree windbreaks are site-specific and depend on rainfall conditions in eastern Canada. Agroforestry System 89:237–246
Sheppard JP, Bohn Reckziegel R, Borrass L, Chirwa PW, Cuaranhua CJ, Hassler SK, Hoffmeister S, Kestel F, Maier R, Mälicke M, Morhart C, Ndlovu NP, Veste M, Funk R, Lang F, Seifert T, du Toit B, Kahle H-P (2020) Agroforestry: an appropriate and sustainable response to a changing climate in southern Africa? Sustainability 12:6796. https://doi.org/10.3390/su12176796
Seserman DM, Veste M, Freese D, Swieter A, Langhof M (2018) Benefits of agroforestry systems for land equivalent ratio—case studies. In: Brandenburg and lower saxony, germany. 4th european agroforestry conferennce—agroforestry as sustainable land use. Proceedings 26–29
Seserman DM, Freese D, Swieter A, Langhof M, Veste M (2019) Trade-off between energy wood and grain production in temperate alley-cropping systems: An empirical and simulation-based derivation of land equivalent ratio. Agriculture 9:147. https://doi.org/10.3390/agriculture9070147
Schulze E-D, Reif A, Küppers M (1984) Die pflanzenökologische Bedeutung und Bewertung von Hecken. Akademie für Naturschutz und Landschaftspflege, Laufen/Salzach
Smith DM, Jarvis PG, Odongo JCW (1998) Management of windbreaks in the Sahel: The strategic implications of tree water use. Agrofor Syst 40:83–96. https://doi.org/10.1023/a:1005962813999
Smith MM, Bentrup G, Kellerman T, MacFarland K, Straight R, Ameyaw L (2021) Windbreaks in the United States: A systematic review of producer-reported benefits, challenges, management activities and drivers of adoption. Agric Syst 187:103032. https://doi.org/10.1016/j.agsy.2020.103032
Stander OPJ, Cronjé P (2016) Important considerations for citrus production under shade nets. Tegnologie CRI (Citrus Research International), 64–67
Stones W, Taylor N, Van Rooyen Z, Kohne J (2019) Growing avocados under shade netting: A Review. Department of Plant and Soil Sciences, University of Pretoria, 1–11
Sudmeyer R, Flugge F (2005) The economics of managing tree–crop competition in windbreak and alley systems. Aust J Exp Agric 45(11):1403–1414.
Sudmeyer RA, Daniels TL, Jones HR, Huxtable D (2012) The extent and cost of mallee–crop competition in unharvested carbon sequestration and harvested mallee biomass agroforestry systems. Crop Pasture Sci 63:555–569. https://doi.org/10.1071/cp12028
Steubing L (1968) Untersuchungen über den Einfluss von Windschutzhecken auf den Wärme- und Wasserhaushalt von Pflanzen und die Rückwirkungen auf Stoffwechselprozesse. In: Tüxen, R. (eds) Pflanzensoziologie und Landschaftsökologie. Berichte über die Internationalen Symposia der Internationalen Vereinigung für Vegetationskunde, vol 7. Springer, Dordrecht. https://doi.org/10.1007/978-94-015-7180-7_20
Swieter A, Langhof M, Lamerre J (2021) Competition, stress and benefits: trees and crops in the transition zone of a temperate short rotation alley cropping agroforestry system. J Agron Crop Sci 208:209–224. https://doi.org/10.1111/jac.12553
Swieter A, Langhof M, Lamerre J, Greef JM (2019) Long-term yields of oilseed rape and winter wheat in a short rotation alley cropping agroforestry system. Agrofor Syst 93(5):1853–1864
Thevs N, Aliev K (2021) Agro-economy of tree wind break systems in Kyrgyzstan, Central Asia. Agrofor Syst 97:319–334. https://doi.org/10.1007/s10457-020-00557-8
Thiel B, Smukler SM, Krzic M, Gergel S, Terpsma C (2015) Using hedgerow biodiversity to enhance the carbon storage of farmland in the Fraser River delta of British Columbia. J Soil Water Conserv 70(4):247–256. https://doi.org/10.2489/jswc.70.4.247
Turner J, Hardy S (2006): Windbreaks for citrus. CITT groups Australia, pp. 1
Ucar T, Hall FR (2001) Windbreaks as a pesticide drift mitigation strategy: a review. Pest Manag Sci Formerly Pestic Sci 57(8):663–675

Veldkamp E, Schmidt M, Markwitz C, Beule L, Beuschel R, Biertümpfel A, Bischel X, Duan X, Gerjets R, Göbel L, Graß R, Guerra V, Heinlein F, Komainda M, Langhof M, Luo J, Potthoff M, van Ramshorst JGV, Rudolf C, Seserman DA, Shao G, Siebicke L, Svoboda N, Swieter A, Carminati A, Freese D, Graf T, Greef JM, Isselstein J, Jansen M, Karlovsky P, Knohl A, Lamersdorf N, Priesack E, Wachendorf C, Wachendorf M (2023) Corre MD (2023): Multifunctionality of temperate alley-cropping agroforestry outperforms open cropland and grassland. Communications Earth & Environment 4:20. https://doi.org/10.1038/s43247-023-00680-1

Veste M, Littmann T, Kunneke A, Du Toit B, Seifert T (2020): Windbreaks as part of climate-smart landscapes reduce evapotranspiration in vineyards, Western Cape Province, South Africa. Plant, Soil and Environment, 66: 119–127. https://doi.org/10.17221/616/2019-pse

Veste M, Böhm C (2018) Agrarholz—Schnellwachsende Bäume in Der Landwirtschaft: Biologie - Ökologie - Management. Springer Spektrum. ISBN 3662499304.

Veste M, Gao J, Sun B, Breckle S-W (2006) The green great wall—combating desertification in China. Geographische Rundschau Internationale Edition 2:14–20

Veste M, Sheppard JP, Abdulai I, Ayisi KK, Borrass L, Chirwa PW, Funk R, Kapinga K, Morhart K, Mwale SE, Ndlovu NP, Nyamadzaw G, Nyoka BI, Sebola P, Seifert T, Senyolo MP, Sileshi GW, Syampungani S, Kahle H-P (2024) The need for sustainable agricultural land-use systems: benefits from integrated agroforestry systems. In G.P. von Maltitz, G.F. Midgley, J. Veitch, C. Brümmer, R.P. Rötter, F.A. Viehberg, M. Veste (Eds.), Sustainability of Southern African Ecosystems under Global Change, Ecological Studies 248. Springer Cham. pp 587–623. https://doi.org/10.1007/978-3-031-10948-5_21

Veste, M, Grey K-A, Gottschalk N, Trachte K, Midgley, GF (2021) Can shelterbelt trees reduce evapotranspiration and ecophysiological stress in irrigated vineyards and citrus orchards? A transcontinental experiment in the Western Cape, South Africa and Lower Lusatia, Germany. Landscape 2021 - Diversity for Sustainable and Resilient Agriculture, Berlin, Germany. https://doi.org/10.13140/RG.2.2.34333.03043

Weninger T, Scheper S, Lackóová L, Kitzler B, Gartner K, King N, Cornelis W, Strauss P, Michel K (2021) Ecosystem services of tree windbreaks in rural landscapes—a systematic review. Environ Res Lett 16:103002. https://doi.org/10.1088/1748-9326/ac1d0d

Zanamwe P (2018) The effect of pre-harvest wind damage on post-harvest quality of 'Triumph' persimmon fruit. Afr J Sci Technol Innov Dev 10:753–758. https://doi.org/10.1080/20421338.2018.1491138

Zhang D, Brandle JR, Hubbard KG, Hodges L, Daningsih E (1999) The Response of muskmelon growth and development to microclimate modification by shelterbelts. HortScience 34(1) 64–68. https://doi.org/10.21273/hortsci.34.1.64

Zheng X, Zhu J, Xing Z (2016) Assessment of the effects of shelterbelts on crop yields at the regional scale in Northeast China. Agric Syst 143:49–60. https://doi.org/10.1016/j.agsy.2015.12.008

Ziegler J, Easter M, Swan A, Brandle J, Ballesteros W, Domke G, Chambers A, Eve M, Paustian K (2016) A A model for estimating windbreak carbon within COMET-Farm™. Agrofor Syst 90:875–887. https://doi.org/10.1007/s10457-016-9977-0

Zubay P, Kunzelmann J, Ittzés A, Németh Zámboriné É, Szabó K (2021) Allelopathic effects of leachates of *Juglans regia* L., *Populus tremula* L. and juglone on germination of temperate zone cultivated medicinal and aromatic plants. Agrofor Syst 95:431–442. https://doi.org/10.1007/s10457-020-00572-9

Chapter 14
Farm Trees in Multifunctional Landscapes in South Africa: Synthesis and Outlook

Maik Veste, Elmi Lötze, and Thomas Littmann

Abstract The development of future multifunctional land use is an important task under the conditions of climate change. The ecological and sustainable agriculture of the future requires more than just the optimization of agronomic objectives. As ecological systems, windbreaks and farm trees make an important contribution to the agricultural landscape. They not only influence the microclimatic site conditions but also give the entire cultivated landscape its structure. Trees can fulfill an important ecological function around farms and settlements. The preservation and promotion of biodiversity in the Western Cape region is of particular importance. With its high diversity of species, the flora and fauna of the Cape Floristic Region is one of the hotspots of biodiversity. It is important to integrate this into the landscape, so that both land use and nature conservation benefit. Integrative solutions can thus lead to a multifunctionality of the cultural landscape and bridge the gap between ecology, economy, and human needs. Holistic sustainability is an important basis for future agriculture, and its implementation requires consideration of all components along the value chain.

Keywords Agricultural landscapes · Biodiversity · Farm trees · Cape winelands · Farms

---

M. Veste (✉)
CEBra—Centre for Energy Technology Brandenburg e.V., Cottbus, Germany
e-mail: veste@cebra-cottbus.de

Institute of Biology, Department Molecular Botany, University of Hohenheim, Stuttgart, Germany

E. Lötze
Department of Horticultural Science, Stellenbosch University, Stellenbosch, South Africa

Agri Technovation South Africa, Wellington, South Africa

T. Littmann
DLC—Dr. Littmann Consulting GmbH, Ennepetal, Germany

M. Veste et al. (eds.), *Trees in South African Agriculture*, Innovations in Landscape Research, https://doi.org/10.1007/978-3-032-00993-7_14

14.1 Introduction

Like in many regions of the world, land will become an increasingly precious resource with a growing human population in Southern Africa. The development targets of a growing economy and the economic empowerment of the people in Southern Africa will add pressure to the agricultural and forestry sectors. Large areas in South Africa for example are already intensively used as cultural landscapes which provide important quantities of food for local consumption, but also for the regional and world market. About 60% of the country is currently under commercial or subsistence farming. Therefore, more and more agricultural fields are converted to intensified, highly productive agricultural production systems, and agricultural production is extended into the drier areas toward the north. However, production conditions are far from optimal in South Africa because 91% of the land surface is classified as dry lands. Moreover, the entire region of Southern Africa is under pressure with considerable water scarcity, whose situation is challenged by substantial bio-physical (biodiversity, climate, and droughts) and socio-economic issues (poverty, a lack of social coherence, and an increasing population) (von Maltitz et al. 2024a). Fluctuations in climatic conditions and water availability are particularly relevant due to their direct impact on agricultural production. In general, there is an urgent need of adaptation of agricultural systems toward a sustainable and integrated landscape development to overcome the current society and climate pressures. Water resources in the region are limited, and a more water efficient landscape and climate-smart agricultural production is needed to mitigate negative effects of climate change. Already a climate change response framework and implementation plan for the agricultural and horticultural sectors was developed for the Western Cape and provides a roadmap for adaptation to climate change (Midgley and Schulze 2021). An integrated approach to water and landscape management is required combining agricultural, forestry, and natural ecosystems to increase ecosystem resilience, to optimize water use efficiency, to ensure a sustainable bioeconomy, and to strengthen conservation of biodiversity over the next decades. This implies the transformation of ecosystems, food, and energy systems toward decarbonized and sustainable production systems (Figs. 14.1, 14.2).

14.1.1 Contribution of Windbreaks in Climate-Smart Agriculture

The current climatic situation in South Africa is characterized by a dynamic and complex ocean-land interaction with a high variability (Chap. 4). Beside the rainfall pattern, the local wind field is highly depending on seasonal variation and the topography. High wind speed with mean annual wind speed of 5–8 m s^{-1} at 10 m above ground in the cultivated areas of the Cape Winelands, and up to 20 m s^{-1} along the ridges are characteristic along the coastal regions of the Western Cape Province. The

Fig. 14.1 Present agricultural landscapes in the Western Cape: **a** Increasing grape cultivation converted agricultural fields to large-scale vineyards at the West Coast region, **b** shading nets span larger fields for berry production in the Overberg region, **c** intensive cereal farming in the Swartland, **d** Rooibos farming at the near Clamwilliam. (Photographs: Maik Veste **a**–**c**, **d** Elbé du Toit)

Fig. 14.2 Arial photograph of landscape mosaic in the Western Cape, South Africa (Photograph Maik Veste)

specific environmental conditions require a rethinking of the agricultural production. ´Climate-smart agriculture´ (CSA) aims to increase agricultural production to support profitability, food security, development, and agriculture resilience, and carbon sinks (Campbell et al. 2014, Thierfelder et al. 2017). Like other agroforestry systems, tree-based windbreaks can make a substantial contribution to the objective of CSA. Windbreaks are important eco-engineering measures influencing directly the microclimate environment (Chaps. 3, 5) and therefore the growth conditions for the crops. The effects of different design windbreaks including shading nets on evapotranspiration and the local water balance at field level have been shown during the FarmImpact project (Chaps. 5, 6, 9 and 10). The benefits of any design must be evaluated at the farm level. In any case, it was shown that there is no universal solution as each design (trees, hedges, and artificial windbreaks) may have individual advantages and disadvantages. Specifically, the recent trend to substitute hedgerows by both vertical and horizontal shading nets will have to realize that the benefits of reducing wind speeds and evapotranspiration in the crop stand are countermeasured by a practical low impermeability to rainfall.

14.2 From Climate-Smart Farms to Climate-Smart Landscapes

The reduction of the wind speed is a well-documented feature and important also for soil conservation and soil development (Sheppard et al. 2020). Therefore, an explicitly integrating adaptation and mitigation into landscape management of agricultural systems is needed. The concept of climate-smart landscapes includes in principle three key features: climate-smart practices at the field and farm level, diversity of land use across the landscape, and management of land-use interactions at landscape scale (Scheer et al. 2012). The design of an integrated landscape is underpinning the climate-smart agriculture goals. Developing an integrative agricultural landscape approach requires more than just the sum of individual fields for agricultural production. Interactions and synergies between agricultural production, climate change adaptation, and environmental aspects are essential to a functional landscape. A climate-smart landscape includes a diversity of different land uses across the landscape (Table 14.1). This must be coordinated at the farm level, but also at the level of the larger landscapes. Agroforestry systems and windbreaks can be part of the climate-smart landscapes, but their detailed planning requires scientifically based analyses in addition to the topographical conditions, as the different climatic conditions also play a key role here (Chaps. 9–12).

More complex multifunctional landscapes need an improved understanding of the water-land-use relations. This is essential to develop advanced strategies for integrating resource-efficient agricultural productivity, sustainable water management, and conservation of the natural resources. Moreover, innovative methodologies and decision-support tools to manage land for a balanced provision of food, timber, fiber, a

Table 14.1 Examples of components for climate-smart practices at the farm level (compiled after Breckle et al. 2001; Esler et al. 2010; Scheer et al. 2012; Sheppard et al. 2020; Veste et al. 2024)

Climate-smart practices	Objectives	Utilization of trees
Building soil organic matter	Minimal tillage and using cover crops and crop residues enhance the organic matter stored in soil, while also supporting biological processes and nutrient and hydrological cycling. Preventing soil erosion	Trees for carbon sequestration wind windbreaks, agroforestry
Integrated nutrient management	Green manures, promoting biological nitrogen-fixation, and incorporating livestock manures into the soil, decrease the amount of nitrogen lost to runoff and emissions of nitrous oxide	Nitrogen-fixing tree and shrubs
Farming with perennials	The development root and woody biomass can substitute for an annual tillage regime, providing year-round ground cover and retaining organic matter and water in soil, preventing erosion	Agroforestry systems fruit, nuts trees
Efficient water management	Best practices for irrigation, water-harvesting technology, and terrace or contour farming systems can contribute to improved water-use efficiency and conservation, reduction of evapotranspiration	Integration of windbreaks and hedges
Climate-friendly livestock systems	Rotational grazing systems, optimal stocking rate, manure management, improved feeds quality, veld management, and preventing overgrazing	Silvo-pastoral agroforestry systems
Restoration of degraded rangelands	Combating land degradation and erosion, rehabilitation of veld flora, grasslands for feed	Windbreaks
Restoration of degraded watersheds	Restoration of watersheds functioning, reducing degradation	Water efficient trees
Protection of natural habitats	Conservation of natural woodlands, grasslands, biodiversity, integration of semi-natural habitats for insects and pest control, fire control	Promoting native tree species

sustainable bioeconomy, and conservation of biodiversity and development of natural ecosystems are urgently needed. The complex interactions in the climate-ocean-land system (Chap. 4) and feedbacks make it necessary to continue the research work in this field, which is more than necessary, especially in view of the fluctuations in the climate system (von Maltitz et al. 2024b).

14.3 Multifunctional Landscapes—The Needs of Integration Different Land-Use Aspects

Land-use systems of the future have to be holistic to support different needs of society. The development of multifunctional agricultural landscapes on farms is a human-centered approach but includes also the inclusion of natural or semi-natural ecosystems. Furthermore, they must be able to balance ecosystem services such as water and food provision, carbon sequestration, and biodiversity. The most diverse needs and demands of humans must be satisfied from the given space. The demands on the landscape can be described from three points of view from the perspective of the land users, which are opposed to each other, but are nevertheless interconnected. Multifunctionality (Fig. 14.3) is therefore an optimal composition of a landscape and is thus initially focused exclusively on human well-being. The cultural landscape in the Western Cape in particular is ideally placed to realize these concepts based on these three pillars (Fig. 14.3). In our case of modern agricultural landscapes, this should integrate economic, environmental, and social aspects (Fig. 14.4).

(a) **Economic:** The primary function is the profitable production of food and processed food products to generate income. Optimized yield and cost efficiency are important factors here.

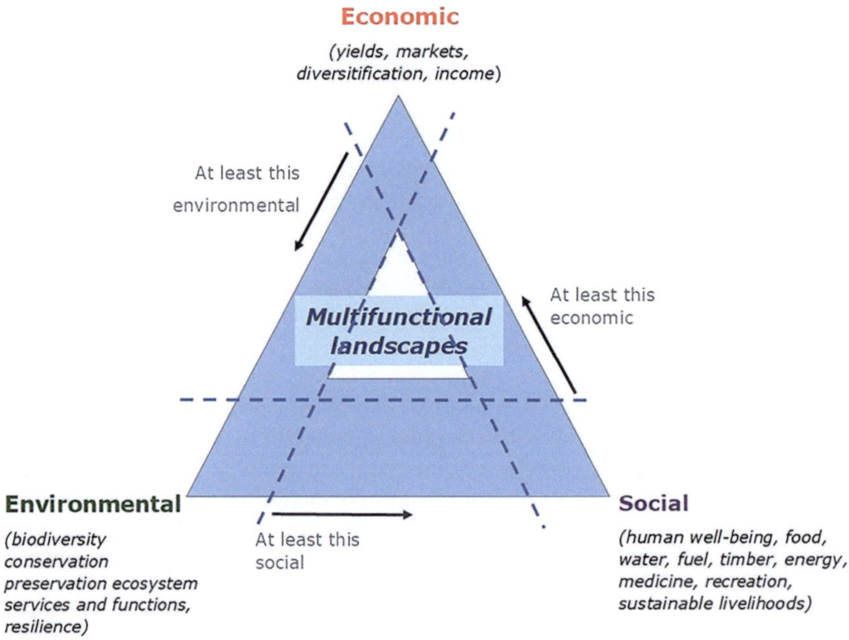

Fig. 14.3 Integration of economic, environmental, and social aspects for multifunctional landscapes

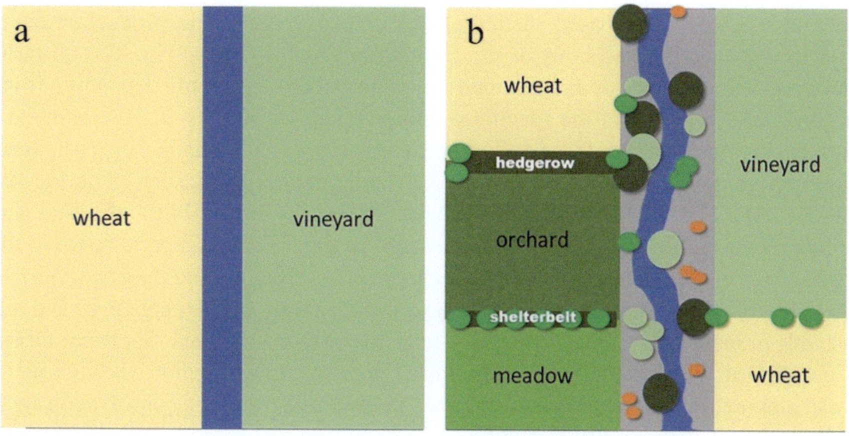

Fig. 14.4 Scheme of the concept of differentiated land use (after Haber 1998) on a farm/landscape level: **a** state of development with large-scale agriculture and **b** ideal image of differentiated and multifunctional land use in agricultural landscapes with the integration of natural habitats (modified after Veste and Böhm 2018)

(b) **Environmental:** Natural habitats are fundamental to the preservation of biodiversity and the associated ecosystem functions and services. This includes also the natural ecosystem, soil, water, and atmosphere. This is the natural basis for sustainable land use and strengthens resilience to climate change.

(c) **Social:** The provision of food, water, fuel, wood, and medicine are basic needs for human well-being. Recreation is gaining in importance in modern societies.

14.3.1 The Basis of the Farm Economy: Agronomy, Yields, and Income

From the economical viewpoint, reduction of evapotranspiration and water saving in agricultural systems are equally important for crop management and sustainable production (Kanzler et al. 2019; Veste et al. 2020; Hofmeister et al. 2024). Meanwhile, it is a well-documented feature of agroforestry systems that tree shelterbelts may have positive effects on the yields of arable crops in different climatic zones (chapter 13) and in many cases the establishment of agroforestry systems is encouraged (Sheppard et al. 2020; Chirwa et al. 2024). On the other hand, selected crops, cultivation, and management are dependent on a variety of internal and external factors. These ultimately influence the economic development of the farms and their income. In the Cape Winelands, the cultivation of fruit and wine plays the central role for the farms, which are favored by the specific climatic conditions. In addition to the yield, the quality of the fruit is of central importance for the marketing of the fruit to the end consumer. As the visual quality of the fruit is damaged by wind (Chap. 6), this has repercussions on the sales prices achieved on the export market. Wind protection measures such as tree-based shelterbelts or nets are reflected in production costs and influencing the decisions of the farm management (Chap. 12). The on-farm decisions for agricultural and horticultural management are highly depending from a good understanding of the specific weather conditions (Chaps. 4, 10), which are influencing finally the yields and economical income. Climate change and prolonged water shortages (Midgley et al. 2005, von Maltitz et al. 2024a), such as during the 2015–2018 "Day Zero" drought, also demonstrated the vulnerability of wheat and fruit producers and the agricultural economy in the Western Cape (Theron et al. 2022, 2023). Following the analysis by Theron et al. (2022), more than two-third of the farmers and producers in the region are already aware on the challenges and are actively preparing for climate change. In response to climate change, apple producers view on-farm water management as the most important strategy, while wheat farmers' strategies are focused on crop management, cultivar selection, and conservation agriculture. Diversification of production systems and incomes are important strategies of commercial farm management to increase their economical resilience (Malherbe et al. 2024). This includes a wide range of activities within industry from crop diversification, introduction of new cover crops, modified livestock management, but also the development of alternative income sources like tourist and recreation activities of the scenic landscapes and game farming (Raaijmakers and Swanepoel 2019; Theron et al. 2022; Talanow et al. 2021; Rötter et al. 2024). During droughts, water supply becomes a critical issue and needs to be discussed further in respect to the different water uses and management goals (Lanari et al. 2021). Furthermore, decisions at farm level are linked in many ways to the market components along the supply chain from the farm to the supermarket and, finally, by the end consumer. Environmental aspects are important for export markets, and South African producers supply the retail market at higher prices for South African European export than that of the wholesale market (van Dijk and Kleynhans 2004). An environmental care product image and social responsibility can be a selling point for the retail market.

14.3.2 Integration of Biodiversity and Preservation Ecosystems

Southern African landscapes harbor a significant part of global biodiversity and unique ecosystems. The Cape Floristic Region (CFR), the smallest of the six recognized floral kingdoms of the world, is a biodiversity hotspot at the southern tip of Africa, extremely rich in endemic plants and arthropods (Mucina and Rutherford 2024; Samways et al. 2024b). The Fynbos belongs to the world's biodiversity hotspots and is recognized as a global priority area for nature conservation. According to the South African "Threatened Plant Species Program" (South African National Biodiversity Institute, SANBI), 67% of all threatened plant species occur in the fynbos biome of the Cape region. Many of these species have a very limited distribution range and only persist in small areas or even in a single location. The major threat to more than 80% of the endangered species is habitat loss. The land-use conversion to agriculture and forestry have occurred at larger spatial scales since the first large settlements were established by European settlers in the seventeenth century. More recently, habitat loss has been accelerated by the ongoing transformation and fragmentation of landscapes. Unique biomes like fynbos, renosterveld and strandveld have been converted into fruit and cereal production, and nowadays, only 5% of the original renosterveld biome remains in the agricultural lowlands (Esler et al. 2014). Therefore, there is an urgent need to reorganize the agricultural landscape and integrate natural or semi-natural structural elements to restore ecosystem functions and the associated ecosystem services for agriculture (Ekroos et al. 2014; Landis 2017; dos Santos et al. 2021). Promoting biodiversity in the Western Cape agricultural landscape can include, e.g., intercropping; the planting and maintenance of trees and hedges, buffer strips and natural corridors; retaining riparian and other areas of high value natural vegetation (Giliomee 2006).

To develop sustainable and complex land use, changes in cultivation concepts and management methods must be considered in addition to the interests of nature and landscape conservation in the agricultural landscape. This concept was developed as early as in the mid-1960s by the landscape ecologist Wolfgang Haber as "*differentiated land use*" (Haber 1968, 1998). In this concept, the monotonous and economically trimmed agricultural landscape is supplemented by a variety of natural structural elements, creating a small-scale mosaic of different land uses that comes close to the original (multifunctional) cultural landscape. These small-scale structures also offer space for natural developments and are usually of high nature conservation value.

There are various reports that in vineyards agroforestry and vineyards located within biologically diverse landscapes show that trees can play a positive role in vineyard arthropod pest management (Altieri and Nicholls 2002; Favor et al. 2024). Furthermore, within-vineyard vegetation and flower cover (Fig. 14.5a) had a consistent positive influence on most arthropods, but the different taxa showed contrasting relationships with configurational heterogeneity (Gaigher et al. 2024). In general, cover crops also play a vital role in reducing evaporation, increasing infiltration, and preventing soil erosion. Also, fruit trees species may have influence on the species compositions (surface-active arthropod, web-building nematodes communities) as a demonstrated for an organic farm in the Western Cape (Birkhofer et al. 2019). Samways et al. (2024b) reviewed the importance of natural and semi-natural patches of vegetation in the agricultural matrix for biodiversity in the Western Cape. In this study, connected small patches have a high value for the arthropods. The authors concluded that the integration of vegetation strips and natural vegetation patches into the farmland will increase the positive effects between different zones and are important for biodiversity conservation of the regions that a heterogenic landscape for maintaining biodiversity is becoming increasingly important for protecting insect populations against climate change by providing a wide range of environmental conditions, resources, and refuges (Pryke and Samways 2015, Harvey et al. 2022,

Samway et al. 2024a). These are essential observations and arguments for transferring the concept of differentiated land use from the agricultural landscapes of Central Europe to the Western Cape.

On the other hand, a far-reaching debate about the spatial organization and size of nature conservation measures in relation to the agricultural landscape is still ongoing. The development of promotion of multifunctional landscapes with biodiversity conservation can take place on different spatial arrangements. The mosaic-shaped arrangement with clearly separated high-yielding farmland from non-crop areas ("land sparing") can more effectively conserve biodiversity than promoting low-yielding, less intensively managed farmland occupying larger areas ("land sharing"). Ekroos et al. (2016) argued that there is no single correct spatial scale for segregating biodiversity protection and commodity production in multifunctional

Fig. 14.5 Integration of biodiversity into farms: **a** Agroforestry and intercropping with cover crops and, Nederburg Wine Estate, Paarl, **b** natural fynbos vegetation at Tierhoek Organic Farm, Robertson (Photographs Maik Veste)

landscapes. The authors concluded that the implementation of biodiversity simultaneously at multiple spatial scales, which are connected, is the best way toward the multifunctional landscape. Connecting the different scales natural corridors are important parts (Pryke and Samways 2015). In this context, trees in agroforestry systems and windbreaks can also contribute to the overall goal (Giliomee 2006, Sheppard. et al. 2020). However, the development of such multifunctional land use requires the combination of fundamental ecological studies on the interaction between biodiversity and ecosystem functions, as well as applied ecological studies on the influence of biodiversity from the field to the cultural landscape.

This would also increase the positive feedback on agricultural production and the preservation of the entire ecosystem toward a climate-smart landscape. Such integration can take place on different scales (Fig. 14.5). Meanwhile, many wine farmers are planting rows of fynbos in between the grapevines which as beneficial effects on growing "healthy" vines: moisture retention, less soil erosion, and minimizing the movement of insects through the vineyard.

The introduction of planted trees alternates the structure of the remaining agricultural and farm landscapes. These may affect biodiversity in different ways, creating new habitats due to new landscapes structures. This is obvious in the dry areas of the Karoo to the north, where trees are rare and concentrated to non-perennial rivers with ground water river connections or rivers—such as the Orange river—(Veste and Jürgens 2004). In these ecosystems, the increase of abundance of woody species (e.g., *Acacia karroo*, syn. *Vachea karroo*) along drainage lines is beneficial for animals, providing a better and constant food source. This can be linked to the extension of distribution of the greater Kudu and vervet monkeys into the Karoo (Walker et al. 2018). Furthermore, alien trees (Fig. 14.6) are planted at farmhouses, town, and along road sites, providing an important nest site for several bird species (Fig. 14.7), which are not part of the original natural regional fauna.

However, the introduction of trees can have also drastic implications for ecosystem functioning and leading finally to ecosystem disservices (Esler et al. 2014, van Willgen et al. 2020), and especially, their implication on water bodies and ground water supply is under consideration (Potgieter et al. 2014; Dzikiti et al. 2016, 2017). In general, it is assumed that indigenous trees use significantly less water, but they grow more slowly than exotic, non-native tree species. Furthermore, invasive trees may alter natural fire regimes through changes to the flammable biomass and are high risks for the fynbos and renosterveld (Esler et al. 2014). The most problematic invasives in fynbos are trees and shrubs, for example, pines and Australian Acacias that are larger than the native shrublands. Fuel loads can increase up to 60% in invaded fynbos, which could then support higher intensity fires. Based on these many negative experiences, there is a major concern about introduced trees in Southern African ecosystems. In this context, large-scale tree plantings for carbon sequences are viewed critically, as the international programs for tree planting ignores the regional ecosystems and their structure and functions (Bond et al. 2019). Landscape transformation by trees-for-carbon projects for Africa includes inappropriate targets. In such cases, the misunderstanding of the ecological principles can lead to

Fig. 14.6 Planted alien trees at farm houses near Darling at the West Coast, Western Cape (Photograph Maik Veste)

the destruction of habitats and thus biodiversity with a one-sided focus on climate effects.

14.3.3 Cultural Landscape and Tree Planting

Far beyond the ecological and economic (agricultural) significance of the landscape, there is also the human component of landscape utilization. In addition to the ecosystem goods (food, water, medical resources, raw material, and energy), the cultural ecosystem services are also important components of the land use for human well-being. Cultural landscapes have developed over centuries under the influence of man and combine different land uses, which are, however, also subject to change. Even if they give the impression of being "natural," they have been created artificially, and natural ecosystems no longer exist apart from small fragments. Nevertheless, they are subject to the basic ecological processes between soil, vegetation, and climate. Agriculturally and horticulturally utilized areas are combined with forests or settlement areas and result in a mosaic of different forms of use—a concept which was mainly developed in Europe (Haber 1998; Zerbe 2022). In this context, farm trees as

Fig. 14.7 Planted trees in the Karoo town Montagu, Western Cape **a** breeding colonies of African sacred ibises, herons, cormorants, and **b** nests of weavers in the Klein Karoo, (Photographs. Maik Veste)

key species and agroforestry can play an important role in biocultural conservation and sustainable development (Zinngerbe et al. 2020; Mobarak et al. 2005).

Also, the landscapes of the Western Cape and the Cape Winelands in particular have been shaped by man since the seventeenth century. Situated in the scenic regions, the wine estates and fruit farms also characterize the contemporary landscape. In addition to agriculture, the landscape also contributes a cultural value for

Fig. 14.8 Trees as part of farmlands in the Western Cape: **a** Orchards and individual trees at Tierhoek Organic Farm, Robertson; **b** Oak trees at Babylonstoren, Paarl; **c** Ornamental garden at Nederburg Wine Estate, Paarl; **d** Helderberg mountains, Stellenbosch. Photographs Maik Veste

tourism and recreation. The "beauty of the landscape" in combination with culture and architecture is seen as a positive objective by land users. The landscape is characterized by positive features and reflects the ideal image of a small, structured landscape such as a landscape garden (Haber 2004). Nowadays, ornamental farms have started to embrace this concept again. Farm trees in the context for the local farm play and important role, either close to the farmhouses as ornamental trees, in the landscape as shelterbelt or solitaire tree or in form of fruit and nut trees often in combination with vineyards and orchards (Fig. 14.8).

The positive image of combined farm use beyond pure agricultural production was recognized decades ago. Initiatives were also launched to promote the possibilities of integrating the rich biodiversity ("flowering landscapes"). The awareness for habitats and biodiversity on agricultural farms has increased, resulting in impressive public initiatives and campaigns for promoting the integration of biodiversity in the Western Cape region (Fig. 14.9a). The "Wine & Biodiversity Initiative," for example, was established in 1998 and developed guidelines for sustainable and eco-friendly wine production to encourage wine growers to integrate natural habitats of high value (mainly fynbos and renosterveld) and even promoting endangered animal species like the Cape leopard. Furthermore, the integration of natural vegetation in combination with sustainable sheep farming and agricultural use was also proposed (Rötter et al. 2024) to promote the high biodiversity of the Succulent Karoo biome (Veste and Jürgens 2004, Mucina 2024) with their annual plants, geophytes, and succulents. The "flowering" Namaqualand and Bokkeveld Plateau attracts many local and overseas tourists every year. Meanwhile, touristic activities and recreation in combination with scenic landscapes and biodiversity is integrated in overall farm management (Fig. 14.9b).

In this way, the landscape experience can be combined with a positive farm image and its agricultural products. This can lead to positive feedback on their higher economic value for the consumer. The beauty and esthetics (Haber 2004) are also an integral part of the motivation for buyers for the acquisition of agricultural land for the combination of agriculture with alternative uses (Reed and Kleynhans 2010). In addition to the classic farm characteristics such as production capacity and soil quality, also the setting of the farm (in valley, against mountain), the natural scenery and biodiversity, and furthermore farm trees as part of the landscape, are rated positively.

Fig. 14.9 **a** Wine management promoting the integration of biodiversity into the wine estate as part of multifunctional land-use concept at Delheim, Stellenbosch, **b** Recreation as part of multifunctional landscapes, Papkuilsfontein, Karoo. Photographs Maik Veste

Thus, solutions for multifunctional land use can be found to bridge the gap between ecology, economy, and human needs in the context of differentiated and multifunctional land use. Holistic sustainability is an important basis for future agriculture and its implementation requires consideration of all components along the value chain (e.g., in viticulture, Ulrich 2022). The implementation of a real multifunctional landscape as a climate-smart system is still a complex challenge for farmers, society, and science (Lanari et al. 2021; Topp et al. 2022). Further preservation and adaptation of the cultural landscape in the Western Cape also requires the networking of land-use stakeholders and secure incentives for their implementation (Stoll-Kleemann and O'Riordan 2002; Manhique and Wätzold 2024). In addition to commercial farmers, smallholder farms must also be included in sustainable land use and the establishment of natural measures—such as tree planting—as practical projects have shown (Evenepoel 2002).

14.4 Lessons Learned and Outlook

Farm trees are integral parts of the Western Cape farmlands, but with the economic pressure on delivering unblemished, visual attractive fruit, the artificial netting of orchards and vineyards to manage fruit damages, such as sunburn and rub marks, is rapidly replacing natural tree-based windbreaks. Similarly, expanding farm boundaries in broad acre crops often result in removal of natural windbreaks to allow for easier cultivation and harvesting of bigger fields. Thus, historical windbreaks are rapidly being replaced on farms which poses a new challenge for the bigger environment in future. In future, natural windbreaks and farm trees will be important in climate-smart landscapes. Especially the benefit for wildlife and insects can play a major role in commercial agriculture with the new focus on regenerative agriculture initiated in Europe. The challenge is to establish a balance between economy, environment, human wellbeing, and commercial agriculture. A different approach needs to be investigated to address this at farm level, but also on the bigger environment on landscape level which is also affected by the same climate conditions as agricultural products, especially when considering future microclimatic challenges. At the farm level, establishment of different species and multiple row-windbreaks with carefully designed positioning can contribute to environmental outcomes, without competing with artificial wind breaks close to crops. In addition, we should also consider unconventional opportunities outside of the agricultural sector and including the local communities as important stakeholder. The bigger landscape lends itself to the introduction of new species as wind breaks, i.e., drought resistant spekboom (*Portulacaria afra*) along public and national roads, sports fields and schools. This should be conducted with support from local authorities or the public to share the responsibility of repopulating open areas to introduce the multiple benefits of windbreaks, which will also address other aspects of human health and well-being well established in literature.

Acknowledgements The research was supported over many years by the Federal Ministry of Education and Research (BMBF), National Research Foundation (NRF) and the Department of Agriculture, Forestry and Fisheries (DAFF) of South Africa, German Academic Exchange Service (DAAD), and Volkswagen Foundation. Thanks to Stellenbosch University for their hospitality to carry out the research and scientific exchange.

References

Altieri MA, Nicholls CI (2002) The simplification of traditional vineyard based agroforests in northwestern Portugal: some ecological implications. Agrofor Syst 56:185–191

Birkhofer K, Addison P, Addison MF, Arvidsson F, Bazelet C, Bengtsson J, Haddad C, Booysen R, Conlong D, Janion-Scheepers C, Kapp C, Lindborg R, Louw S, Malan AP, Storey SG, Swart WJ (2019) Effects of ground cover management on biotic communities, ecosystem services and disservices in organic deciduous fruit orchards in South Africa. Front Sustain Food Syst 3:107. https://doi.org/10.3389/fsufs.2019.00107

Breckle S-W, Veste M, Wucherer W (eds.) (2001): Sustainable Land-Use in Deserts, Springer, Heidelberg, pp. 465. https://doi.org/10.1007/978-3-642-59560-8

Bond WJ, Stevens N, Midgley GF, Lehmann CER (2019) The trouble with trees: afforestation plans for Africa. Trends Ecology Evolution 34:963–965. https://doi.org/10.1016/j.tree.2019.08.003

Chirwa PW, Syampungani S, Mwamba TM (2024) Trees in a sub-saharan multi-functional landscape: research, management, and policy. Springer Cham. https://doi.org/10.1007/978-3-031-69812-5

Dos Santos JS, Dodonov P, Oshima JE, Martello F, de Jesus AS, Ferreira ME, Silva-Neto CM, Ribeiro MC, Collevatti RG (2021) Landscape ecology in the Anthropocene: an overview for integrating agroecosystems and biodiversity conservation. Persp Ecol Conserv 19:21–32. https://doi.org/10.1016/j.pecon.2020.11.002

Dzikiti S, Ntshidi Z, Le Maitre DC, Bugan RDH, Mazvimavi D, Schachtschneider K, Jovanovic NZ, Pienaar HH (2017) Assessing water use by *Prosopis* invasions and *Vachellia karroo* trees: Implications for groundwater recovery following alien plant removal in an arid catchment in South Africa. For Ecol Manage 398:153–163

Dzikiti S, Gush MB, Le Maitre DC, Maherry A, Jovanovic NZ, Ramoelo A, Cho MA (2016) Quantifying potential water savings from clearing invasive alien Eucalyptus camaldulensis using in situ and high resolution remote sensing data in the Berg River catchment, Western Cape, South Africa. For Ecol Manage 361:69–80

Ekroos J, Olsson O, Rundlöf M, Wätzold F, Smith HG (2014) Optimizing agri-environment schemes for biodiversity, ecosystem services or both? Biol Conserv 172:65–71

Ekroos J, Ödman AM, Andersson GKS, Birkhofer K, Herbertsson L, Klatt BK, Olsson O (2016) Sparing land for biodiversity at multiple spatial scales. Front Ecolology Evol 3:145. https://doi.org/10.3389/fevo.2015.00145

Esler KJ, Milton SJ, Dean WRJ (2010) Karoo Veld: ecology and management. Briza Publications, Pretoria

Esler KJ, Pierce SM, de Villiers C (2014) Fynbos: ecology and management. Briza Publications, Pretoria

Evenepoel H (2002) Meerdoelige bome vir windbreak en skuiling. Department of Forestry, Faculty of Agriculture, Stellenbosch University.

Favor K, Gold M, Halsey S, Hall M, Vallone R (2024) Agroforestry for enhanced arthropod pest management in Vineyards. Agrofor Syst 98:213–227. https://doi.org/10.1007/s10457-023-00900-9

Gaigher R, Pryke JS, Kratschmer S, Zaller JG, Winter S, Samways MJ, Pachinger B, Geldenhuys M, Van Schalkwyket J. (2024) Area and configuration of both semi-natural habitats and vineyards influence arthropods in South African and central European agricultural landscapes. Insect Conservation and Diversity 17(4): 702–718. Available from: https://doi.org/10.1111/icad.12741

Giliomee JH (2006) Conserving and increasing biodiversity in the large-scale, intensive farming systems of the Western Cape, South Africa. S Afr J Sci 102:375–378

Haber W (1965) Landschaftsökologie in der Flurbereinigung. In: Tüxen, R. (eds) Pflanzensoziologie und Landschaftsökologie. Berichte Über die Internationalen Symposia der Internationalen Vereinigung für Vegetationskunde 7: 381–396 Springer, Dordrecht. https://doi.org/10.1007/978-94-015-7180-7_35.

Haber W (1998) Das Konzept der differenzierten Landnutzung – Grundlage für Naturschutz und nachhaltige Naturnutzung. In: Bundesministerium für Umwelt, Naturschutz und Reaktorsicherheit (BMU) (Hrsg) Ziele des Naturschutzes und einer nachhaltigen Naturnutzung in Deutschland, S 57–64

Haber W (2004) Landscape ecology as a bridge from ecosystems to human ecology. Ecol Res 19:99–106. https://doi.org/10.1111/j.1440-1703.2003.00615.x

Harvey JA, Tougeron K, Gols R, Heinen R, Abarca M, Abram PK et al (2023) Scientists' warning on climate change and insects. Ecol Monogr. https://doi.org/10.1002/ECM.1553

Hoffmeister S, Bohn Reckziegel R, du Toit B, Hassler SK, Kestel F, Maier R, Sheppard JP, Zehe E (2024) Hydrological and pedological effects of combining Italian alder and blackberries in an agroforestry windbreak system in South Africa. Hydrol Earth Syst Sci 28:3963–3982. https://doi.org/10.5194/hess-28-3963-2024

Kanzler M, Böhm C, Mirck J, Schmitt D, Veste M (2019) Microclimate effects on evapotranspiration and winter wheat (*Triticum aestivum* L.) yield within a temperate agroforestry system. Agrofor Syst 93:1821–2184. https://doi.org/10.1007/s10457-018-0289-4

Lanari N, Bek D Timms, J Simkin L (2021) In whose interests? Water risk mitigation strategies practiced by the fruit industry in South Africa's Western Cape. Geoforum 126: 105–114

Malherbe W, Biggs R, Sitas N (2024) Comparing apples and pears: linking capitals and capacities to assess the resilience of commercial farming operations. Agricultural Systems217: Article 103934, https://doi.org/10.1016/j.agsy.2024.103934

Manhique H, Wätzold F (2024) Effects of institutional distrust on value estimates of stated preference surveys in developing countries: A choice experiment on conserving biodiversity within agricultural landscapes in a biodiversity hotspot, Q Open 4 (1): qoae014. https://doi.org/10.1093/qopen/qoae014

Mobarak C, Hatoum L, Kmoch L, Torralba M, Lieblein G, Wezel A, Plieninger T (2025) Farm trees as cultural keystone species: Bridging biocultural conservation and sustainable development in the Morocco High Atlas Mountains. Mt Res Dev 45(1):R11–R22. https://doi.org/10.1659/mrd.2024.00024

Mucina L, Rutherford MC (2024) Zonal Biomes of Southern Africa. Biome Ecology, vol 2. Springer, Cham. https://doi.org/10.1007/978-3-031-63731-5_5

Midgley GF, Chapman RA, Hewitson B, Johnston P, De Wit M et al (2005) A status quo, vulnerability and adaptation assessment of the physical and socioeconomic effects of climate change in the Western Cape. Report to the Western Cape Government, Cape Town, South Africa. CSIR Report No. ENV-S-C 2005–073, Stellenbosch

Midgley SJE, Davis N, Schulze RE (2021) Scientific and practical guide to climate change and pome/stone fruit production in South Africa. Part 1: Atlas of key climate-related variables for historical and future climate conditions as relevant to pome and stone fruit production. Report submitted to Hortgro Pome and Hortgro Stone, Stellenbosch.

Potgieter LJ, Richardson DM, Wilson JRU (2014) *Casuarina cunninghamiana* in the Western Cape, South Africa: Determinants of naturalisation and invasion, and options for management. S Afr J Bot 92:134–146

Pryke JS, Samways MJ (2015) Conserving natural heterogeneity is crucial for designing effective ecological networks. Landscape Ecol 30:595–607. https://doi.org/10.1007/s10980-014-0096-x

Raaijmakers S, Swanepoel PA (2019) Vulnerability, institutional arrangements and the adaptation choices made by farmers in the Western Cape of South Africa. South African Journal of Plant and Soil 37(1):51–59. https://doi.org/10.1080/02571862.2019.1645219

Rötter RP, Nkomo, M, Meyer zu Drewer J, Veste M (2024) Agricultural Land-Use Systems and Management Challenges. In: von Maltitz GP, Midgley GF, Veitch J, Brümmer C, Rötter RP, Viehberg, FA, Veste M (eds) Sustainability of southern African ecosystems under global change: Science for management and policy interventions. Ecological Studies 248, Springer Nature, Cham, 551–586. https://doi.org/10.1007/978-3-031-10948-5_20

Samways MJ, Gaigher R, Deacon C, Pryke JS (2024a) Heterogeneous conservation corridors of remnant vegetation protect biodiversity in South African timber mosaics. Biodivers Conserv. https://doi.org/10.1007/s10531-024-02993-9

Samways MJ, Pryke JS, Gaigher R, Deacon C (2024b) Biosphere reserves in the megadiverse cape floristic region are effective in conserving arthropod diversity. Integr Conserv, 118. https://doi.org/10.1002/inc3.72

Scherr, SJ, Shames S, Friedman R (2012) From climate-smart agriculture to climate-smart landscapes. *Agriculture & Food Security 1 (*12): https://doi.org/10.1186/2048-7010-1-12

Sheppard JP, Bohn Reckziegel R, Borrass L, Chirwa PW, Cuaranhua CJ, Hassler SK, Hoffmeister S, Kestel F, Maier R, Mälicke M, Morhart C, Ndlovu NP, Veste M, Funk R, Lang Seifert F, du Toit TB, Kahle HP (2020) Agroforestry: an appropriate and sustainable response to a changing climate in Southern Africa? Sustainability 12(17):6796. https://doi.org/10.3390/su12176796

Stoll-Kleemann S, O'Riordan T (2002) From participation to partnership in biodiversity protection: experience from Germany and South Africa. Soc Nat Resour 15:161–177

Talanow K, Topp EN, Loos J, Martín-López B (2021) Farmers' perceptions of climate change and adaptation strategies in South Africa's Western Cape. Journal Rural Study 81:203–219. https://doi.org/10.1016/j.jrurstud.2020.10.026

Thierfelder C, Chivenge P, Mupangwa W, Rosenstock TS, Lamana C, Eyre JX (2017) How climate-smart is conservation agriculture (CA)?—its potential to deliver on adaptation, mitigation and productivity on smallholder farms in southern Africa. Food Sec 9:537–560. https://doi.org/10.1007/s12571-017-0665-3

Theron SN, Midgley S, Hochrainer-Stigler S, Archer E, Tramberand S, Walker S (2023) Agricultural resilience and adaptive capacity during severe drought in the Western Cape. South Africa. Reg Environ Change 23:98

Theron SN, Archer ERM, Midgley SJE, Walker S (2022) Exploring farmers' perceptions and lessons learned from the 2015–2018 drought in the Western Cape, South Africa. J Rural Stud 95:208–222. https://doi.org/10.1016/j.jrurstud.2022.09.002

Topp EN, Loos J, Martín-López B (2022) Decision-making for nature's contributions to people in the Cape Floristic Region: the role of values, rules and knowledge. *Sustain Sci* **17**, 739–760 (2022). ttps://doi.org/https://doi.org/10.1007/s11625-020-00896-6

van Dijk T, Kleynhans TE (2004) An evaluation of the environmental care orientation of deciduous fruit producers in the Western Cape. Agrekon 43(3):317–330. https://doi.org/10.1080/03031853.2004.9523652

Ulrich K (ed) (2022) Ganzheitliche Nachhaltigkeit in der Weinwirtschaft. Ulmer, Stuttgart

van Wilgen BW, Measey J, Richardson DM, Wilson JR, Zengeya TA (2020). Biological invasions in South Africa: an overview. In: Measey J, editor. Biological invasions in South Africa. Invading nature, Springer series in invasion ecology. Springer, Cham. https://doi.org/10.1007/978-3-030-32394-3

Von Maltitz GP, Midgley GF, Veitch J, Brümmer C, Rötter RP, Viehberg FA, Veste M (eds) (2024). Sustainability of Southern African Ecosystems under Global Change, Ecological Studies 248, Springer, Cham, https://doi.org/10.1007/978-3-031-10948-5

Veste M, Böhm C (2018) Agrarholz—Schnellwachsende Bäume in der Landwirtschaft: Biologie - Ökologie - Management. Springer Spektrum, Heidelberg. https://doi.org/10.1007/978-3-662-49931-3

Veste M, Jürgens N (2004) Zonobiom III: Die Karoo. In: Walter, H., Breckle, S.-W., Ökologie der Erde, Band 2 Spezielle Ökologie der tropischen und subtropischen Zonen, Spektrum Akademischer Verlag, pp. 514–537

Veste M, Littmann T, Kunneke A, Du Toit B, Seifert T (2020) Windbreaks as part of climate-smart landscapes reduce evapotranspiration in vineyards, Western Cape Province, South Africa. Plant Soil Environ 66:119–127. https://doi.org/10.17221/616/2019-pse

Veste M, Sheppard JP, Abdulai I, Ayisi KK, Borrass L, Chirwa PW, Funk R, Kapinga K, Morhart K, Mwale SE, Ndlovu NP, Nyamadzaw G, Nyoka BI, Sebola P, Seifert T, Senyolo MP, Sileshi GW, Syampungani S, Kahle H-P (2023) The need for sustainable agricultural land-use systems: benefits from integrated agroforestry systems. In: von Maltitz GP, Midgley GF, Veitch J, Brümmer C, Rötter RP, Viehberg, FA, Veste M (eds) Sustainability of southern African ecosystems under global change: Science for management and policy interventions. Ecological Studies 248, Springer Nature, Heidelberg, pp 587–623. https://doi.org/10.1007/978-3-031-10948-5_21

Walker C, Milton SJ, O'Connor TG, Maguire JM, Dean WRJ (2018) Drivers and trajectories of social and ecological change in the Karoo, South Africa. Afr J Range & Forage Sci 35:157–177

Zerbe S (2022) Restoration of multifunctional cultural landscapes. Merging tradition and innovation for a sustainable future. Springer. Landscape Series 30:1–716

Zinngrebe Y, Borasino, E, Chiputwa B, Dobie P, Garcia E, Gassner, A, Kihumuro P, Komarudin, H, Liswanti, N, Makui P, Plieninger T, Winter E, Hauck J (2020). Agroforestry governance for operationalising the landscape approach: connecting conservation and farming actors. Sustainbility Science 15:1417–1434. https://doi.org/10.1007/s11625-020-00840-8

Author Index

M. Veste et al. (eds.), *Trees in South African Agriculture*, Innovations in Landscape Research, https://doi.org/10.1007/978-3-032-00993-7

MIX
Papier aus verantwortungsvollen Quellen
Paper from responsible sources
FSC® C105338

If you have any concerns about our products,
you can contact us on
ProductSafety@springernature.com

In case Publisher is established outside the EU,
the EU authorized representative is:
Springer Nature Customer Service Center GmbH
Europaplatz 3, 69115 Heidelberg, Germany

Printed by Libri Plureos GmbH
in Hamburg, Germany